TOPOLOGICAL ALGEBRAS

NORTH-HOLLAND
MATHEMATICS STUDIES **24**

Notas de Matemática (60)
Editor: Leopoldo Nachbin
Universidade Federal do Rio de Janeiro
and University of Rochester

Topological Algebras

EDWARD BECKENSTEIN
St. John's University, Notre Dame College, Staten Island, New York

LAWRENCE NARICI
St. John's University, Jamaica, New York

CHARLES SUFFEL
Stevens Institute of Technology, Hoboken, New Jersey

1977

NORTH-HOLLAND PUBLISHING COMPANY - AMSTERDAM • NEW YORK • OXFORD

North-Holland ISBN: 0 7204 0724 9

PUBLISHERS:
NORTH-HOLLAND PUBLISHING COMPANY
AMSTERDAM, NEW YORK, OXFORD

SOLE DISTRIBUTORS FOR THE U.S.A. AND CANADA:
ELSEVIER / NORTH HOLLAND, INC.
52 VANDERBILT AVENUE, NEW YORK, N.Y. 10017

Library of Congress Cataloging in Publication Data

Beckenstein, Edward, 1940-
Topological algebras.

(Notas de matemática ; 60) (North-Holland mathematics studies ; 24)
Includes index.
1. Topological algebras. I. Narici, Lawrence, joint author. II. Suffel, Charles, joint author. III. Title. IV. Series.
QA1.N86 no. 60 [QA326] 510'.8s [512'.55] 77-1127
ISBN 0-7204-0724-9

PRINTED IN THE NETHERLANDS

Kilimanjaro is a snow-covered mountain 19,710 feet high, and is said to be the highest mountain in Africa. Its western summit is called by the Masai "Ngaje Ngai," the House of God. Close to the western summit there is the dried and frozen carcass of a leopard. No one has explained what the leopard was seeking at that altitude.

From "The Snows of Kilimanjaro," by
Ernest Hemingway

For Paul and Maria, Dori and Chuck, and Marshall and Cheryl.

PREFACE

Let T be a completely regular Hausdorff space and let $\underset{\sim}{F}$ stand for the real numbers R or the complex numbers $\underset{\sim}{C}$ without specifying either.

Three main subjects are dealt with in this book: (1) general topological algebras; (2) the space $C(T,\underset{\sim}{F})$ of continuous functions mapping T into $\underset{\sim}{F}$ as an algebra only (with pointwise operations); and (3) $C(T,\underset{\sim}{F})$ endowed with compact-open topology as a topological algebra $C(T,\underset{\sim}{F},c)$. We wish to characterize the maximal ideals and homomorphisms of $C(T,\underset{\sim}{F})$ and the <u>closed</u> maximal ideals and <u>continuous</u> homomorphisms of topological algebras in general and $C(T,\underset{\sim}{F},c)$ in particular. In addition a considerable inroad is made into the properties of $C(T,\underset{\sim}{F},c)$ as a topological vector space in Chapter 2. Naturally enough, many of the results about $C(T,\underset{\sim}{F},c)$ serve to illustrate and motivate results about general topological algebras.

Attention is restricted to the algebra $C(T,\underset{\sim}{R})$ of real-valued continuous functions in Chapter 1 and to the pursuit of the maximal ideals and real-valued homomorphisms of such algebras. The clue to their identity and capture is found in the case when T is compact. The collection M_t of functions $x \in C(T,\underset{\sim}{R})$ which vanish at the point t is a maximal ideal whether T is compact or not, but when T is compact, every maximal ideal of $C(T,\underset{\sim}{R})$ is of this type. For noncompact T, the maximal ideals of $C(T,\underset{\sim}{R})$ are tied to the points of the Stone-Čech compactification βT of T in a very similar way. When T is compact, all homomorphisms of $C(T,\underset{\sim}{R})$ are evaluation maps, maps t^* taking functions $x \in C(T,\underset{\sim}{R})$ into their values $x(t)$ at t. (Note that the kernel of t^* is M_t.) By a quirk of nature, this remains essentially true even for noncompact T, but the whole story is a little more complicated. Gnerally the homomorphisms of $C(T,\underset{\sim}{R})$ are evaluation maps but associated with the points of the repletion υT of T, a certain subspace of βT, rather than just T. The quirk mentioned above by which the homomorphisms of $C(T,\underset{\sim}{R})$ are usually given by just the points of T is that $\upsilon T = T$ for most spaces.

These things and others are discussed in Chapter 1. Rather than deal with them from the z-ultrafilter point of view however, as Gillman and Jerison do for example, we have used uniform spaces as the habitat for the development of the theory. The idea that such an environment provides a felicitous setting for the development of the theory of rings of continuous functions is due to Nachbin and Chapter 1 owes a great deal to the way in which Warner carried such a development through in a set of lectures given at Reed College. Some background results on uniform spaces are given without proofs in Chapter 0 with references given for the details.

As mentioned above, the Stone-Čech compactification βT and repletion υT of T plays an important role in the development of the algebraic properties of $C(T,\underset{\sim}{R})$. Thus, significance attaches to obtaining them and viewing T as a uniform space enables a simple and direct realization of each. The theory of uniform spaces provides that every Hausdorff uniform (= completely regular) space T has a unique completion. With no further fuss this fact produces one form of the Stone-Čech compactification of T, as the completion of T with respect to the weakest uniformity with respect to which each bounded continuous function is uniformly continuous. (Mercurial entity that it is, βT emerges as a space of measures in Section 1.7 and as a space of homomorphisms of a Banach algebra in Section 4.10.) The repletion is obtained similarly — as a completion of T with respect to a different uniformity.

An interest that is always present when studying $C(T,\underset{\sim}{F})$ is the correlation of algebraic properties of $C(T,\underset{\sim}{F})$ with purely topological properties of T. For examples: (1) If consideration is restricted to compact spaces, S and T say, then $C(S,\underset{\sim}{R})$ and $C(T,\underset{\sim}{R})$ are isomorphic as algebras if and only if S is homeomorphic to T; (2) T is connected if and only if 0 and 1 are the only idempotents in $C(T,\underset{\sim}{F})$. When $C(T,\underset{\sim}{F})$ takes on the compact-open topology to become $C(T,\underset{\sim}{F},c)$, the scope for possible interactions broadens as one now seeks interplay between topological properties of T and topologico-algebraic properties of $C(T,\underset{\sim}{F},c)$. In this spirit, in Chapter 2, the famous theorems of Nachbin and Shirota are presented which settled Dieudonné's question: Must a barreled topological vector space be bornological? Nachbin and Shirota independently obtained necessary and sufficient conditions on T for $C(T,\underset{\sim}{R},c)$ to be barreled and for it to be bornological. The condition for bornologicity is especially simple: $C(T,\underset{\sim}{R},c)$ is bornological if and only if $T = \upsilon T$. Referring to the necessary and sufficient conditions on T which makes $C(T,\underset{\sim}{R},c)$ barreled as "condition NS", one can investigate the question: Is there a T which satisfies condition NS but for which $T \neq \upsilon T$? I.e. is there a T for which $C(T,\underset{\sim}{R},c)$ is barreled but for which $T \neq \upsilon T$? There are such spaces (of ordinals, predictably enough) and so bornologicity is not implied by barreledness. All of Chapter 2 is devoted to correlating topological properties of T with topological vector space properties of $C(T,\underset{\sim}{F},c)$. In particular, in addition to barreledness and bornologicity, conditions which guarantee or characterize metrizability, completeness, and separability of $C(T,\underset{\sim}{F},c)$ are obtained.

Another compactification, the Wallman compactification, plays an important role in characterizing the maximal ideals of certain topological algebras (Chapter 5). To develop the Wallman compactification however some knowledge of lattice theory is required. What is needed, together with the Wallman compactification itself, is presented in Chapter 3. In Chapter 4 the general subject of commutative topological algebras (with identity) is introduced and developed. To be more accurate, it is the theory of commutative locally m-convex algebras that is developed there. For just as topological vector spaces display an almost disappointing similarity to topological groups without the added assumption of local convexity, general topological algebras are similar to the point of disinterest to topological rings without "local m-convexity". With this added property, scalar multiplication plays an important role.

In Chapter 6 a special type of algebra is dealt with which we call an LB-algebra. The reason for the "LB" is that they are essentially just inductive Limits of Banach algebras.

In dealing with algebras of continuous functions as algebras we have stuck to real-valued functions in the text. As pointed out in Section 1.4 tho, most of what appears in Chapter 1 remains true for algebras of complex-valued functions. In dealing with topological algebras, where whether the underlying field is $\underset{\sim}{R}$ or $\underset{\sim}{C}$ can make a significant difference, we have tried to treat the real and complex cases on an equal footing wherever possible, and pointed out where it is not possible. In excursions in the exercises we consider algebras of $\underset{\sim}{K}$-valued functions where $\underset{\sim}{K}$ is a topological field or, more specially, a nonarchimedean valued field.

What sort of background should one have to read the book? Basically some algebra (one should certainly know what prime and maximal ideals are), some topology (having taken a year course in it somewhere along the line should suffice), and some functional analysis including some things about Banach algebras. As to the functional analysis, the elementary properties of locally convex spaces plus their duality theory should do; as for Banach algebras, not much is required per se, but the generalization of Banach algebra results to locally m-convex algebras will be more meaningful if something about Banach algebras is known. Essentially it is the same argument that would be given as to the desirability of some acquaintance with metric spaces before studying general topological spaces: It's not absolutely necessary, but it's nice to have.

Our notational conventions do not require any special comment. The only one that is unusual in any way is the use of ∇ to mark the end of a proof, but the reason for that choice had best remain a mystery. Except for Chapter 6, each chapter has a large number of exercises attached to it. The earliest ones are routine and meant for practice as well as information; the reverse is true of the later ones. References are given as well as extensive hints, many of which are really proofs written in telegraphic style. These later exercises are meant mainly to provide additional information about or information tangential to topics developed in the text. As simply "exercises", they should be approached with extreme caution.

Before commending our lucubrations to you, there are many people we would like to thank for many different reasons: St. John's University for providing facilities and a reduced teaching load to Narici; the National Science Foundation for providing a summer grant to Suffel; Elizabeth Suffel for doing some onerously difficult typing; to many other friends, some for having made comments about the book which were directly helpful to it, others for just being there. As for errors that might appear in the text, we shudder, apologize now, and vigorously stipulate that any that might remain are all the fault of the first-named author.

CONTENTS

ZERO

Fundamentals

THIS SHORT CHAPTER contains some things which are basic for what follows and makes certain things explicit, such as "completely regular" not including "Hausdorff." Mainly it deals with topologies and uniformities determined by families of functions and how the two are related; a few facts about uniform spaces are listed, with references to Bourbaki for proofs, in Sec. 0.2. These latter facts are put to use right away in Sec. 1.1 where the Stone-Cech compactification βT of a completely regular Hausdorff space T is obtained as a uniform space completion of T with respect to the uniformity induced by the space $C(T,\underset{\sim}{R})$ of real-valued continuous functions on T. In Sec. 1.5 the repletion (real compactification) υT of T is obtained similarly.

0.1 Topologies defined by families of functions. We assume familiarity with topology and the theory of uniform spaces and choose Bourbaki's General Topology, Parts 1 and 2, hereinafter referred to as Bourbaki 1966a and 1966b respectively, as our standard reference on these subjects.

In dealing with topological spaces, Hausdorff separation is not included in any instance unless specifically indicated. By saying that two subsets A and B of a topological space T are separated by open sets, we mean that disjoint open sets U and V exist containing A and B respectively. If a continuous function $x:T \to [0,1]$ exists which maps A into $\{0\}$ and B into $\{1\}$, we say that A and B are separated by a continuous function. Thus a completely regular space is one in which each point t and the complement of any neighborhood of t may be separated by a continuous function. Occasionally "Tihonov space" is used as a synonym for "completely regular Hausdorff space."

Unlike Bourbaki, Hausdorff separation is not included in "compact." By "locally compact" we mean that each point in the space possesses a neighborhood whose closure is compact. A space is σ-compact if it is a countable union of compact subsets, Lindelöf if every open cover contains a countable subcover.

Call a set clopen if it is closed and open and a topological space zero-dimensional if it possesses a base of clopen sets. An example of a zero-dimensional space follows.

Example 0.1-1 VALUED FIELDS A field K together with a real-valued map $|\,|:K \to \underset{\sim}{R}$ such that for all a, $b \in K$

(a) $|a| \geq 0$ and $=0$ iff $a=0$;

(b) $|ab| = |a||b|$

(c) $|a+b| \leq |a| + |b|$

is called a valued field; the map $||$ is called a valuation on K. If $|\ |$ satisfies (c') below instead of (c), then $|\ |$ is a nonarchimedean valuation and K a nonarchimedean valued field:

(c') $|a+b| \leq \max(|a|, |b|)$.

In either case $d(a,b) = |a-b|$ is a metric on K and when K carries the metric topology, K is a topological field. If the valuation is nonarchimedean, it is straightforward to verify that spheres, open or closed, in K, $\{a\in K \mid |a| < r \text{ or } \leq r\}$, $r > 0$, are clopen in K. Thus any nonarchimedean valued field is a zero-dimensional topological space.

NOTATIONS $\underset{\sim}{R}$ and $\underset{\sim}{C}$ stand for the real and complex numbers respectively carrying their usual topologies. $\underset{\sim}{F}$ denotes $\underset{\sim}{R}$ or $\underset{\sim}{C}$ without specifying either. $\underset{\sim}{N}$, $\underset{\sim}{Z}$, and $\underset{\sim}{Q}$ denote the natural numbers, integers, and rationals respectively.

If S and T are topological spaces, C(S,T) stands for the set of all continuous maps of S into T. If there is a notion of "bounded" set in T, if T was a topological vector space, for example, then $C_b(S,T)$ denotes the collection of all bounded continuous maps from S into T, i.e. all continuous maps whose range is a bounded subset of T. If T is a topological field then C(S,T) and $C_b(S,T)$ are each T-algebras with respect to the pointwise operations: $(x+y)(t)=x(t)+y(t)$, $(xy)(t)=x(t)y(t)$, and $(ax)(t)=ax(t)$ for x and y in C(S,T) or $C_b(S,T)$ and $a\in T$. It is always to these operations that we refer when we speak of spaces of continuous functions as algebras, rings, or linear spaces.

(0.1-1) INITIAL TOPOLOGIES (Bourbaki 1966a, p. 30, Prop. 4). Let T be a set, $((T_\mu, \mathcal{T}_\mu))_{\mu\in M}$ a family of topological spaces, and, for each $\mu\in M$, x_μ a map from T into T_μ. The topology $\mathcal{T}$ generated by the sets $\bigcup_{\mu\in M} x_\mu^{-1}(\mathcal{T}_\mu)$, i.e. the topology having $\bigcup_{\mu\in M} x_\mu^{-1}(\mathcal{T}_\mu)$ as a subbase, is the coarsest topology for T with respect to which each of the maps x_μ is continuous. $\mathcal{T}$ is called the initial topology determined by the maps $(x_\mu)_{\mu\in M}$ and a base for it is given by finite intersections of sets of the form $x_\mu^{-1}(G)$ where $\mu\in M$ and $G\in\mathcal{T}_\mu$.

Initial topologies are a means to transport topologies in ranges of functions back to the domain. Final topologies, discussed next, are a vehicle for the reverse direction.

(0.1-2) FINAL TOPOLOGIES (Bourbaki 1966a, p. 32, Prop. 6). If T is set,

$(T_\mu)_{\mu\in M}$ a family of topological spaces and x_μ a map from T_μ into T for each $\mu\in M$ then there is a finest topology $\mathcal{J}$ for T with respect to which each x_μ is continuous. $\mathcal{J}$ is called the final topology for T (determined by the maps (x_μ)) and $\mathcal{J}$ consists of those subsets U of T such that $x_\mu^{-1}(U)$ is open in T_μ for each $\mu\in M$.

If $(T,\mathcal{J})$ is a topological space and $C(T,\mathbb{R})$ the class of all continuous real-valued functions on T then certainly the initial topology $\mathcal{J}_c$ determined by $C(T,\mathbb{R})$ on T is coarser than $\mathcal{J}$. The same is true of the initial topology $\mathcal{J}_b$ determined by $C_b(T,\mathbb{R})$ on T. Moreover $\mathcal{J}_b\subset\mathcal{J}_c\subset\mathcal{J}$. When T is a completely regular Hausdorff space, however, the three topologies coincide ((0.2-5)).

Example 0.1-2 COMPACT-OPEN AND POINT-OPEN TOPOLOGIES ON $C(T,\mathbb{F})$ If K is a compact subset of the topological space T then the map

$$p_K : C(T,\mathbb{F}) \to \mathbb{F} \ (=\mathbb{R} \text{ or } \mathbb{C})$$
$$x \to \sup |x(K)|$$

is a seminorm on the linear space $C(T,\mathbb{F})$. Viewing $C(T,\mathbb{F})$ as an algebra, p_K is a multiplicative seminorm on $C(T,F)$ in the sense that $p_K(xy) \le p_K(x)p_K(y)$. The initial topology determined by the maps p_K as K runs through the compact subsets of T is the compact-open topology for $C(T,\mathbb{F})$. When $C(T,\mathbb{F})$ carries the compact-open topology it is denoted by $C(T,\mathbb{F},c)$. The compact-open topology is a locally convex Hausdorff topology for the linear space $C(T,\mathbb{F})$ and a locally m-convex Hausdorff topology for the algebra $C(T,\mathbb{F})$ (see (4.3-2)). A neighborhood base at 0 in $C(T,\mathbb{F},c)$ is given by the collection of all positive multiples of sets of the form

$$V_{p_K} = \{x\in C(T,\mathbb{F}) \mid p_K(x) < 1\} \text{ or } \overline{V}_{p_K} = \{x\in C(T,\mathbb{F}) \mid p_K(x) \le 1\}$$

as K runs through the class of compact subsets of T. If T is compact, then a base at 0 is given by positive multiples of just V_{p_T} and in this case $C(T,\mathbb{F},c)$ is a Banach algebra, its norm being simply the sup norm.

Another name for the compact-open topology is the topology of uniform convergence on compact sets, the reason being that a net $\mu \to x$ in the compact-open topology iff $\mu \to x$ uniformly on each compact subset of T.

If, instead of taking the class of all compact subsets of T, we take the collection of all singletons, the ensuing weaker locally m-convex Hausdorff topology for $C(T,\mathbb{F})$ is the point-open topology (topology of pointwise convergence or simple convergence). $C(T,\mathbb{F},p)$ denotes $C(T,\mathbb{F})$

endowed with the point-open topology.

Example 0.1-3 $\sigma(X,X')$ AND TVS CONVENTIONS "Topological vector space" (TVS) does not include Hausdorff; when desired we will say Hausdorff TVS (HTVS). "Locally convex topological vector space" and "locally convex Hausdorff topological vector space" will occasionally be abbreviated to LCS and LCHS respectively. All TVS's are assumed to have $\underset{\sim}{R}$ or $\underset{\sim}{C}$ as their underlying field unless otherwise specified. If X is a TVS, X' denotes the linear space of all continuous linear functionals on X. The weak (or weakened) topology $\sigma(X,X')$ for X is the initial topology generated by the maps

$$p_{x'}:X \to \underset{\sim}{F}$$
$$x \to |\langle x,x' \rangle|$$

as x' runs through X'. It is clearly the coarsest topology for X with respect to which each $x' \in X'$ is continuous, i.e. also the initial topology generated by the family of maps X' on X. $\sigma(X',X)$ is the initial topology determined on X' by the maps $x' \to |\langle x,x' \rangle|$ as x runs through X. These weak topologies, being determined by families of seminorms, are clearly locally convex topologies.

The polar S^o of a subset S of a TVS X is the collection of all $x' \in X'$ such that $\sup|x'(S)| \leq 1$; the dual consideration applies to polars of subsets of X'.

Example 0.1-4 $\mathfrak{S}$-TOPOLOGIES, $\tau(X,X')$ AND $\beta(X,X')$ If $\mathfrak{S}$ is a collection of $\sigma(X,X')$-bounded subsets of the LCHS X, then the collection $\mathfrak{S}^o$ of polars S^o of sets $S \in \mathfrak{S}$ is a set of absorbent balanced convex subsets of X'. Hence the collection of positive multiples of finite intersections of sets from $\mathfrak{S}^o$ forms a neighborhood base at 0 for a locally convex topology for X' called the $\mathfrak{S}$-topology. Dual considerations apply to $\mathfrak{S}$-topologies for X.

Another way to view $\mathfrak{S}$-topologies is as follows. If $\mathfrak{S}$ is a collection of $\sigma(X',X)$-bounded subsets of X', then $\langle x,S \rangle$ is a bounded set of scalars for each $S \in \mathfrak{S}$ and each $x \in X$. The maps

$$p_S:X \to \underset{\sim}{F}$$
$$x \to \sup|\langle x,S \rangle|$$

are seminorms (p_S is actually the gauge of S^o) on X. The (initial) topology determined by the seminorms $(p_S)_{S \in \mathfrak{S}}$ is the $\mathfrak{S}$-topology mentioned above.

Taking $\mathfrak{S}$ to be the class of all balanced convex $\sigma(X',X)$-compact subsets of X', or all $\sigma(X',X)$-bounded subsets of X', the $\mathfrak{S}$-topologies generated are respectively the Mackey topology $\tau(X,X')$ and the strong topology $\beta(X,X')$.

0.2 Uniformities defined by families of functions

(0.2-1) INITIAL UNIFORMITIES (Bourbaki 1966a, p. 176, Prop. 4). Let T be a set, $(T_\mu)_{\mu\in M}$ a family of uniform spaces, and, for each $\mu\in M$, x_μ a map from T into T_μ. Then there is a coarsest uniformity $\mathcal{U}$ for T, called the initial uniformity determined by the maps $(x_\mu)_{\mu\in M}$, with respect to which each x_μ is uniformly continuous. A fundamental system of entourages for $\mathcal{U}$ is given by sets of the form

$$U(V_1,\ldots,V_n) = \{(t,t')\in T\times T \mid (x_{\mu_i}(t),x_{\mu_i}(t'))\in V_i,\ 1 \le i \le n\}$$

where $\{\mu_1,\ldots,\mu_n\}$ is a finite subset of M and each V_i is an entourage in T_{μ_i}. If y is a map from a uniform space S into the uniform space $(T,\mathcal{U})$, then y is uniformly continuous iff $x_\mu \circ y:S \to T_\mu$ is uniformly continuous for each $\mu\in M$.

If each T_μ is a Hausdorff uniform space and if the family $(x_\mu)_{\mu\in M}$ separates points in T (i.e. for each pair (t,t') of points from T, if $t \neq t'$ then there is some $\mu\in M$ such that $x_\mu(t) \neq x_\mu(t')$) then $\mathcal{U}$ is a Hausdorff uniform structure.

As stated in (0.2-2) below, the topology determined by $\mathcal{U}$ is just the initial topology determined by the maps $(x_\mu)_{\mu\in M}$.

(0.2-2) INITIAL UNIFORMITIES VS. INITIAL TOPOLOGIES (Bourbaki 1966a, p. 177, Corollary). The topology determined by an initial uniformity determined by maps $(x_\mu)_{\mu\in M}$ is the coarsest topology with respect to which each x_μ is continuous, i.e. it is the initial topology determined by the maps $(x_\mu)_{\mu\in M}$.

A uniformity $\mathcal{U}$ is compatible with a topology $\mathcal{T}$ if the topology determined by $\mathcal{U}$ is just $\mathcal{T}$. A topological space T is uniformizable if a compatible uniformity $\mathcal{U}$ exists on T. The central characterization of uniformizability is:

(0.2-3) UNIFORMIZABILITY (Bourbaki 1966b, p. 144, Theorem 2). A topological space T is uniformizable iff it is completely regular.

(0.2-4) THE UNIFORMITY OF A COMPACT HAUSDORFF SPACE (Bourbaki 1966a, p. 199, Theorem 1). On a compact Hausdorff space T there is exactly one uniformity compatible with the topology of T. The entourages of this uniformity are all neighborhoods of the diagonal in $T\times T$.

Two uniformities of special interest are the initial uniformities determined by the real-valued continuous functions $C(T,\underset{\sim}{R})$ and $C_b(T,\underset{\sim}{R})$ on a topological space denoted by $\mathcal{C}(T,\underset{\sim}{R})$ and $\mathcal{C}_b(T,\underset{\sim}{R})$ respectively and sometimes shortened to simply $\mathcal{C}$ and $\mathcal{C}_b$. If T is completely regular and Hausdorff, $C(T,\underset{\sim}{R})$ and $C_b(T,\underset{\sim}{R})$ each separate points in T and so determine Hausdorff uniform structures for T. Letting $\mathcal{J}_c$ and $\mathcal{J}_b$ denote the topologies determined by $\mathcal{C}$ and $\mathcal{C}_b$ on T and letting $\mathcal{J}$ be T's original topology, it is clear that $\mathcal{J}_b \subset \mathcal{J}_c \subset \mathcal{J}$. When T is completely regular, the three topologies coincide.

(0.2-5) COMPLETE REGULARITY AND THE INITIAL TOPOLOGY DETERMINED BY $C(T,\underset{\sim}{R})$ ON T (T,$\mathcal{J}$) is completely regular iff $\mathcal{J}=\mathcal{J}_b=\mathcal{J}_c$.

Proof Suppose that T is completely regular and let V be a neighborhood of $s\in T$. There is some $x\in C_b(T,\underset{\sim}{R})$ such that $x(s)=0$ and $x(CV)=\{1\}$, and $\{t\in T \mid |x(t)| < 1/2\}$ is a basic neighborhood of s in the topology determined by $\mathcal{C}_b$ that is contained in V. It follows that $\mathcal{J}\subset\mathcal{J}_b$.

Conversely, if $\mathcal{J}=\mathcal{J}_b$, then T is uniformizable, hence completely regular by (0.2-3). ∇

Concerning completions of uniform spaces, we need the following results, the upshot of which is that Hausdorff uniform spaces are densely embedded in an essentially unique complete Hausdorff uniform space.

(0.2-6) COMPLETION (Bourbaki 1966a, p. 191, Theorem 3).

(a) For any uniform space T there is a complete Hausdorff uniform space $\hat{T}$, called the Hausdorff completion of T, and a uniformly continuous map $i:T \to \hat{T}$ which the property:

(P) Given any uniformly continuous map f of T into a complete Hausdorff uniform space S, there is a unique uniformly continuous map $g:T \to S$ such that $f=g\circ i$.

If (i_1,T_1) is another pair consisting of a complete Hausdorff uniform space T_1 and a uniformly continuous map $i_1:T \to T_1$ having property (P) then there is a unique uniform space isomorphism $h:T \to T_1$ such that $i_1=h\circ i$.

(b) (Bourbaki 1966a, p. 194, Corollary). If T is a Hausdorff uniform space then the canonical map $i:T \to \hat{T}$ is a uniform space isomorphism of T onto a dense subspace of $\hat{T}$. In this case, $\hat{T}$ is called the completion of T.

ONE

Algebras of Continuous Functions

AS P. SAMUEL has remarked, there are two principal methods of investigation in point set topology. The first — the "internal" method — refers to the topological space alone. Separation axioms, compactness, and connectedness, for example, are usually expressed solely in terms of the topological space. The second method — the "external" — uses the real numbers as an analytic tool. Here they usually appear via the channel of real-valued continuous functions on the topological space T. At times the entire class of such functions is called upon, as in the definition of complete regularity; in other instances a subclass such as the continuous pseudometrics is singled out, as with uniform spaces. From the external standpoint, if one wants the topological space T described accurately by its continuous functions, $C(T,\underset{\sim}{R})$, one must have enough of them. And to guarantee a good supply of nontrivial continuous functions, T is assumed to be completely regular and Hausdorff throughout.

A goal of this chapter is to develop an algebraic external method — interplay between the topological structure of T and the algebraic structure of the algebra $C(T,\underset{\sim}{R})$. In Chapter 2 the more complex interactions between the topological space T and the topological vector space (or algebra) $C(T,\underset{\sim}{R},c)$, $C(T,\underset{\sim}{R})$ with compact-open topology, are investigated. But here consideration is restricted to topologicoalgebraic interactions. Generally what can be said of the algebraic structure of algebras of continuous functions $C(T,\underset{\sim}{R})$? How sensitive is $C(T,\underset{\sim}{R})$ to T? I.e. can essentially different T's determine the same space of continuous functions? The converse question is clearly trivial: homeomorphic spaces do have the same space of continuous functions. The former question is answered (negatively) in Section 1.6. The machinery to answer it, and others like it, is developed along the way. To answer that particular question, the idea of a "repletion" υT, a certain type of completion of a completely regular Hausdorff space T, is needed. This leads to the related issue of "replete" spaces. (If replete spaces produce the same algebra of continuous functions, they must be homeomorphic.)

The Stone-Čech compactification βT of T is also cultivated as a completion (Sec. 1.1) and later, in Section 1.7, as a space of finitely additive 0 - 1 measures. The kinship between βT and υT, both being completions, emerges in the measure-theoretic setting with υT appearing as the countably additive members of βT.

Because of our interest in replete spaces, for reasons which will become apparent as the chapter develops, questions concerning which particular spaces are replete ($\underset{\sim}{R}$ is, for example) or the broader issue of what classes of spaces are replete are unavoidable. In trying to answer the latter question we are led to certain fundamental set-theoretic questions — a prospect apt at first to strike terror in the hearts of many mathematicians (us, in particular) — such as: Can Ulam cardinals exist within the framework of Zermelo-Fraenkel set theory? Some discussion of these matters is given in Section 1.7 before going on to Shirota's theorem in Section 1.8 governing repleteness of complete uniform spaces.

1.1 The Stone-Čech Compactification

Very early on, with respect to the definition of uniform spaces, Weil recognized that every compact Hausdorff space — hence every subspace of a compact Hausdorff space — could be viewed as a uniform space. He did this (Weil 1937, p. 24) using only internal methods. In order to prove a converse statement — namely that every uniform space could be viewed as a subspace of a compact Hausdorff space — he had to use the real numbers. An internal construction of this fact, similar to the way in which the Wallman compactification is obtained in Chapter 3 as a space of ultrafilters, was given by Samuel (1948).

That every uniform (= completely regular) Hausdorff space T possesses a compactification* to which every bounded continuous function on the space may be extended is the subject of this section. The reason for our interest in it lies in the close connection between the maximal ideals of $C(T,\underset{\sim}{R})$ and the points of the compactification, the Stone-Čech compactification, as elaborated on further in Section 1.4. The construction of the Stone-Čech compactification given here, as a uniform space completion — hence as a space of ultrafilters — is due to L. Nachbin.

Existence and uniqueness of the Stone-Čech compactification βT were first proved by Stone (1937), using the methods of Boolean rings. Kakutani (1941) has a construction using Banach lattices. Čech (1937) simplified Stone's original proof while Wallman-type compactifications (cf. Chapter 4

---

* A compactification S of a topological space T is a compact space containing a dense homeomorphic image of T. A Hausdorff compactification is a compactification which is a Hausdorff space.

and Wallman 1938) yield βT if T is normal. An approach using Banach algebra techniques, due to Gelfand and Silov, is discussed in Section 4.12 A realization as a space of measures due to Varadarajan is given in Section 1.7 while still other avenues to βT appear in the exercises.

In this section T is a completely regular Hausdorff space, $C(T,\underset{\sim}{R})$ is the algebra (with pointwise operations on the functions of $C(T,\underset{\sim}{R})$) of continuous functions taking T into $\underset{\sim}{R}$, and $C_b(T,\underset{\sim}{R})$ will be the subalgebra of $C(T,\underset{\sim}{R})$ consisting of all bounded functions.

The protean Stone-Cech compactification of T appears as the maximal ideals of $C_b(T,\underset{\sim}{R})$ (appropriately topologized), as a subspace of a product of closed unit intervals, as the z-ultrafilters of T (appropriately topologized), a space of measures (Sec. 1.7) and as the completion of T with respect to a certain uniform structure. We examine the last construction and show (Theorem 1.1-1) that in some sense βT is unique.

Let $x \in C_b(T,\underset{\sim}{R})$ and consider the sets $V(x,\varepsilon) = \{(s,t) \in T\times T \mid |x(s) - x(t)| < \varepsilon\}$ for $\varepsilon > 0$. Then the collection of entourages $\{V(x,\varepsilon) \mid x \in C_b(T,\underset{\sim}{R}), \varepsilon > 0\}$ form a subbase for a uniform structure $\mathcal{C}_b$ on T compatible with the topology of T.

<u>Definition 1.1-1. THE STONE-CECH COMPACTIFICATION</u>. The completion βT of the completely regular Hausdorff space T with respect to the uniform structure $\mathcal{C}_b$ is called the <u>Stone-Cech compactification</u> of T.

<u>Theorem 1.1-1. ELEMENTARY PROPERTIES OF βT</u>. Let T be a completely regular Hausdorff space. Then:

(a) βT is a compact Hausdorff space.

(b) Each $x \in C_b(T,\underset{\sim}{R})$ can be (uniquely)* extended to $\hat{x} \in C(\beta T,\underset{\sim}{R}) = C_b(\beta T,\underset{\sim}{R})$.

(c) If $\overline{\beta}T$ is a compactification of T which is a Hausdorff space and each $x \in C_b(T,\underset{\sim}{R})$ can be (uniquely)* extended to $\hat{x} \in C(\overline{\beta}T,\underset{\sim}{R})$, then βT and $\overline{\beta}T$ are equivalent compactifications of T (i.e. βT and $\overline{\beta}T$ are homeomorphic under a mapping which extends the "identity" on T).

<u>Proof</u> (a) To show that βT is compact, it is sufficient to show that T with the uniform structure $\mathcal{C}_b$ is totally bounded for then βT is complete and totally bounded and (Bourb. 1966a, p. 202) therefore compact. For each $x \in C_b(T,\underset{\sim}{R})$, there exists a closed interval $[a,b] \subset R$ such that

---

* As $\underset{\sim}{R}$ is a Hausdorff space, the extension $\hat{x}$ must be unique.

$x(T) \subset [a,b]$. Consider a finite set of points $s_o = a < s_1 < \ldots < s_n < b = s_{n+1}$ such that $|s_{i+1} - s_i| < \varepsilon/2$.

Now if $[s_i, s_{i+1}] \cap x(T) \neq \emptyset$, there exists $t_i \in T$ such that $x(t_i) \in [s_i, s_{i+1}]$ and it follows that $x^{-1}[s_i, s_{i+1}] \subset V(x,\varepsilon)[t_i]$ where $V(x,\varepsilon)[t_i] = \{t \in T \mid (t_i, t) \in V(x,\varepsilon)\}$. If $[s_i, s_{i+1}] \cap x(T) = \emptyset$, then $x^{-1}[s_i, s_{i+1}] = \emptyset$. Hence, since $T = \bigcup_{i=0}^{n} x^{-1}[s_i, s_{i+1}]$, it follows that $T = \bigcup V(x,\varepsilon)[t_i]$ where the union is taken over those i such that $[s_i, s_{i+1}] \cap x(T) \neq \emptyset$.

(b) Since $x(V(x,\varepsilon)[t]) \subset S_\varepsilon(x(t))$ for each $t \in T$, $S_\varepsilon(x(t))$ denoting the open sphere of radius ε about $x(t)$, it follows that x is uniformly continuous on T and, since R is complete, x may be uniquely continuously extended to $\hat{x} : \beta T \to R$ (Bourbaki 1966a, p. 190, Th. 2).

(c) Let $\overline{\beta}T$ be a Hausdorff compactification of T with the property that every $x \in C_b(T,\underset{\sim}{R})$ can be extended to $\hat{x} \in C(\overline{\beta}T,\underset{\sim}{R})$. As $\overline{\beta}T$ is a completely regular Hausdorff space, a subbase of entourages for a uniform structure generating the topology on $\overline{\beta}T$ is given by the sets $V(\hat{x},\varepsilon) = \{(s,t) \subset \overline{\beta}T \times \overline{\beta}T \mid |\hat{x}(s) - \hat{x}(t)| < \varepsilon\}$. Since $V(\hat{x},\varepsilon) \cap T \times T = V(x,\varepsilon)$, we see that we may view T as a subspace of the uniform space $\overline{\beta}T$. Since $\overline{\beta}T$ is complete, we see that $\overline{\beta}T$ is a completion of T as is βT, and by uniqueness of completions, $\overline{\beta}T$ is equivalent to βT. ∇

1.2 Zero Sets

In this section, as in the preceeding, T is a completely regular Hausdorff space. Letting x denote a function belonging to the algebra $C(T,\underset{\sim}{R})$, $z(x)$ will denote the set of points in T at which the function x vanishes. It is not difficult to see that the set $z(x)$ does not determine the function x. Yet the collection $z(M) = \{z(x) \mid x \in M\}$, where M is a maximal ideal in $C(T,\underset{\sim}{R})$, uniquely determines M. Moreover the properties of $z(M)$ will enable us to tell whether M is the kernel of a homomorphism of $C(T,\underset{\sim}{R})$. In this section we begin the study of the zero sets $z(x)$ of functions $x \in C(T,\underset{\sim}{R})$ and their relationship to the algebraic structure of the ring $C(T,\underset{\sim}{R})$. In particular we go from T to $C(T,\underset{\sim}{R})$ back to T again in showing that the zero sets $z(x)$ of continuous functions x on T determine the topology of T.

A subset of T of the form $x^{-1}(0)$, $x \in C(T,\underset{\sim}{R})$, is called a <u>zero set</u> and is also denoted by $z(x)$. For collections E of functions from $C(T,\underset{\sim}{R})$, $z(E)$ stands for $\{z(x) \mid x \in E\}$. Some elementary properties of zero sets follow.

<u>(1.2-1) ELEMENTARY PROPERTIES OF ZERO SETS</u>. For continuous real-valued functions on the completely regular Hausdorff space T:

(a) $z(x^2+y^2) = z(x) \cap z(y)$;

(b) $z(x) = z(|x|)$;

(c) $z(xy) = z(x) \cup z(y)$;

(d) For any $x \in C(T,\underset{\sim}{R})$, there exists $y \in C(T,\underset{\sim}{R})$ such that $0 \leqq y \leqq 1$ and $z(x) = z(y)$;

(e) $\bigcap_{n=1}^{\infty} z(x_n) = z(x)$ for some $x \in C(T,\underset{\sim}{R})$;

(f) $x^{-1}([r,\infty))$ is a zero set for any real number r and any $x \in C(T,\underset{\sim}{R})$.

<u>Proof</u> (a), (b), and (c) are evident. To prove (d) we let y be the continuous function defined by $y(t) = \min\{|x(t)|,1\}$. To prove (e) we assume $0 \leqq x_n \leqq 1$ for each n and consider $x = \Sigma\, 2^{-n} x_n$. Since $\Sigma\, 2^{-n} x_n$ is uniformly convergent, $x \in C(T,\underset{\sim}{R})$. Clearly $z(x) = \bigcap_{n=1}^{\infty} z(x_n)$. To prove (f), we consider $z(y)$ where $y = \min\{x,r\} - r$. ∇

The result established in (1.2-3) below generalizes the result of (1.2-1)(f). First we need the following topological result.

<u>(1.2-2) G_δ SETS AND ZERO SETS</u>. Let H be a closed subset of the normal space T. A necessary and sufficient condition that H be a zero set is that H be G_δ.

<u>Remark</u> Thus if T is a metric space, then every closed set is a zero set.

<u>Proof</u> If $H = z(x)$, then $H = \bigcap\{t \in T \mid |x(t)| < 1/n\}$. Conversely suppose $H = \bigcap U_n$, U_n open. By Urysohn's lemma there exist functions $x_n \in C(T,\underset{\sim}{R})$, $0 \leqq x_n \leqq 1$, such that $x_n(H) = \{0\}$ and $x_n(CU_n) = \{1\}$. Similar to the proof of (1.2-1)(e), we consider $x = \Sigma\, 2^{-n} x_n$. By the uniform convergence of this series, $x \in C(T,\underset{\sim}{R})$. A simple argument shows that $z(x) = \bigcap U_n = H$. ∇

<u>(1.2-3) A CHARACTERIZATION OF ZERO SETS</u>. For any completely regular Hausdorff space T, the collection of zero sets $z(C(T,\underset{\sim}{R}))$ is given by $\{x^{-1}(K) \mid x \in C(T,\underset{\sim}{R}), K \subset \underset{\sim}{R} \text{ is closed}\}$.

<u>Proof</u> Since $K \subset \underset{\sim}{R}$ and K is a closed set, by the remark after (1.2-2) there exists $g : \underset{\sim}{R} \to \underset{\sim}{R}$ such that $z(g) = K$. Then for each $x \in C(T,\underset{\sim}{R})$, $gx \in C(T,\underset{\sim}{R})$ and $z(gx) = \{t \in T \mid g(x(t)) = 0\} = \{t \in T \mid x(t) \in K\} = x^{-1}(K)$. ∇

We now prove that the zero set neighborhoods of points in T generate the topology of T. This result is useful in proving the Gelfand-Kolmogorov theorem (Theorem 1.4-1), in proving that βT can be realized as a lattice

(Wallman) compactification of T (Chapter 3), and is generally useful in the study of algebras of continuous functions.

(1.2-4) ZERO SET BASES. In any completely regular Hausdorff space T, the zero set neighborhoods of any $t \in T$ form a base for the neighborhood filter at t.

Proof Let N be an open neighborhood of t in T. There exists an open neighborhood $\hat{N}$ of t in βT such that $N = \hat{N} \cap T$. As βT is a regular space there exists a neighborhood $\hat{U}$ of t in βT whose closure is contained in $\hat{N}$. By the normality of βT there exists $\hat{x} \in C(\beta T, \underset{\sim}{R})$ such that $\hat{x}(\text{cl } \hat{U}) = \{0\}$ and $\hat{x}(C\hat{N}) = \{1\}$. Let x be the restriction of $\hat{x}$ to T. Then $\text{cl}\hat{U} \cap T \subset \hat{x}^{-1}(0) \cap T = x^{-1}(0) \subset \hat{N} \cap T = N$. ∇

Definition 1.2-2. COMPLETELY SEPARATED SETS. Two sets $A, B \subset T$ are completely separated if there exists $x \in C_b(T,\underset{\sim}{R})$, $0 \leq x \leq 1$, such that $x(A) = \{1\}$ while $x(B) = \{0\}$.

In the result that follows, we show that the sets which are completely separated in T are those whose closures in βT are disjoint.

Theorem 1.2-1. COMPLETELY SEPARATED SETS. For any two subsets A and B of the completely regular Hausdorff space T, the following conditions are equivalent.

(a) A and B are completely separated.

(b) $\text{cl}_\beta A \cap \text{cl}_\beta B = \emptyset$.

(c) A and B are contained in disjoint zero-sets. In particular note that if A and B are zero sets, then they are disjoint if and only if they are completely separated.

Proof To prove that (a) implies (b) suppose that A and B are completely separated. Then for some $x \in C_b(T,\underset{\sim}{R})$, $x(A) = \{0\}$ while $x(B) = \{1\}$. Letting $\hat{x}$ denote the continuous extension of x to βT, $\hat{x}\ (\text{cl}_\beta A) = \{0\}$ while $\hat{x}\ (\text{cl}_\beta B) = \{1\}$. Thus $\text{cl}_\beta A \cap \text{cl}_\beta B = \emptyset$. Conversely, if $\text{cl}_\beta A \cap \text{cl}_\beta B = \emptyset$, by Urysohn's lemma there exists $\hat{x} \in C(\beta T,\underset{\sim}{R})$, $0 \leq x \leq 1$, such that $\hat{x}(\text{cl}_\beta A) = \{0\}$ while $\hat{x}(\text{cl}_\beta B) = \{1\}$. The function $\hat{x}$ when restricted to T clearly separates A and B. Thus (a) and (b) are equivalent.

To prove that (a) implies (c), let $x \in C_b(T,\underset{\sim}{R})$ be such that $x(A) = \{0\}$ while $x(B) = \{1\}$. Then $A \subset x^{-1}(0)$, $B \subset x^{-1}(1)$. By (1.2-3) $x^{-1}(1)$ is a zero set and (c) follows. To prove that (c) implies (a) let $A \subset z(x)$, $B \subset z(y)$ for $x,y \in C_b(T,\underset{\sim}{R})$ and $z(x) \cap z(y) = \emptyset$. Taking $w = x^2/(x^2+y^2)$, we see that $w \in C_b(T,\underset{\sim}{R})$, $w(A) = \{0\}$, $w(B) = \{1\}$, and therefore that A and B are completely separated. ∇

In general, if A and B are subsets of T, $cl_\beta(A \cap B) \subset cl_\beta A \cap cl_\beta B$. We now show that if A and B are zero sets, then equality prevails. This result will be useful in proving the Gelfand-Kolmogorov theorem (Theorem 1.4-1).

Theorem 1.2-2. CLOSURES OF INTERSECTIONS OF ZERO SETS. Let T be a completely regular Hausdorff space. If $x, y \in C(T, \underset{\sim}{R})$, then $cl_\beta z(x) \cap cl_\beta z(y) = cl_\beta (z(x) \cap z(y))$.

Proof Clearly we need only show that $cl_\beta z(x) \cap cl_\beta z(y) \subset cl_\beta(z(x) \cap z(y))$. Let $p \in cl_\beta z(x) \cap cl_\beta z(y)$. Applying (1.2-4) let $\hat{V}$ be a zero set neighborhood of p in βT and $V = \hat{V} \cap T$. Then $V \cap z(x)$ and $V \cap z(y)$ are zero sets in T. Since $\hat{V}$ is a neighborhood of p in βT, it follows that $p \in cl_\beta (V \cap z(x))$. Similarly $p \in cl_\beta (V \cap z(y))$. Thus by Theorem 1.2-1(b) the zero sets $V \cap z(x)$ and $V \cap z(y)$ are not completely separated. By Theorem 1.2-1(c), it follows that $V \cap z(x) \cap z(y) \neq \emptyset$ and thus $\hat{V} \cap z(x) \cap z(y) \neq \emptyset$. Since by (1.2-4) the neighborhoods $\hat{V}$ form a base for the neighborhoods of p, $p \in cl_\beta(z(x) \cap z(y))$. ∇

1.3 Maximal Ideals and z-Filters

In this section we consider a notion which lies at the foundation of an approach to the study of algebras of continuous functions — the z-filter. The principal result, Theorem 1.3-1, establishes a connection between z-filters and maximal ideals.

Definition 1.3-1. z-FILTERS. A z-filter F is a nonempty collection of nonempty zero sets, closed under the formation of finite intersections, and such that if a zero set A contains some $B \in F$, then $A \in F$. A z-ultrafilter is a z-filter which is not properly contained in any other z-filter.

Thus to define z-filter, we simply cast "filter" in the setting of zero sets (of a completely regular Hausdorff space) rather than in the class of all subsets. The usual Zorn's lemma argument confirms the following result.

(1.3-1) z-ULTRAFILTERS. Every z-filter may be embedded in a z-ultrafilter.

Some of the standard properties of filters are shared by z-filters, as the following result shows.

(1.3-2) z-FILTERS. (a) If the zero set A meets each member of the z-filter F then there is a z-filter containing F and A.

(b) The z-filter F is a z-ultrafilter if and only if the only zero sets which meet every member of F actually belong to F.

(c) If a union, $A \cup B$, of zero sets belongs to a z-ultrafilter F, then $A \in F$ or $B \in F$.

Proof (a) The collection $\mathcal{H}$ of zero sets which contain an element of $\mathcal{F} \cap A$ or $\mathcal{F}$ is a z-filter containing $\mathcal{F}$ and A.

(b) This follows directly from (a) and the definition of z-ultrafilter.

(c) Suppose that neither A nor B belongs to $\mathcal{F}$. Then, by (b), there are sets $A', B' \in \mathcal{F}$ such that $A' \cap A = B' \cap B = \emptyset$. Thus $A \cup B$ fails to meet $A' \cap B'$ which is contradictory. ∇

Theorem 1.3-1. z-ULTRAFILTERS AND MAXIMAL IDEALS. The map

$$z : \mathcal{I} \to Z$$
$$I \to z(I) = \{z(x) \mid x \in I\}$$

establishes a (not necessarily 1-1) correspondence between the collection $\mathcal{I}$ of proper ideals I of $C(T,\underset{\sim}{R})$, T a completely regular Hausdorff space, and the set $\bar{Z}$ of all z-filters. The restriction of this map to $\mathcal{M}$, the collection of all maximal ideals, produces a 1-1 correspondence between $\mathcal{M}$ and the collection of all z-ultrafilters. Specifically, if $\mathcal{F}$ is a z-ultrafilter, then $z^{-1}(\mathcal{F})$ is the maximal ideal associated with $\mathcal{F}$.

Proof First we show that z(I) is a z-filter. Since I contains no units, $\emptyset \notin z(I)$. Since $z(x) \cap z(y) = z(x^2+y^2)$ for each pair $x,y \in I$, z(I) is stable under finite intersections. If $x \in I$ and $z(x) \subset z(w)$ for some $w \in C(T,\underset{\sim}{R})$ then $xw \in I$ and, by (1.2-1),

$$z(w) = z(x) \cup z(w) = z(xw) \in z(I) .$$

Next suppose that M is a maximal ideal. To see that z(M) is a z-ultrafilter, suppose that z(w) is a zero set which meets every element of z(M) and consider the ideal J generated by M and w. A typical element of J is of the form xw+m where $x \in C(T,\underset{\sim}{R})$ and $m \in M$. Since $z(w) \cap z(m) \neq \emptyset$, it follows that $z(xw+m) \neq \emptyset$. Thus no element of J is invertible, J is proper, and it follows from the maximality of M that $w \in M$. By (1.3-2)(b) it now follows that z(M) is a z-ultrafilter.

Conversely suppose that $\mathcal{F}$ is a z-filter. We contend that $I = z^{-1}(\mathcal{F}) = \{x \mid z(x) \in \mathcal{F}\}$ is an ideal. Utilizing the relations $z(xy) = z(x) \cup z(y)$ and $z(x+y) \supset z(x) \cap z(y)$, it is clear that I is closed under addition and multiplication by elements from $C(T,\underset{\sim}{R})$. Thus I is an ideal and clearly $z(I) = \mathcal{F}$. It remains to be shown that $Z|_{\mathcal{M}}$ is 1-1 and maps $\mathcal{M}$ onto the class of all z-ultrafilters. To demonstrate the surjectivity just mentioned, we show the ideal $z^{-1}(\mathcal{F})$ to be maximal whenever $\mathcal{F}$ is a z-ultrafilter. If $y \notin z^{-1}(\mathcal{F})$, then $z(y) \notin \mathcal{F}$; if $\mathcal{F}$ is a z-ultrafilter, there exists $z(x) \in \mathcal{F}$ such that $z(x) \cap z(y) = \emptyset$. Thus $z(x^2+y^2) = \emptyset$ and

$x^2 + y^2$ is invertible in $C(T,\underset{\sim}{R})$. It follows that the ideal generated by I and y is $C(T,\underset{\sim}{R})$ which proves that $z^{-1}(\mathcal{F})$ is a maximal ideal.

For any ideal I, $z^{-1}(z(I))$ is a proper ideal containing I. Thus $z^{-1}(z(M)) = M$ for maximal ideals M, and $Z|_M$ is seen to be injective. ∇

The main characteristics of the Stone-Čech compactification are summarized in Theorem 1.3-2.

Theorem 1.3-2. CHARACTERIZING βT. Let S be a Hausdorff compactification of the completely regular Hausdorff space T. Then the following statements are equivalent.

(1) S and βT are equivalent compactifications.

(2) For any pair $x,y \in C(T,\underset{\sim}{R})$, if $z(x) \cap z(y) = \emptyset$ then $cl_S z(x) \cap cl_S z(y) = \emptyset$.

(3) For any pair $x,y \in C(T,\underset{\sim}{R})$, $cl_S(z(x) \cap z(y)) = cl_S z(x) \cap cl_S z(y)$.

(4) Each $x \in C_b(T,\underset{\sim}{R})$ can be extended to some $x^S \in C(S,\underset{\sim}{R})$.

Proof By Theorems 1.1-1 and 1.2-2 we already know that (1) and (4) are equivalent and that (1) ⇒ (3) ⇒ (2). Thus it only remains to show that (2) ⇒ (3) and (3) ⇒ (4).

(2) ⇒ (3). Clearly $cl_S(z(x) \cap z(y)) \subset cl_S z(x) \cap cl_S z(y)$. Suppose that $s \in cl_S z(x) \cap cl_S z(y)$. As S is a compact Hausdorff space, there is a base of zero set neighborhoods V_μ^S at s. Certainly each $V_\mu = V_\mu^S \cap T$ is a zero set in T so that $z(x) \cap V_\mu$ and $z(y) \cap V_\mu$ are zero sets. Let V_α^S be an arbitrary zero set neighborhood of s in S and note that

$$\left(z(x) \cap V_\mu\right) \cap V_\alpha^S = z(x) \cap \left(V_\mu^S \cap V_\alpha^S\right).$$

Since $s \in cl_S z(x)$ however, $z(x) \cap \left(V_\mu^S \cap V_\alpha^S\right) \neq \emptyset$. Hence $s \in cl_S(z(x) \cap V_\mu)$. In the same way $s \in cl_S(z(y) \cap V_\mu)$ and therefore

$$cl_S(z(x) \cap V_\mu) \cap cl_S(z(y) \cap V_\mu) \neq \emptyset .$$

Thus, by (2), $z(x) \cap z(y) \cap V_\mu \neq \emptyset$ and, as V_μ^S is an arbitrary zero set neighborhood of s, it follows that $s \in cl_S(z(x) \cap z(y))$.

(3) ⇒ (4). To prove (4) we must extend an arbitrary $x \in C_b(T,\underset{\sim}{R})$ up to S so that the resulting function x^S is continuous. For each $s \in S$ we wish to define $x^S(s)$. To accomplish this we show that there is a unique z-ultrafilter $\mathcal{F}_s$ on T which converges to s and then unambiguously define $x^S(s)$ as the limit of a z-filter on $\underset{\sim}{R}$ derived from $\mathcal{F}_s$. To that end let $\mathcal{F}_s$ be any z-ultrafilter on T containing the z-filter base (V_μ) where $\left(V_\mu^S\right)$ is

a base of zero set neighborhoods of s in S and $V_\mu = V_\mu^S \cap T$ for each μ. Since $(V_\mu) \subset F_s$ it follows that the filterbase $F_s \to s$.

If F is another z-ultrafilter convergent to s and distinct from F_s we may choose $z(w) \in F$ and $z(y) \in F_s$ such that $z(w) \cap z(y) = \emptyset$. By (3), or more fundamentally by (2), $cl_S z(w) \cap cl_S z(y) = \emptyset$. Since $s \in cl_S z(y)$ — i.e. s is an adherence point of F_s — $s \notin cl_S z(w)$ and s is not an adherence point of F thus contravening the convergence of F to s.

Now, to obtain the definition of $x^S(s)$, let [a,b] be any closed interval containing x(T) and B_s be the class of all closed subsets E of [a,b] such that $x^{-1}(E) \in F_s$. Clearly B_s, a z-filter of subsets of [a,b] is a filterbase and, as [a,b] is compact, there is an adherence point $x^S(s)$ of B_s.

To see that B_s actually converges to s we show first that B_s is prime, i.e. if A and B are closed subsets of [a,b] whose union belongs to B_s then one or the other of A and B belongs to B_s. Indeed if $A \cup B \in B_s$ then $x^{-1}(A) \cup x^{-1}(B) = x^{-1}(A \cup B) \in F_s$. But F_s is a z-ultrafilter and it follows that $x^{-1}(A)$ or $x^{-1}(B)$ belongs to F_s and B_s is seen to be prime.

To show that $B_s \to x^S(s)$, let V be an arbitrary zero set neighborhood of $x^S(s)$ in [a,b]. As [a,b] is completely regular there is a zero set $U \subset [a,b]$ such that $x^S(s) \in [a,b] - U \subset V$. Now $U \cup V = [a,b] \in B_s$ and, since B_s is prime, either U or V belongs to B_s. But $x^S(s) \notin U$ and therefore $U \notin B_s$. Hence $V \in B_s$ and it follows that $B_s \to x^S(s)$.

All that now remains is to show that x^S is continuous. To this end let V be any zero set neighborhood of $x^S(s)$. The job now is to exhibit a neighborhood W of $s \in S$ such that $x^S(W) \subset V$. By the complete regularity of [a,b] there exists a zero set $V' \subset [a,b]$ such that $x^S(s) \in [a,b] - V'$. Thus $V \cup V' = [a,b]$ and $x^{-1}(V) \cup x^{-1}(V') = T$, so $cl_S x^{-1}(V) \cup cl_S x^{-1}(V') = S$.

We contend that $s \notin cl_S x^{-1}(V')$. If it did, if $s \in cl_S x^{-1}(V')$, then, as s adheres to each set in F_s, $cl_S x^{-1}(V') \cap cl_S z(w) \neq \emptyset$ for each $z(w) \in F_s$. By (3) it now follows that $x^{-1}(V') \cap z(w) \neq \emptyset$ for all $z(w) \in F_s$. Since F_s is a z-ultrafilter and $x^{-1}(V')$ is a zero set, $x^{-1}(V') \in F_s$. By the definition B_s, $V' \in B_s$. On the other hand $B_s \to x^S(s)$ and $x^S(s) \notin V'$ — a contradiction. Thus $s \notin cl_S x^{-1}(V')$.

The foregoing argument shows that s belongs to the open set $S - cl_S x^{-1}(V') = W \subset cl_S x^{-1}(V)$. If $p \in W \subset cl_S x^{-1}(V)$ then as W is a neighborhood of p, every neighborhood of p will intersect $x^{-1}(V)$ and therefore $x^{-1}(V) \in \mathcal{F}_p$. Thus $V \in \mathcal{B}_p \to x^S(p)$ so, as V is closed in [a,b], $x^S(p) \in V$. This proves that $x^S(W) \subset V$. ∇

1.4 Maximal Ideals and the Stone-Cech Compactification

It is clear that if t is a point of the completely regular Hausdorff space T and $M_t = \{x \in C(T,\underset{\sim}{R}) \mid x(t) = 0\}$, then M_t is a maximal ideal of $C(T,\underset{\sim}{R})$. In fact it is the kernel of the nontrivial homeomorphism $x \to x(t)$ from $C(T,\underset{\sim}{R})$ onto $\underset{\sim}{R}$. As is verified in (1.4-2) below, if T is compact, then these "fixed" ideals M_t constitute the set of all maximal ideals of $C(T,\underset{\sim}{R})$. Even if T is not compact tho, each of the maximal ideals of $C(T,\underset{\sim}{R})$ is "fixed" on a point p of the Stone-Cech compactification βT of T, a fact which is the essential content of Theorem 1.4-1, the main result of the section.

Definition 1.4-1. FIXED MAXIMAL IDEALS. Let T be a completely regular Hausdorff space. A maximal ideal M of $C(T,\underset{\sim}{R})$ is fixed if there is some $t \in T$ such that $M = M_t = \{x \in C(T,\underset{\sim}{R}) \mid x(t) = 0\}$. Otherwise M is called free.

(1.4-1) FIXED IDEALS. The maximal ideal M is fixed iff $\bigcap_{x \in M} z(x) \neq 0$. In this case $\bigcap_{x \in M} z(x) = \{t\}$ for some t in the completely regular Hausdorff space T.

Proof If $M = M_t$ then surely $t \in \bigcap_{x \in M} z(x)$. If $s \neq t$ then, since T is completely regular, there is some $x \in C(T,\underset{\sim}{R})$ such that $x(t) = 0$ — hence $x \in M_t$ — while $x(s) = 1$ so $s \notin \bigcap_{x \in M} z(x)$. Thus $\bigcap_{x \in M} z(x) = \{t\}$.

Conversely suppose that M is maximal and $\bigcap_{x \in M} z(x) \neq \emptyset$. If $t \in \bigcap_{x \in M} z(x)$, then $x(t) = 0$ for all $x \in M$ and M is seen to be a subset of M_t. The maximality of M then implies that $M = M_t$. ∇

For any T, completely regular and Hausdorff as usual, $C(T,\underset{\sim}{R})$ possesses fixed maximal ideals — namely those of the form $M_t = \{x \mid x(t) = 0\}$. As (1.4-2) shows, if T is compact, then all maximal ideals of $C(T,\underset{\sim}{R})$ are of this type. Conversely if T is not compact, $C(T,\underset{\sim}{R})$ always possesses free ideals. ∇

(1.4-2) MAXIMAL IDEALS ALL FIXED WHEN T IS COMPACT. If T is a compact Hausdorff space then the maximal ideals of $C(T,\underset{\sim}{R})$ are all fixed.

Proof We show that there are no free ideals in $C(T,\underset{\sim}{R})$ when T is compact: we show that if M is free then M contains a unit.

If M is free then for each $t \in T$ there is some $x_t \in M$ such that $x_t(t) \neq 0$. Since each x_t is continuous there must be open neighborhoods U_t

for each t on which x_t does not vanish. Since T is compact, finitely many of the U_t's, $U_{t_1},\ldots,U_{t_n}$ say, must cover T. It is now easy to see that the function

$$y = x_{t_1}^2 + \cdots + x_{t_n}^2 \in M$$

never vanishes on T and is therefore invertible. ∇

As a consequence of (1.4-2) we may determine the form of the maximal ideals of $C_b(T,\underset{\sim}{R})$ whether T is compact or not. First consider the map

$$\varphi : C_b(T,\underset{\sim}{R}) \to C(\beta T,\underset{\sim}{R})$$
$$x \to \hat{x}$$

where $\hat{x}$ denotes the unique extension of x to βT. Through the algebra isomorphism φ the maximal ideals of $C_b(T,\underset{\sim}{R})$ are seen to be in 1-1 correspondence with those of $C(\beta T,\underset{\sim}{R})$. Since βT is compact, the maximal ideals of $C(\beta T,\underset{\sim}{R})$ are all of the form

$$\hat{M}_p = \{\hat{x} \in C(\beta T,R) \mid \hat{x}(p) = 0\} \qquad p \in \beta T .$$

And $\hat{M}_p$ corresponds (i.e. $\varphi(M_p) = \hat{M}_p$) to the maximal ideal M_p in $C_b(T,\underset{\sim}{R})$ consisting of those $x \in C_b(T,\underset{\sim}{R})$ such that $\hat{x}(p) = 0$.

In the case when T is not necessarily compact, Theorem 1.4-1 below characterizes the maximal ideals of $C(T,\underset{\sim}{R})$. Before proving it, note that (1.4-2) may evidently be recast as follows:

If T is compact then all maximal ideals of C(T,R) are of the form

$$M_p = \{x \in C(T,\underset{\sim}{R}) \mid p \in cl_\beta z(x)\} \qquad p \in \beta T .$$

<u>Theorem 1.4-1. βT AND MAXIMAL IDEALS.</u> (Gelfand-Kolmogorov) The mapping

$$\beta T \to \mathcal{M}$$

$$p \to \{x \in C(T,\underset{\sim}{R}) \mid p \in cl_\beta z(x)\} = M_p$$

establishes a 1-1 correspondence between the class $\mathcal{M}$ of maximal ideals of $C(T,\underset{\sim}{R})$ and the points p of βT.

<u>Proof</u> First we show that M_p is an ideal in $C(T,\underset{\sim}{R})$. If $x \in M_p$ and $y \in C(T,\underset{\sim}{R})$, then $z(x) \subset z(xy)$, so $p \in cl_\beta z(x) \subset cl_\beta z(xy)$. For $x,y \in M_p$, $p \in cl_\beta z(x) \cap cl_\beta z(y) = cl_\beta[z(x) \cap z(y)] \subset cl_\beta z(x+y)$, the equality concerning closures of zero sets following from Theorem 1.2-2. Thus M_p is an ideal.

To prove that M_p is maximal we first show that $z(M_p)$ is a z-ultrafilter and then that $z^{-1}(z(M_p)) = M_p$ (Theorem 1.3-1). Let y be such that $z(y) \cap z(x) \neq \emptyset$ for all $x \in M_p$. We show that $z(y) \in z(M_p)$ by showing that $y \in M_p$, i.e. that $p \in cl_\beta z(y)$. If $z(\hat{x})$ is any zero set neighborhood of p, we claim that $x = \hat{x}|_T$ belongs to M_p. To see this let $p \in int_\beta z(\hat{x})$. Then for each neighborhood V of p, $V \cap int_\beta z(\hat{x})$ is a nonempty neighborhood of p. As T is dense in βT, $V \cap int_\beta z(\hat{x}) \cap T$ is also nonempty. Thus V meets $z(x) = z(\hat{x}) \cap T$ for each neighborhood V of p and it follows that

$$p \in int_\beta z(\hat{x}) \subset cl_\beta z(x) .$$

Hence $x \in M_p$. We have shown that every member of a basic family of neighborhoods of p (i.e. the zero set neighborhoods; see (1.2-4)) meets $z(y)$, so $p \in cl_\beta z(y)$ and $y \in M_p$. It only remains to show that $z^{-1}(z(M_p)) = M_p$. To see this let $w \in z^{-1}(z(M_p))$. As such there must be some $m \in M_p$ such that $z(w) = z(m)$. Hence $p \in cl_\beta z(m) = cl_\beta z(w)$ and therefore $w \in M_p$. The desired equality now follows.

Now let M be a maximal ideal in $C(T,\underset{\sim}{R})$. We wish to show that $M = M_p$ for some $p \in \beta T$. To prove this we show that $M \subset M_p$ for some $p \in \beta T$.

If no M_p contains M then for each $p \in \beta T$ there is some $x_p \in M$ such that $p \notin cl_\beta z(x_p)$. Thus $\beta T = \bigcup_{p \in \beta T} C(cl_\beta z(x_p))$. Since βT is compact,

$$\beta T = \bigcup_{i=1}^{n} C(cl_\beta z(x_{p_i}))$$

for some $p_1,\ldots,p_n \in \beta T$. Thus

$$\bigcap_{i=1}^{n} cl_\beta z(x_{p_i}) = cl_\beta\left(\bigcap_{i=1}^{n} z(x_{p_i})\right) = \emptyset$$

since closure and intersection may be interchanged for zero sets by Theorem 1.2-2; it follows that $\bigcap_{i=1}^{n} z(x_{p_i}) = \emptyset$. But $x_{p_1},\ldots,x_{p_n} \in M$ and therefore $\bigcap_{i=1}^{n} z(x_{p_i}) = \emptyset$ is impossible by Theorem 1.3-1. It now follows that there is some $p \in \beta T$ for which $M = M_p$.

Finally we show that if $p \neq q$ then $M_p \neq M_q$. If p and q are distinct, there are disjoint zero set neighborhoods $\hat{U}$ and $\hat{V}$ of p and q respectively in βT and functions $x,y \in C(T,\underset{\sim}{R})$ such that $z(x) = \hat{U} \cap T = U$ and $z(y) = \hat{V} \cap T = V$. As $U \cap V = \emptyset$, x and y cannot belong to the same maximal ideal. ∇

Theorem 1.4-1 enables us to characterize compactness externally, namely: The completely regular Hausdorff space T is compact if and only if each maximal ideal in $C(T,\underset{\sim}{R})$ is fixed. To see the sufficiency of the condition, we need only note that if T is not compact, then for any $p \in \beta T - T$, M_p is not fixed.

The algebra $C(T,\underset{\sim}{C})$ of complex-valued continuous functions does not differ markedly from $C(T,\underset{\sim}{R})$ if one takes preservation of the results of this chapter as the datum. The $C(T,\underset{\sim}{R})$-completion of T yields βT; so does the $C(T,\underset{\sim}{C})$-completion. The maximal ideals of $C(T,\underset{\sim}{C})$ are in 1-1 correspondence with the maximal ideals of $C(T,\underset{\sim}{R})$ under the mapping $M \to \text{re}\, M + i\, \text{re}\, M$ where re M denotes the collection of real parts re x of functions x in M. The inverse of this map is the map sending the maximal ideal J of $C(T,\underset{\sim}{R})$ into $J + iJ$. The Gelfand-Kolmogorov theorem then pairs the point $p \in \beta T$ with the maximal ideal $M_p + iM_p$ of $C(T,\underset{\sim}{C})$ where $M_p = \{x \in C(T,\underset{\sim}{R}) \,|\, p \in \text{cl}_\beta z(x)\}$. By similar considerations, a marked propinquity is exhibited between the subalgebras of bounded continuous functions, whether real- or complex-valued. Even if $\underset{\sim}{R}$ is replaced by the topological division ring $\underset{\sim}{H}$ of quaternions with euclidean (i.e. $\underset{\sim}{R}^4$) topology, and "ideal" by "two-sided ideal," the similarities persist: The collection $H_b(T,\underset{\sim}{H})$ of hom omorphisms of $C_b(T,\underset{\sim}{H})$ onto $\underset{\sim}{H}$ endowed with the initial topology generated by $C_b(T,\underset{\sim}{H})$ again yields βT, for example, so the maximal two-sided ideals of $C(T,\underset{\sim}{H})$ are in 1-1 correspondence with the maximal ideals of $C(T,\underset{\sim}{R})$. It is worth noting here that $\underset{\sim}{R}$, $\underset{\sim}{C}$ and $\underset{\sim}{H}$ are the only connected locally compact topological fields; if a locally compact field is not connected, then it must be totally disconnected (Bourbaki 1964, Ch. 6, §9, no. 3, Cor. 2).

Some discussion is devoted to algebras of continuous functions taking values in a topological field or a topological ring in the exercises at the end of this chapter and scattered in a few other places throughout the book.

1.5 Replete Spaces

The Stone-Cech compactification βT of T enabled us to characterize the maximal ideals of $C(T,\underset{\sim}{R})$ in a certain way (Theorem 1.4-1, the Gelfand-Kolmogorov theorem): Each $p \in \beta T$ corresponded to the maximal ideal

$$M_p = \{x \in C(T,\underset{\sim}{R}) \,|\, p \in \text{cl}_\beta z(x)\} .$$

When is M_p the kernel of a hom omorphism of $C(T,\underset{\sim}{R})$ into $\underset{\sim}{R}$? A sufficient condition is that $p \in T$ for then M_p is the kernel of the hom omorphism $\hat{p}$ sending x into x(p). As we shall see later ((1.6-1) and (Theorem

1.5-1), the "super adherence points" of T-those points $p \in \beta T$ such that for all countable families (V_n) of neighborhoods of p in βT, $\cap V_n$ must meet T — determine all maximal ideals which are kernels of homomorphisms. The collection υT of all such points is the <u>repletion</u> of T and it is to this collection that this section is devoted. Once the maximal ideals which are kernels of homomorphisms have been characterized as above, we also know which spaces T have evaluation maps as the only homomorphisms of $C(T,\underset{\sim}{R})$: namely those and only those for which $\upsilon T = T$. Spaces T with this latter property are called replete and constitute the only spaces for which $C(T,\underset{\sim}{R},c)$ — $C(T,\underset{\sim}{R})$ with compact-open topology — is bornological, as is shown in Chapter 2. A surprising fact about replete spaces is that nonreplete spaces are so hard to come by. The space $[0,\Omega)$ of ordinals less than the first uncountable ordinal Ω is not replete (Example 1.5-1) but any subspace of $\underset{\sim}{R}^n$ is (Theorem 1.5-3), as is any complete uniform space T, provided only that T is of non-Ulam (nonmeasurable) cardinal (Theorem 1.8-1), to name just two broad classes of replete spaces.

Before starting the definition of repletion we recall the following result on extension by continuity; for a proof see Kelley 1955, page 153 or Dugundji 1966, page 243.

<u>(1.5-1) EXTENSION BY CONTINUITY</u>. Let T be a completely regular Hausdorff space and $\underset{\sim}{R} \cup \{\infty\}$ be the one-point compactification of $\underset{\sim}{R}$. Any function $x \in C(T,\underset{\sim}{R})$ can be uniquely extended to a function $\hat{x} \in C(\beta T, \underset{\sim}{R} \cup \{\infty\})$.

Using the extensions whose existence were just noted, we now define that subspace of βT called the <u>repletion</u> of T.

<u>Definition 1.5-1. REPLETION</u>. The <u>repletion</u> (<u>real compactification</u>) υT of the completely regular Hausdorff space T consists of those $p \in \beta T$ for which each $\hat{x}(p)$ is finite, i.e. $\hat{x}(p) \neq \infty$, for every $x \in C(T,\underset{\sim}{R})$. If $T = \upsilon T$, then T is called <u>replete</u>.

Since T is dense in βT, the only way to extend a function $x \in C(T,\underset{\sim}{R})$ to a continuous real-valued function on any subspace S, $T \subset S \subset \beta T$, is by continuity. Thus υT is the largest subspace of βT to which each $x \in C(T,\underset{\sim}{R})$ possesses a real-valued continuous extension. Any time a subspace T of a topological space S is such that each $x \in C(T,\underset{\sim}{R})$ may be extended to a continuous real-valued function on S, T is said to be <u>C-embedded</u> in S. A similar meaning is attached to the expression <u>C_b-embedded</u>, meaning that all <u>bounded</u> continuous functions are continuously extendible. Thus in this nomenclature, υT is the largest subspace of βT in which T is C-embedded.

Note also that βT is the only compact Hausdorff space in which T is C_b-embedded by Theorem 1.1-1.

It is easier to exhibit spaces which are replete than spaces which are not. The reals, for example, are replete, as is any separable metric space (Theorem 1.5-3); so is any compact space and, more generally, any Lindelöf space. As Shirota's theorem (Theorem 1.8-1) shows, any complete uniform space whose cardinal is non-Ulam is replete. The space $[0,\Omega)$ of ordinals less than the first uncountable ordinal Ω is not replete as is established in Example 1.5-1.

In our next theorem we show that υT may be obtained as the completion of T with respect to the initial uniformity $\mathcal{C}$ determined by $C(T,\underset{\sim}{R})$ on T, paralleling the way in which βT was defined. Among other things this shows repletions υT to be themselves replete: $\upsilon(\upsilon T) = \upsilon T$.

<u>Theorem 1.5-1. CHARACTERIZATIONS OF THE REPLETION.</u> Let T be a completely regular Hausdorff space. Then:

(a) (Nachbin) υT is the completion of T with respect to the initial uniformity $\mathcal{C}$ determined by $C(T,\underset{\sim}{R})$ (i.e. $\mathcal{C}$ is the weakest uniform structure for T with respect to which each $x \in C(T,\underset{\sim}{R})$ is uniformly continuous).

(b) υT consists of those points $p \in \beta T$ such that for all sequences (V_n) of neighborhoods of p in βT, $(\bigcap V_n) \cap T \neq \emptyset$. Thus $p \notin \upsilon T$ if and only if there is some G_δ set (in βT) containing p which does not meet T.

<u>Proof</u> (a) Let $\mathcal{C}_b$ denote the initial uniformity determined on T by $C_b(T,\underset{\sim}{R})$ and $\hat{\mathcal{C}}$ the initial uniformity determined on υT by $C(\upsilon T,\underset{\sim}{R})$, i.e. the initial uniformity determined on υT by the functions $\hat{x}$ for $x \in C(T,\underset{\sim}{R})$.

We show that υT is the $\mathcal{C}$-completion of T. We do this by observing that $(T,\mathcal{C})$ is a dense uniform subspace of $(\upsilon T,\hat{\mathcal{C}})$ and proving that υT is $\hat{\mathcal{C}}$-complete. (That T is dense in υT follows from the fact that T is dense in βT.)

To show that υT is $\hat{\mathcal{C}}$-complete, it suffices (see for example Bourbaki 1966a, Prop. 9, p. 186) to show that any $\mathcal{C}$-Cauchy filterbase $\mathcal{B}$ on the dense subset T converges to a point of υT. Since $\mathcal{C}_b$ is weaker than $\mathcal{C}$, any such $\mathcal{B}$ will also be $\mathcal{C}_b$-Cauchy. Hence there is some $p \in \beta T$ such that $\mathcal{B} \to p$. It remains to show that $p \in \upsilon T$, i.e. that each $\hat{x}(p)$ is finite. By the continuity of any such $\hat{x}$, $x(\mathcal{B}) = \hat{x}(\mathcal{B})$ converges to $\hat{x}(p)$ in the Hausdorff space $\underset{\sim}{R} \cup \{\infty\}$. The fact that $\mathcal{B}$ is $\mathcal{C}$-Cauchy however implies that $x(\mathcal{B})$ is Cauchy in $\underset{\sim}{R}$, hence convergent to a point of $\underset{\sim}{R}$ so that its unique limit, namely $\hat{x}(p)$, is finite.

(b) We show that $p \notin \upsilon T$ if and only if there is a denumerable family (V_n) of neighborhoods of p in βT such that $(\bigcap V_n) \cap T = \emptyset$.

If $p \notin \upsilon T$ then there is some $x \in C(T,\underset{\sim}{R})$ such that $\hat{x}(p) = \infty$. The neighborhoods

$$V_n = \{t \in \beta T \mid |\hat{x}(t)| > n\} \qquad (n = 1,2,\ldots)$$

have the desired property: indeed

$$\bigcap V_n = \hat{x}^{-1}(\{\infty\}) \subset \beta T - T .$$

Conversely suppose that p possesses a sequence (V_n) of neighborhoods in βT whose intersection does not meet T. Clearly we may assume that (V_n) is a nested sequence of closed neighborhoods. Since βT is normal there is a nested sequence of neighborhoods U_n of p in βT such that $U_n \subset \overline{U}_n \subset \text{int } V_n$ and continuous functions y_n on βT such that $0 \leqq y_n \leqq 1$, $y_n(\overline{U}_n) = 1$ and $y_n(C(\text{int } V_n)) = 0$. As we verify next, the function

$$x : T \to \underset{\sim}{R}$$

$$t \to \sum_{n=1}^{\infty} y_n(t)$$

is continuous on T and $\hat{x}(p) = \infty$. Since (V_n) is nested and $\bigcap V_n$ does not meet T, $(T-V_n)$ is an ascending open cover of T. If $t \in T - V_n$ then $y_m(t) = 0$ for all $m \geqq n$ so

$$x(t) = \sum_{n=1}^{m-1} y_n(t) .$$

A straightforward argument now reveals the continuity of x on T.

Before showing that $\hat{x}(p) = \infty$, note that in any topological space, if A is a dense subset and V a neighborhood of t then $\overline{V \cap A}$ is also a neighborhood of t: $\text{int } V = \text{int } V \cap \overline{A}$ and $\text{int } V \subset \overline{\text{int } V \cap A} \subset \overline{V \cap A}$ for if W is any neighborhood of $s \in \text{int } V$ then $W \cap (\text{int } V \cap A) = (W \cap \text{int } V) \cap A \neq \emptyset$.

Since each of the functions y_m, $m \leq n$, is continuous and equal to 1 everywhere on U_n, then $x(t) \geqq n$ at any $t \in U_n \cap T$. Therefore $\hat{x}(t) \geqq n$ everywhere on the neighborhood $\overline{U_n \cap T}$ of p and since this holds for every n, the desired result follows. ∇

Clearly $C(T,\underset{\sim}{R})$ and $C(\upsilon T,\underset{\sim}{R})$ are indistinguishable as algebras. And, as part (a) of the above theorem shows, $(T,\mathcal{C})$ is a dense uniform subspace of

$(\upsilon T, \hat{\mathcal{C}})$. Moreover, since υT is $\hat{\mathcal{C}}$-complete, it is seen that repletions are replete — i.e. that $\upsilon(\upsilon T) = \upsilon T$.

If S and T are homeomorphic completely regular Hausdorff spaces under a homeomorphism $h : T \to S$ say, then the map

$$f : C(T,\underset{\sim}{R}) \to C(S,\underset{\sim}{R}) \qquad x \to x' \tag{1}$$

where $x'(s) = x \cdot h^{-1}(s)$, $s \in S$, shows $C(S,\underset{\sim}{R})$ and $C(T,\underset{\sim}{R})$ to be algebra-isomorphic, a hardly surprising fact whose converse for replete spaces S and T we shall establish shortly. Using this we can demonstrate the topological invariance of repleteness, something a bit more striking — a bit more striking because it is closely related to a particular uniform structure of a space and homeomorphisms do not necessarily preserve uniform properties. The homeomorphism $t \to t/(1+|t|)$, for example, maps the complete uniform space $\underset{\sim}{R}$ onto the incomplete space (0,1). Note too that repleteness was only defined for completely regular — hence uniformizable — spaces.

<u>(1.5-2) REPLETENESS IS A TOPOLOGICAL INVARIANT</u>. Let S and T be homeomorphic completely regular Hausdorff spaces. Then if S is replete, so is T.

<u>Proof</u> Let f be as in (1) above and let $\mathcal{C}(S)$ and $\mathcal{C}(T)$ denote the initial uniformities induced on S and T by $C(S,\underset{\sim}{R})$ and $C(T,\underset{\sim}{R})$ respectively. Our goal is to show that T is $\mathcal{C}(T)$-complete and to this end let $\mathcal{B}$ be a $\mathcal{C}(T)$-Cauchy filterbase in T, and note that this is equivalent to saying that $x(\mathcal{B})$ is a Cauchy filterbase in $\underset{\sim}{R}$ for each $x \in C(T,\underset{\sim}{R})$. With h as in (1), consider $h(\mathcal{B})$. For any $x' \in C(S,\underset{\sim}{R})$

$$x' \cdot h(\mathcal{B}) = x(\mathcal{B})$$

by the way x' was defined, so $h(\mathcal{B})$ is $\mathcal{C}(S)$-Cauchy — hence convergent to some $s \in S$. It follows that $\mathcal{B} \to h^{-1}(s)$. ∇

Before getting to more specialized examples, the $\mathcal{C}(T,\underset{\sim}{R})$-completeness characterization of replete spaces T yields the following embeddability theorem for them, a property which is useful in proving many subsequent theorems.

<u>Theorem 1.5-2. EMBEDDABILITY IN PRODUCTS OF $\underset{\sim}{R}$</u>. A completely regular Hausdorff space T is replete if and only if it is homeomorphic to a closed subspace of a cartesian product of real lines. Thus closed subspaces, intersections, and products of replete spaces are replete.

Proof Necessity. Let T be replete — i.e. $\mathcal{C}$-complete — and consider the evaluation map

$$e : T \to \underset{\sim}{R}^{C(T,\underset{\sim}{R})}$$
$$t \to (x(t))_{x\in C(T,\underset{\sim}{R})}$$

Since T is completely regular and Hausdorff, e is injective and e×e (i.e. the map $(s,t) \to (e(s),e(t))$ maps the subbasic $\mathcal{C}$-entourage $V(x,\varepsilon) = \{(s,t)\in T\times T \mid |x(s) - x(t)| < \varepsilon\}$ into the relative subbasic product entourage

$$\{((y(s)),(y(t)))_{y\in C(T,\underset{\sim}{R})} \mid |x(s) - x(t)| < \varepsilon\}.$$

Hence e and e^{-1} are uniformly continuous (e is a unimorphism). Thus the $\mathcal{C}$-completeness of T implies the completeness of e(T) in the product uniformity. Completeness is a Hausdorff uniform space however implies closedness, so the necessity of the condition is established.

Sufficiency. Since repleteness is a topological invariant by (1.5-2) it suffices to prove that closed subspaces T of a product $\underset{\sim}{R}^A$ are replete. Since products of complete spaces are complete (Bourbaki 1966a, p. 186, Prop. 10), any closed subspace T is complete in the product uniformity. By the definition of the product uniformity each projection pr_a $(a\in A)$ of T into $\underset{\sim}{R}$ is uniformly continuous, hence certainly continuous, i.e.

$$\{pr_a \mid a\in A\} \subset C(T,\underset{\sim}{R})$$

so the product uniformity is coarser than $\mathcal{C}$, the initial uniformity induced by $C(T,\underset{\sim}{R})$ on T. Since $\mathcal{C}$-Cauchy filters must therefore also be Cauchy filters with respect to the product uniformity, the desired result is seen to follow. ∇

The above result shows closed subspaces of replete spaces to be replete but the property of repleteness is obviously not passed on to every subspace — provided there are nonreplete spaces, i.e. provided there is a space T for which $T \neq \upsilon T$ for then T is a nonreplete subspace of the replete space υT. But where to look for a nonreplete space? Theorem 1.5-3 below eliminates a wide class of spaces. Among other things it shows every subspace of $\underset{\sim}{R}$ and $\underset{\sim}{R}^n$ to be replete.

Theorem 1.5-3. CLASSES OF REPLETE SPACES (Cf. Theorem 1.8-1) If the completely regular Hausdorff space T possesses any of the following properties, then T is replete.

(a) Lindelöf (i.e. open covers possess countable subcovers);

(b) σ-compact (T possesses a countable cover consisting of compact sets);

(c) second countable (the topology possesses a countable base);

(d) separable metric.

Proof Certainly every σ-compact space is second countable and every second countable space is Lindelöf, as is any separable metric space. So, the only thing that needs to be shown is (a). To do this we use the characterization of repleteness of part (b) of Theorem 1.5-1. For $p \in \beta T - T$, let $U(p)$ denote the collection of closed neighborhoods of p. Since $\bigcap U(p) = \{p\}$, βT being Hausdorff, then

$$T \subset C(\bigcap U(p)) = \bigcup_{U \in U(p)} CU .$$

If T is Lindelöf $(CU)_{U \in U(p)}$ possesses a countable subcover (CU_n) or, equivalently, $(\bigcap U_n) \cap T = \emptyset$. Thus $p \notin \upsilon T$. ∇

Though completely regular Hausdorff Lindelöf spaces are replete, the converse is false. To see this, let T denote $\underset{\sim}{R}$ with the topology generated by the half-open intervals (a,b]. T is completely regular, Hausdorff and Lindelöf (see below), hence replete, hence so is $T \times T$. $T \times T$ is not Lindelöf however, since the closed subspace $\{(t,-t) \mid t \in T\}$ is discrete and nondenumerable.

Proof that T is Lindelöf. Let $\mathcal{G} = (G_\mu)$ be an open cover of T and let I denote the set of all points $x \in \underset{\sim}{R}$ belonging to an open interval (a,b) such that (a,b) is contained in some covering set G_μ. As I is second countable in the Euclidean topology, a countable collection of the intervals (a,b) covers I; thus a countable subset of $\mathcal{G}$ covers I. Next we claim that $E = \underset{\sim}{R} - I$ is a countable set. If $x \in E$ then, since some G_μ contains x, there is some $a_x \in \underset{\sim}{R}$ such that $(a_x,x] \subset G_\mu$. Furthermore distinct x's in E yield disjoint intervals $(a_x,x]$: indeed if $x < y$ and $(a_x,x] \cap (a_y,y] \neq \emptyset$, then $x \in (a_y,y]$ which contradicts the fact that $x \notin I$. Now by associating a rational number $r_x \in (a_x,x]$ with each $x \in E$, it follows that E is countable. It remains to add one covering set G_μ for each $x \in E$ to the countable set of $\mathcal{G}$ which covers I to obtain a countable subcover of T. ∇

Having just generated a replete discrete space, it is natural to inquire about conditions under which discrete spaces are replete. In Section 1.7 it is seen that precisely those not of Ulam cardinal are replete (see discussion before and after Definition 1.7-1).

Before giving an example of a nonreplete space we set forth two topological notions: <u>countable compactness</u> — every countable open cover has a finite subcover — and <u>pseudocompactness</u> — every real-valued continuous function is bounded $\Leftrightarrow C(T,\underset{\sim}{R}) = C_b(T,\underset{\sim}{R})$. That the former implies the latter is clear since the open cover $\{t \in T \mid |x(t)| < n\}$, $n \in \underset{\sim}{N}$, has a finite subcover for each continuous function x. Tho the two notions are generally distinct, they coincide in normal Hausdorff spaces (Dugundji 1966, XI, 3.7 and XI, 3.9, pp. 231 - 2).

Generally we can make the following identifications:

$$C_b(T,\underset{\sim}{R}) = C(\beta T,\underset{\sim}{R}) \quad \text{and} \quad C(T,\underset{\sim}{R}) = C(\upsilon T,\underset{\sim}{R}) .$$

If T is pseudocompact, all four may be identified for then $C(T,\underset{\sim}{R}) = C_b(T,\underset{\sim}{R})$. Moreover:

<u>(1.5-3) T PSEUDOCOMPACT $\Rightarrow$ COMPACT = REPLETE</u>. If T is a completely regular Hausdorff space, then T is pseudocompact if and only if $\beta T = \upsilon T$. Thus in the class of pseudocompact spaces T, compactness is equivalent to repleteness.

<u>Proof</u> If there is a point $p \in \beta T - \upsilon T$ then, as constructed in the proof of Theorem 1.5-1 (b), there is some $x \in C(T,\underset{\sim}{R})$ which is unbounded on T. Thus if T is pseudocompact, $\beta T = \upsilon T$. Conversely if $\upsilon T = \beta T$ then

$$C(T,\underset{\sim}{R}) = C(\upsilon T,\underset{\sim}{R}) = C(\beta T,\underset{\sim}{R}) = C_b(T,\underset{\sim}{R})$$

and T is seen to be pseudocompact. ∇

The problem of exhibiting a nonreplete space is now reduced to that of producing a space which is pseudocompact, but not compact.

<u>Example 1.5-1. $[0,\Omega)$ IS NOT REPLETE</u>. The completely regular Hausdorff ordinal space $[0,\Omega)$, where Ω is the first uncountable ordinal, is countably compact — hence pseudocompact — but not compact. To see that it is not compact, view it as a subspace of the Hausdorff space $[0,\Omega]$ and note that it is not closed. We prove countable compactness by showing each sequence to have a cluster point. If (t_n) is a sequence from $[0,\Omega)$, it can be rearranged so that it is increasing. The rearranged sequence then has its supremum, t say, as a limit. Clearly t is countable — i.e. $t \in [0,\Omega)$ — and is a cluster point of the original sequence. It now follows from (1.5-3) that $[0,\Omega)$ is not replete.

Later conditions under which spaces $C(T,\underset{\sim}{R},c)$ of continuous functions with compact-open topology are barreled and bornological are obtained:

(Th. 2.5-1) $C(T,\underset{\sim}{R},c)$ is barreled if and only if unbounded continuous functions $x|_E$ exist on each closed noncompact subset E of T.

(Th. 2.6-1) $C(T,\underset{\sim}{R},c)$ is bornological if and only if $T = \upsilon T$. Taking <u>relative pseudocompactness</u> to mean that the restriction of each $x \in C(T,\underset{\sim}{R})$ to E is bounded, the barreledness result becomes:

$C(T,\underset{\sim}{R},c)$ is barreled if and only if each closed relatively pseudocompact subset of T is compact.

This version together with the characterization of relative pseudocompactness of our next result is helpful in proving that bornological spaces of continuous functions must be barreled (see (2.6-1)). Note too that saying a subset E of T is relatively pseudocompact is different from saying that E is pseudocompact. If E is closed and T is a normal Hausdorff space, however, then each $x \in C(T,\underset{\sim}{R})$ possesses a continuous extension to $C(T,\underset{\sim}{R})$ and the two notions do coincide.

<u>(1.5-4) RELATIVE PSEUDOCOMPACTNESS</u>. Let T be a completely regular Hausdorff space. Then $E \subset T$ is relatively pseudocompact if and only if $cl_\beta E \subset \upsilon T$. (Thus as already noted in (1.5-3) T being pseudocompact is equivalent to the equality $\upsilon T = \beta T$.)

<u>Proof</u> We show that E is not relatively pseudocompact if and only if $cl_\beta E \not\subset \upsilon T$.

Suppose that $p \in cl_\beta E - \upsilon T$, so that there is some $x \in C(T,\underset{\sim}{R})$ such that $\hat{x}(p) = \infty$. Thus for each $n \in \underset{\sim}{N}$ there is a point $t \in E$ at which x is $\geq n$.

Conversely if E is not relatively pseudocompact, then a positive function $x \in C(T,\underset{\sim}{R})$ exists which is unbounded on E. Thus there is a sequence (t_n) of points from E such that $x(t_n) \geq n$. Consequently at any cluster point p of (t_n) in the compact space βT, $\hat{x}(p) = \infty$.

As for the last statement we need only note that $cl_\beta T = \beta T$. ∇

1.6 <u>Characters and υT</u>

<u>Homomorphism</u> of $C(T,\underset{\sim}{R})$ means algebra homomorphism of $C(T,\underset{\sim}{R})$ into $\underset{\sim}{R}$; <u>character</u> means <u>nontrivial</u> homomorphism and each character h determines an ideal, its kernel ker h, of codimension one, hence a maximal ideal. (Note that being of codimension 1 is only a sufficient condition for maximality of an ideal. For maximal ideals M of $C(T,\underset{\sim}{R})$, if codim $M \neq 1$, then codim $M = \infty$.(Gillman and Jerison, 1960, p. 173).) Codim M = 1 is a necessary and sufficient condition for a maximal ideal to be the kernel of a character, the canonical map $x \to x + M \in X/M = \underset{\sim}{R}$ being the character going with the maximal ideal M. This latter type of maximal ideal is also called

a <u>real</u> maximal ideal. We denote the set of characters of $C(T,\underset{\sim}{R})$ by H or H(T), where necessary, and often identify characters and their kernels.

It has already been established (Theorem 1.4-1, the Gelfand-Kolmogorov theorem) that points $p \in \beta T$ determine maximal ideals

$$M_p = \{x \in C(T,\underset{\sim}{R}) \mid p \in cl_\beta z(x)\}$$

and that all maximal ideals of $C(T,\underset{\sim}{R})$ are of this type. Here we show that

$$\operatorname{codim} M_p = 1 \quad \text{iff} \quad p \in \upsilon T,$$

in other words that M_p is the kernel of a character if and only if $p \in \upsilon T$. Once we have this, another property of real maximal ideals emerges — namely that M_p is real if and only if $z(M_p)$ is stable with respect to countable intersection — which is useful in proving Shirota's remarkable result (Theorem 1.8-1) that every complete uniform space of non-Ulam cardinal is replete.

<u>(1.6-1) υT AND CHARACTERS</u>. Let T be a completely regular space and let x^υ be the unique extension of $x \in C(T,\underset{\sim}{R})$ to $C(\upsilon T,\underset{\sim}{R})$, and let M_p be as in the Gelfand-Komolgorov theorem. Then there is a 1-1 correspondence between υT and H, determined by

$$\sigma : \upsilon T \to H$$
$$p \to p^*$$

with

$$p^* : C(T,\underset{\sim}{R}) \to \underset{\sim}{R}$$
$$x \to x^\upsilon(p)$$

and $\ker p^* = M_p$. In other words M_p is the kernel of a character if and only if $p \in \upsilon T$.

<u>Proof</u> Two things are clear from outset: for $p \in \upsilon T$, the evaluation map p^* is a character, and different p's determine different characters. To show that $\ker p^* = M_p$, it suffices to show that $\ker p^* = M_p$. In turn this is equivalent to showing that for each $x \in C(T,\underset{\sim}{R})$

$$x^\upsilon(p) = 0 \Rightarrow p \in cl_\beta z(x) .$$

Since there is a neighborhood base of zero sets at p, by (1.2-4), it suffices to show that any neighborhood $z(y^\beta)$, $y^\beta \in C(\beta T,\underset{\sim}{R})$, of p meets z(x) if $x^\upsilon(p) = 0$. Clearly $y = y^\beta|_T \in \ker p^*$ so $x^2 + y^2 \in \ker p^*$. Consequently there exists $t \in z(x^2+y^2) = z(x) \cap z(y)$.

Let e denote the identity of $C(T,\underset{\sim}{R})$, the mapping sending each $t \in T$ into 1. To see that the map $p \to p^*$ is onto H, let h be any character. There exists $p \in \beta T$ such that ker $h = M_p$ and it remains only to show that $p \in \upsilon T$, for then ker h = ker $p^* = M_p$, whence $h = p^*$. To see that $p \in \upsilon T$, consider any $x \in C(T,\underset{\sim}{R})$ and its continuous extension $x^\beta \in C(\beta T,\underset{\sim}{R}\cup\{\infty\})$; we show that $x^\beta(p) \in \underset{\sim}{R}$. As $x - h(x)e \in$ ker $h = M_p$

$$p \in cl_\beta z(x-h(x)e).$$

Hence there must be a net (t_μ) from $(z(x-h(x)e)$ converging to p. By continuity $x^\beta(t_\mu) \to x^\beta(p)$, but for all μ, $x^\beta(t_\mu) = h(x)$, so $x^\beta(p) = h(x) \in \underset{\sim}{R}$. ∇

It is now clear that the spaces T for which evaluation maps $x \to x(t)$ are all the characters of $C(T,\underset{\sim}{R})$ — i.e. $H = T^*$ — are precisely the replete spaces. As a special case, one sees that compact spaces have this property. The spaces T for which all the maximal ideals of $C(T,\underset{\sim}{R})$ are of codimension one are the pseudocompact spaces, i.e. those for which $\beta T = \upsilon T$ ((1.5-4)).

In the discussion immediately preceding (1.5-2) the hardly surprising fact that homeomorphic spaces S and T produce the same algebra of continuous functions was noted. The converse is certainly false for υT can differ† from T (Example 1.5-1), yet $C(\upsilon T,\underset{\sim}{R})$ is always the same as $C(T,\underset{\sim}{R})$. Still, there is a result going in the converse direction. If $C(S,\underset{\sim}{R})$ is isomorphic to $C(T,\underset{\sim}{R})$, then the characters of the algebras are also identifiable, hence so are υS and υT — at least in the sense that they are in 1-1 correspondence. But this is rather weak. Are they homeomorphic? Is the 1-1 correspondence alluded to above a homeomorphism? As it happens, it is. Part of the vehicle we use to show it is the following result which provides for a topological identification of υT and H.

<u>(1.6-2) υT IS HOMEOMORPHIC TO H(T)</u>. Let T be a completely regular Hausdorff space and H and $\mathcal{M}$ the sets of characters and maximals ideals respectively of $C(T,\underset{\sim}{R})$. In virtue of the Gelfand-Kolmogorov theorem, $\mathcal{M}$ may be topologized by reflecting βT's topology onto it. Viewing H as a subspace of $\mathcal{M}$ then the bijection of (1.6-1)

---

† Since repleteness is a topological invariant by (1.5-2), $T \neq \upsilon T$ precludes the possibility of T being homeomorphic to υT.

$$\sigma : \upsilon T \to H$$
$$p \to p^*$$

($p^*(x) = x^{\upsilon}(p)$ for each $x \in C(T,\underset{\sim}{R})$) is a homeomorphism. Moreover the topology H inherits as a subspace of M is the initial topology generated by $C(T,\underset{\sim}{R})$ on H. A typical subbasic neighborhood of $p^* \in H$ is given by

$$V(p^*,x,\varepsilon) = \{q^* \in H \,\big|\, |q^*(x) - p^*(x)| = |x^{\upsilon}(p) - x^{\upsilon}(q)| < \varepsilon\}$$

$x \in C(T,\underset{\sim}{R})$, $\varepsilon > 0$.

Proof of (1.6-2) is routine and is omitted. We turn next to settling the questions raised prior to it concerning the "sensitivity" — in Gillman and Jerison's phrase — of the space of continuous functions to the underlying topological space. Generally the space of continuous functions is not sensitive enough to distinguish between spaces, but in the broad class of replete spaces, it is.

<u>(1.6-3) RECOVERING T FROM $C(T,\underset{\sim}{R})$</u>. Let S and T be completely regular Hausdorff spaces. If the algebra $C(S,\underset{\sim}{R})$ is isomorphic to $C(T,\underset{\sim}{R})$ then the repletions of S and T are homeomorphic. Consequently in the class of replete spaces, the spaces are homeomorphic if and only if their algebras of continuous functions are isomorphic.

<u>Proof</u> If $f : C(T,\underset{\sim}{R}) \to C(S,\underset{\sim}{R})$ is an isomorphism, then

$$f' : H(T) \to H(S)$$
$$p^* \to p^* \cdot f^{-1}$$

establishes a 1-1 correspondence between the characters of $C(T,\underset{\sim}{R})$ and $C(S,\underset{\sim}{R})$. With σ as in (1.6-1), then

$$\upsilon T \xrightarrow{\sigma} H(T) \xrightarrow{f'} H(S) \xrightarrow{\sigma^{-1}} \upsilon S$$

establishes the desired homeomorphism: with notation as in (1.6-2), the subbasic neighborhood $V(p,x,\varepsilon)$ of a point $p \in \upsilon T$ is transformed first into $V(p^*,x,\varepsilon)$, then into $V(p^* \cdot f^{-1}, f(x), \varepsilon)$ by f', and finally into the subbasic neighborhood $V(\sigma^{-1}(p^* \cdot f^{-1}), f(x), \varepsilon)$ of $\sigma^{-1}(p^* \cdot f^{-1}) = (\sigma^{-1} \cdot f' \cdot \sigma)(p) \in \upsilon S$. ∇

As sets, $C(T,\underset{\sim}{R})$ and $C_b(T,\underset{\sim}{R}) = C(\beta T,\underset{\sim}{R})$ obviously differ if and only if T is not pseudocompact. But, tho different, could they be isomorphic for non-pseudocompact T? If so, then $\upsilon T = \upsilon(\beta T) = \beta T$ — hence T is pseudocompact.

A forerunner of (1.6-3) is Banach's result (Banach 1932, p. 170, Theorem 3) that in the class of compact metric spaces S and T, S and T are homeomorphic if and only if $C(S,\underset{\sim}{R})$ is isometric to $C(T,\underset{\sim}{R})$ when each space carries its sup norm metric. Stone (1937, p. 469) then generalized this isometry-homeomorphism to compact Hausdorff spaces. Gelfand and Kolmogorov (1940) turned their attention to $C(S,\underset{\sim}{R})$ and $C(T,\underset{\sim}{R})$, S and T compact Hausdorff, as rings alone and showed them to be ring-isomorphic if and only if S and T were homeomorphic. By taking $x \geq y$ if and only if $x(t) \geq y(t)$ for all t, $C(T,\underset{\sim}{R})$ becomes a distributive lattice, and Stone (1941) showed that, as a lattice-ordered group, $C(T,\underset{\sim}{R})$ characterizes the compact Hausdorff space T; Kaplansky (1947) showed that as a lattice alone, $C(T,\underset{\sim}{R})$ characterizes T. Kaplansky's result has been extended to noncompact T in Shirota 1952b and Henriksen 1956.

Theorem 1.6-1. COUNTABLE INTERSECTIONS AND CHARACTERS. Let T be a completely regular Hausdorff space and $p \in \beta T$. Then the following are equivalent:

(a) M_p is the kernel of a character (namely p^*);

(b) $z(M_p)$ is stable under countable intersections;

(c) $z(M_p)$ has the countable intersection property.

Proof It is clear that (b) implies (c). To show that (a) implies (b) we first show that (a) implies (c).

Let (x_n) be a countable family of functions from M_p. Certainly the function $\inf(|x_n|, 2^{-n})$ has the same zero set as x_n so we may assume that $0 \leq x_n \leq 2^{-n}$. Thus $x = \sum x_n \in C(T,\underset{\sim}{R})$ and $z(x) = \bigcap_n z(x_n)$. To show that not (c) $\Rightarrow$ not (a), suppose that $\bigcap_n z(x_n) = \emptyset$ so that x is invertible in $C(T,\underset{\sim}{R})$. We show that $p \notin \upsilon T$ by showing that $(x^{-1})^{\beta}(p) = \infty$.

As $p \in \bigcap_{n=1}^{m} \mathrm{cl}_{\beta} z(x_n) = \mathrm{cl}_{\beta} \bigcap_{n=1}^{m} z(x_n)$ by Theorem 1.2-2, for each $m \in \underset{\sim}{N}$, there is a net $(t_\mu) \subset \bigcap_{n=1}^{m} z(x_n)$ converging to p in βT. For $\varepsilon > 0$ then by choosing m sufficiently large we may guarantee that $\left|\sum_{n=m+1}^{\infty} x_n\right| \leq \sum_{n=m+1}^{\infty} 2^{-n} < \varepsilon$ on T. Hence $\left|x^{-1}(t_\mu)\right| > 1/\varepsilon$ for each index μ and, by the continuity of $(x^{-1})^{\beta}$, $|(x^{-1})^{\beta}(p)| \geq 1/\varepsilon$. The desired result that $(x^{-1})^{\beta}(p) = \infty$ now follows.

To prove (a) $\Rightarrow$ (b) it remains to show that $z(x) \in z(M_p)$. Suppose that $z(y) \in z(M_p)$. Then $z(y) \cap z(x_n) \in z(M_p)$ for each n and so, by the previous argument,

$$z(y) \cap z(x) = \cap_n z(y) \cap z(x_n) \neq \emptyset .$$

As $z(x)$ meets each set in the z-ultrafilter $z(M_p)$, it follows by (1.3-2)(b) that $z(x) \in z(M_p)$.

Last we show that (c) $\Rightarrow$ (a). If $p \notin \upsilon T$ then there is some $x \in C(T,\underset{\sim}{R})$ such that $x^\beta(p) = \infty$. By (1.2-1)(f), $x^{-1}([n,\infty))$ is a zero set for each $n \in \underset{\sim}{N}$. As x^β is continuous at p and $x^\beta(p) = \infty$, $p \in cl_\beta x^{-1}([n,\infty))$ for each $n \in \underset{\sim}{N}$. Thus each $x^{-1}([n,\infty)) \in z(M_p)$ and $\cap_n x^{-1}([n,\infty)) = \emptyset$. ∇

1.7 <u>0-1 Measures, βT, and Ulam Cardinals</u>

There is a 1-1 correspondence between z-ultrafilters on T and points of βT, as well as with maximal ideals of $C(T,\underset{\sim}{R})$. Associated with each z-ultrafilter F is a 0-1 measure (on the algebra A_z of sets determined by the zero sets Z) defined by taking $m(E) = 1$ if $E \in F$, 0 otherwise. Moreover associated with each $t \in T$ is the 0-1 measure m_t concentrated at t ($m_t(E \in A_z) = 1$ if and only if $t \in E$), thus providing a natural embedding of T in the space M_o of 0-1 measures on T. By suitably topologizing M_o, the map $t \to m_t$ actually embeds T homeomorphically in M_o, as shown in Theorem 1.7-1. So topologized, not only is M_o compact, but functions $x \in C_b(T,\underset{\sim}{R})$ may be continuously extended from T to M_o. The vehicle by which this is done is defining the extension $\hat{x}$ to be $\int x dm$ at m, i.e.

$$\hat{x} : M_o \to \underset{\sim}{R}$$
$$m \to \int x dm$$

This is what Varadarajan (1965) did and this result is presented here. Converting the z-ultrafilters into a space of measures — a topological vector space of measures in fact — has the advantage of making a host of results from measure theory available.

A difficulty which this leads us into is confrontation with some problems in set theory. The measures m_t concentrated at points of T are countably additive set functions which may be defined on the class $P(T)$ of all subsets of T, with $m_t(T) = 1$. But are these the only such set functions? Could there be a countably additive $\{0,1\}$-valued set function μ defined on $P(T)$ such that $\mu(T) = 1$ but $\mu(\{t\}) = 0$ for each $t \in T$? The question of whether such a function — an <u>Ulam measure</u> — can exist on a set T has antecedents as far back as 1904 when Lebesgue asked: Does there exist a measure m defined on $P([0,1])$ such that $m([0,1]) = 1$ and such that congruent sets* have the same measure? Since all one-point sets must have the same measure under these conditions, each singleton must have measure 0. The question was answered in the negative in 1905 by Vitali: No such measure could exist. Moreover, as shown by Banach and Kuratowski in 1929, even if the congruence requirement is dropped, no such measure can exist on [0,1] if the continuum hypothesis is assumed. If the generalized continuum hypothesis (GCH) is assumed, then no such measure can exist on any set E, as Banach showed in 1930. Calling a cardinal $|T|$ an Ulam cardinal if an Ulam measure can be defined on $\mathcal{P}(T)$, the question may be phrased: Are there sets of Ulam cardinal? Within the framework of Zermelo-Fraenkel (ZF) set theory, the existence of such cardinals cannot be proved — at least if one assumes ZF to be consistent, for their nonexistence has been shown to be consistent with ZF. Their nonexistence is even consistent with ZF + Axiom of Choice (ZFC) (Shepherdson 1952, Ulam 1930). In ZF + GCH, as previously mentioned, their nonexistence can be demonstrated. Possibly their nonexistence can even be proved in ZF or ZFC.

For us a finitely additive measure on T is a finitely additive nonnegative real-valued function m defined on the algebra $\mathcal{A}_z$ of sets generated by the zero sets $\mathcal{Z}$ of the completely regular Hausdorff space T for which the following "regularity" condition holds:

$$\text{For each } A \in \mathcal{A}_z, \quad m(A) = \sup\{m(Z) \mid Z \in \mathcal{Z},\ Z \subset A\}\,.$$

(Thus knowledge of m on $\mathcal{Z}$ is sufficient to determine its behavior on $\mathcal{A}_z$.) The difference of two finitely additive measures on T is called a finitely additive signed measure on T. In the event that the set function is countably additive, i.e. $m(\bigcup E_n) = \sum m(E_n)$ whenever $(E_n)_{n \in \mathbb{N}}$ is a pairwise disjoint collection from $\mathcal{A}_z$ with union in $\mathcal{A}_z$, it is called either a measure or signed measure, as the case may be. A 0 - 1 measure has $\{0,1\}$ as its range. The collection of all finitely additive 0 - 1 measures is denoted by $\mathcal{M}_o$, the finitely additive measures by $\mathcal{M}^+$, and the finitely additive signed measures by $\mathcal{M}$. For each $m \in \mathcal{M}$, there are $m^+, m^- \in \mathcal{M}^+$ such that $m = m^+ - m^-$ where $m^+(A) = \sup\{m(B) \mid B \subset A,\ B \in \mathcal{A}_z\}$ and $m^-(A) = -\inf\{m(B) \mid B \subset A, B \in \mathcal{A}_z\}$ for $A \in \mathcal{A}_z$. The total variation of m is $|m|(A) = m^+(A) + m^-(A)$. The function $m \to \|m\| = |m|(T)$ defines a norm on the linear space $\mathcal{M}$ with pointwise

---

* X and Y are congruent if there is some $r \in [0,1]$ such that for all $y \in Y$ there is some $x \in X$ such that $y = x + r \pmod 1$, and for each $x \in X$ there is some $y \in Y$ such that $y = x + r \pmod 1$.

operations with respect to which M is a Banach space. It is not difficult to see that

$$|m|(A) = \sup\{ \sum_{i=1}^{n} |m(E_i)| \,\big|\, E_i \in A_z,\ E_i \subset A, (E_i) \text{ pairwise disjoint} \}$$

(see Dunford and Schwartz 1958, p. 137). Moreover a bounded finitely additive real-valued set function m is a finitely additive signed measure if and only if for each $A \in A_z$ there are zero sets Z_1, Z_2 such that $Z_1 \subset A \subset CZ_2$ and $|m|(CZ_2-Z_1) < \varepsilon$ so that the elements of M are just the "regular" finitely additive bounded real-valued functions on T, "regular" in the sense that for each $A \in A_z$, and $\varepsilon > 0$, there is a closed set $C \subset A$ in A_z and an open set $G \supset A$ in A_z such that $|m|(G-C) < \varepsilon$ (Dunford and Schwartz 1958, p. 137). In (2.4-5) it is established that the space M is the continuous dual of the Banach space $C_b(T,\underset{\sim}{R})$ equipped with the uniform norm — a fact that we will have need for later in this section (Theorem 1.7-1 (a)).*

Using the definition of integral given in Dunford and Schwartz 1958, Chapter III, or Alexandrov 1940, we now define a topology on M, the <u>vague topology</u>, by specifying a subbasis at each $m_o \in M$. A typical subbasic neighborhood of m_o is given by

$$\{ m \in M \,\Big|\, \left|\int x\,dm - \int x\,dm_o\right| < \epsilon \} = V(m_o, x, \varepsilon)$$

where $x \in C_b(T,\underset{\sim}{R})$ and $\varepsilon > 0$. In other words, the vague topology is the initial topology determined by the maps $\{ \int x d\cdot \mid x \in C_b(T,\underset{\sim}{R}) \}$ on M. A net (m_μ) from M converges to $m \in M$ in the vague topology if and only if $\int x dm_\mu \to \int x dm$ for each $x \in C_b(T,\underset{\sim}{R})$. In our first result we present an alternate characterization of "vague" convergence to be used in the sequel.

<u>(1.7-1) VAGUE CONVERGENCE</u>. A net (m_μ) from M^+ converges to $m \in M^+$ in the vague topology if and only if

(1) $m_\mu(T) \to m(T)$ and

(2) $\limsup_\mu m_\mu(Z) \leq m(Z)$ for each zero set Z.

Condition (2) may be replaced by

(2') $\liminf_\mu m_\mu(CZ) \geq m(CZ)$ for each zero set Z.

Thus a net converges in the vague topology of M^+ whenever it converges in

---

* The proof of (2.4-5) of course does not depend on anything following Theorem 1.7-1.

the product topology of M^+ (a collection of functions mapping A_Z into $\underset{\sim}{R}$) so that the product topology of M^+ is at least as strong as the vague topology.

<u>Proof</u> First suppose that $(m_\mu)_{\mu\in L}$ is a net from M^+ converging to $m \in M^+$ in the vague topology and let $Z \in \mathcal{Z}$, $\mathcal{Z}$ denoting the zero sets of the completely regular Hausdorff space T. By the regularity of m, given $\varepsilon > 0$, there exists $Z' \in \mathcal{Z}$ such that $Z \subset CZ'$ and $m(CZ') < m(Z) + \varepsilon$. By Theorem 1.2-1 (c), we may choose $x \in C_b(T,\underset{\sim}{R})$ such that $0 \leq x \leq 1$, $x(Z) = \{1\}$ and $x(Z') = \{0\}$. For this x and each $\mu \in L$

$$m_\mu(Z) = \int_Z x dm_\mu \leq \int x dm_\mu .$$

On the other hand

$$\int x dm = \int_{CZ'} x dm \leq m(CZ') < m(Z) + \varepsilon .$$

Since $m_\mu \to m$, $\int x dm_\mu \to \int x dm$ so that

$$\limsup_\mu m_\mu(Z) = \limsup_\mu \int_Z x dm_\mu \leqq \lim \int x dm_\mu = \int x dm < m(Z) + \varepsilon .$$

As ε is arbitrary, $\limsup_\mu m_\mu(Z) \leq m(Z)$. It is clear that $m_\mu(T) \to m(T)$.

Conversely, suppose that the conditions hold. We claim that for each $x \in C_b(T,\underset{\sim}{R})$, $\limsup_\mu \int x dm_\mu \leq \int x dm$. Furthermore it suffices to prove this inequality for those x's for which $0 < x < 1$, for by appropriately choosing scalars $a > 0$ and b we can always force $ax + b$ to satisfy the condition $0 < ax + b < 1$. Assuming the result to hold for $0 < y < 1$ then yields

$$\limsup_\mu \int (ax+b) dm_\mu \leq \int (ax+b) dm .$$

But $\int (ax+b) dm = a \int x dm + bm(T)$ and

$$\limsup_\mu \int (ax+b) dm_\mu = a\cdot\limsup_\mu \int x dm_\mu + b\cdot\lim m_\mu(T) .$$

Since $b\cdot\lim m_\mu(T) = bm(T)$, it follows that

$$\limsup_\mu \int x dm_\mu \leq \int x dm .$$

To prove the inequality for $0 < x < 1$, note first that for each $k \in \underset{\sim}{N}$,

$$T = \bigcup_{i=1}^{k} x^{-1}([(i-1)/k, i/k)) .$$

Letting Z_i be the zero set $x^{-1}([i/k,\infty))$, this becomes

$$T = \bigcup_{i=1}^{k} (Z_{i-1}-Z_i) .$$

Now the simple functions $\underline{x}$ and $\bar{x}$ defined at $t \in Z_{i-1} - Z_i$ by

$$\underline{x}(t) = (i-1)/k \quad \text{and} \quad \bar{x}(t) = i/k \quad (i = 1,\ldots,k)$$

satisfy the inequality $\underline{x} \leq x \leq \bar{x}$. Thus

$$(1) \qquad \int x dm_\mu \leq \int \bar{x} dm_\mu = 1/k + 1/k \cdot \sum_{i=1}^{k} m_\mu(Z_i)$$

and

$$(2) \qquad 1/k \sum_{i=1}^{k} m(Z_i) = \int \underline{x} dm \leq \int x dm .$$

As $\lim\sup_\mu m_\mu(Z_i) \leq m(Z_i)$ for each i, it follows that, taking superior limits in (1) and replacing $(1/k) \sum_{i=1}^{k} \lim\sup_\mu m_\mu(Z_i)$ by $(1/k) \sum_{i=1}^{k} m(Z_i)$, and using (2),

$$\lim\sup_\mu \int x dm_\mu \leq 1/k + \int x dm$$

for each $k \in \mathbb{N}$. Letting $k \to \infty$, establishes the desired inequality. Applying the inequality for $-x$ yields

$$\lim\inf_\mu \int x dm_\mu \geq \int x dm$$

and we may conclude that $\int x dm_\mu \to \int x dm$ for each $x \in C_b(T,\mathbb{R})$. ∇

Now the measures m_t concentrated at the points t in T, defined to be 1 at the sets $A \in \mathcal{A}_z$ to which t belongs and 0 otherwise, evidently constitute an injective image of T in $\mathcal{M}_o$ via the map

$$\varphi: T \to \mathcal{M}_o \subset \mathcal{M}^+$$

$$t \to m_t$$

Indeed if $t \neq s$ and $x \in C(T,\mathbb{R})$ maps t into 0 and s into 1 then $m_t(z(x)) = 1$ while $m_s(z(x)) = 0$. It is equally evident that the elements of $\varphi(T)$ are countably additive.

Theorem 1.7-1. $\beta T = M_o$ Let M_o denote the collection of all finitely additive 0 - 1 measures on the completely regular Hausdorff space T. Then:

(a) M_o with vague topology is the Stone-Cech compactification of T, and

(b) the repletion υT of T is the collection of all countably additive elements of M_o, i.e. the measures in M_o.

Proof (a) First we shall show that the 1-1 mapping φ given above is a homeomorphism. Using the fact that there is a base of zero sets at each point $t \in T$, one may show that for each $x \in C_b(T,\underset{\sim}{R})$ and $t \in T$ that $\int x dm_t = x(t)$. Since $\int x dm_t = x(t)$,

$$\varphi(\{t \in T \mid |x(t)-x(t_o)| < \varepsilon\}) = \left\{ m_t \,\middle|\, \left| \int x dm_t - \int x dm_{t_o} \right| < \varepsilon \right\}$$

$$= M_o \cap V(m_{t_o}, x, \varepsilon)$$

and φ is seen to be a homeomorphism.

Next it is shown that $\varphi(T)$ is dense in M_o by showing that, given $m \in M_o$, there is a net from $\varphi(T)$ which converges to m (in the vague topology).

Let Z_m denote the collection of all zero sets Z for which $m(Z) = 1$. Z_m becomes a directed set with respect to the ordering

$$Z_1 \le Z_2 \quad \text{iff} \quad Z_2 \subset Z_1 .$$

Since each $Z \in Z_m$ is nonempty, we may choose some element t_Z from Z. We contend that the net $\left(m_{t_Z}\right)_{Z \in Z_m}$ converges to m in the product topology, therefore also in the vague topology by (1.7-1).

Suppose that $A \in A_z$. If $m(A) = 1$ then there is a zero set $Z_o \subset A$ such that $m(Z_o) = 1$. Thus for each $Z \in Z_m$, if $Z \ge Z_o$ (i.e. $Z \subset Z_o$) then $m_{t_Z}(A) = 1$ so that $m_{t_Z}(A) \to m(A)$. If $m(A) = 0$, then $m(CA) = 1$ and some zero set $Z_o \subset CA$ must exist whose measure is 1. Hence each $Z \in Z_m$ greater than or equal to Z_o, being a subset of Z_o, fails to meet A and $m_{t_Z}(A) = 0$. Thus $m_{t_Z}(A) \to m(A)$ in both cases and $\varphi(T)$ is seen to be dense in M_o.

The next step in verifying that M_o is a compactification of T is to show that M_o is compact and Hausdorff in its vague topology. As for the separation, let m_1 and m_2 be distinct points of M_o. Then $m_1 - m_2$ is a nontrivial finitely additive regular set function. Clearly the total variation $|m_1 - m_2|$ of $m_1 - m_2$ is either 1 or 2.

Choose pairwise disjoint sets $A_1,\ldots,A_n \in \mathcal{A}_z$ with the properties that $\Sigma_i |(m_1-m_2)|(A_i) = |m_1 - m_2|(T)$ and $m_1(A_1) \neq m_2(A_1)$. Assuming that $m_1(A_1) = 1$ and $m_2(A_1) = 0$, choose zero sets Z and F such that Z and F are each subsets of A_1, $m_1(Z) = 1$, and $m_2(F) = 0$. Then for the zero set $Z_1 = Z \cup F$ we have $Z, F \subset Z_1 \subset A_1$ and it follows that $m_1(Z_1) = m_1(A_1)$ and $m_2(Z_1) = m_2(A_1)$. For each i, $2 \leq i \leq n$, choose zero sets $Z_i \subset A_i$ such that $(m_1-m_2)(Z_i) = (m_1-m_2)(A_i)$ in the same manner as Z_1 was chosen. As the Z_i's are pairwise disjoint, so are their closures in βT. Thus there is an $x^\beta \in C(T,\underset{\sim}{R})$ such that $x^\beta(\mathrm{cl}_\beta Z_1) = \{1\}$ and $x^\beta(\bigcup_{i=2}^{n} \mathrm{cl}_\beta(Z_i)) = 0$. Letting x denote $x^\beta|_T$, it follows that

$$\int x\,d(m_1-m_2) = (m_1-m_2)(Z_1) = 1 .$$

Finally then $\{m \in M_o \mid |\int x\,d(m-m_1)| < 1/2$ and $\{m \in M_o \mid |\int x\,d(m-m_2)| < 1/2\}$ are disjoint neighborhoods of m_1 and m_2 in the vague topology.

Since a typical continuous linear functional x' on $C_b(T,\underset{\sim}{R})$ with the uniform norm is of the form $x'(x) = \int x\,dm$ for some unique finitely additive signed measure $m \in M$ by (2.4-5), the vague topology of M is in fact the weak-topology $\sigma(C_b(T,\underset{\sim}{R})', C_b(T,\underset{\sim}{R}))$. Thus, by Alaoglu's theorem, a subset of M is compact in the vague topology whenever it is closed in the vague topology and norm-bounded. As the norm, i.e. the total variation, of each element of M_o equals 1, it only remains to show that M_o is (vague) closed in M. To this end let $(m_\mu)_{\mu \in L}$ be a net from M_o convergent to $m \in M$.

First we claim that $m \in M^+$. If not, then there is some $A \in \mathcal{A}_z$ such that $m(A) < 0$. Let $d = -m(A)/3 > 0$ and choose $Z_1, Z_2 \in \mathcal{Z}$ such that $Z_1 \subset A \subset CZ_2$ and $|m|(CZ_2 - Z_1) < d$. It follows that $m(Z_1)$ and $m(CZ_2)$ are both less than $-2d$. Since Z_1 and Z_2 are disjoint, so are their closures in βT and we may choose $x^\beta \in C(\beta T,\underset{\sim}{R})$ such that $0 \leq x^\beta \leq 1$, $x^\beta(\mathrm{cl}_\beta Z_2) = \{0\}$, and $x^\beta(\mathrm{cl}_\beta Z_1) = \{1\}$. Let x denote the restriction of x^β to T and consider

$$\int x\,dm = \int_{Z_2} x\,dm + \int_{CZ_2-Z_1} x\,dm + \int_{Z_1} x\,dm = \int_{CZ_2-Z_1} x\,dm + m(Z_1)$$

As $m(Z_1) < -2d$ and

$$\left|\int_{CZ_2-Z_1} x\,dm\right| \leq |m|(CZ_2-Z_1) < d$$

it follows that $\int x\,dm < -d$. On the other hand, for each $\mu \in L$, $\int x\,dm_\mu$ is nonnegative since it is the integral of a nonnegative function with respect to a finitely additive measure. Thus $\int x\,dm_\mu \not\to \int x\,dm$ which contravenes the choice of (m_μ) and m. We conclude that $m \in M^+$. Furthermore as each m_μ and m belong to M^+ and $m_\mu \to m$, the net $m_\mu(T) \to m(T)$ by (1.7-1) so that $0 \leq m(A) \leq m(T) = 1$ for each $A \in A_z$. To see that $m \in M_o$ we make the contrary assumption that $0 < m(A) < 1$ for some $A \in A_z$. Then there must be zero sets Z_1 and Z_2 such that $Z_1 \subset A \subset CZ_2$ and $0 < m(Z_1) \leq m(CZ_2) < 1$. Once again, by the fact that the β-closures of disjoint sets are disjoint, we may choose $x^\beta \in C(\beta T, \underset{\sim}{R})$ such that $x^\beta(cl_\beta Z_1) = \{0\}$ and $x^\beta(cl_\beta Z_2) = \{1\}$.

Now let x denote the restriction of x^β to T, and choose zero sets Z^* Z^{**} (by (1.2-3)) such that

$$CZ^* = \{t \in T \mid |x(t)| < 1/4\}$$

and

$$Z^{**} = \{t \in T \mid |x(t)| \leq 1/4\}.$$

It follows that $Z_1 \subset CZ^* \subset Z^{**} \subset CZ_2$. By (1.7-1) $\limsup m_\mu(Z^{**}) \leq m(Z^{**}) < 1$, so that an index $\mu_o \in L$ exists such that $m_\mu(Z^{**}) < 1$ for each $\mu \geq \mu_o$. As each $m_\mu \in M_o$, $m_\mu(Z^{**}) = 0$ for each $\mu \geq \mu_o$. Using condition (2') of (1.7-1) and a similar argument it follows that an index $\mu_1 \in L$ exists such that $m_\mu(CZ^*) = 1$ for each $\mu \geq \mu_1$. Choosing $\mu \geq \mu_o, \mu_1$ we see that $m_\mu(Z^{**}) = 0$ and $m_\mu(CZ^*) = 1$ even though $CZ^* \subset Z^{**}$, a contradiction. Hence $m \in M_o$, M_o is closed, and therefore compact in the vague topology.

Having shown that M_o is a compactification of T the one thing remaining to do is to prove that each $x \in C_b(T,\underset{\sim}{R})$ has a continuous extension to M_o (Theorem 1.1-1 (c)). We effect the extension of $x \in C_b(T,\underset{\sim}{R})$ to M_o by taking

$$\hat{x}(m) = \int x\,dm \qquad (m \in M_o).$$

Clearly $\hat{x}$ is continuous on M_o by the very definition of the vague topology, but does $\hat{x}$ extend x? Since there is a base of zero set neighborhoods at each $t \in T$, we may conclude that $\hat{x}(m_t) = \int x\,dm_t = x(t)$ at each $t \in T$, and it is seen that $\hat{x}$ is an extension of x. It now follows that $\mathcal{M}_o$ with vague topology is the Stone-Cech compactification of T.

(b) Recall (Theorem 1.6-1) that a point m in $\mathcal{M}_o$ (= βT) belongs to the repletion υT of T iff (1) its associated maximal ideal $M_m = \{x \in C(T,\underset{\sim}{R}) \mid m \in cl_\beta z(x)\}$ has the property that $z(M_m)$ is stable under countable intersections. [Or (2) the codimension of M_m is 1, i.e. $C(T,\underset{\sim}{R})/M_m$ is isomorphic to $\underset{\sim}{R}$.]

Prior to showing that each m ∈ υT is countably additive via statement (1) above, we establish the following technical fact:

$$z(M_m) = \{Z \in \mathcal{Z} \mid m(Z) = 1\} = \mathcal{Z}_m .$$

As $\mathcal{Z}_m$ is clearly a z-filter and $z(M_m)$ is a z-ultrafilter, it is only necessary to show that $z(M_m) \subset \mathcal{Z}_m$ to establish the equality of the two sets. A point in $z(M_m)$ is of the form z(x), $x \in C(T,\underset{\sim}{R})$, where $m \in cl_\beta(z(x))$. Thus there must be a net (t_μ) from z(x) such that $m_{t_\mu} \to m$. Therefore, by (1.7-1), $\limsup m_{t_\mu}(z(x)) \le m(z(x))$. By the way the m_{t_μ} are defined, each $m_{t_\mu}(z(x)) = 1$ however. Hence $m(z(x)) = 1$, and the desired inclusion is established.

Suppose now that m ∈ υT, so that $\mathcal{Z}_m$ is stable under countable intersections, and that m is not countably additive. This being the case, there must be a sequence (A_n) of sets from $\mathcal{A}_z$ decreasing to ∅ such that each $m(A_n) = 1$. And we may choose zero sets $F_n \subset A_n$ such that each $m(F_n) = 1$.

Now each of the zero sets $Z_n = \bigcap_{m=1}^{n} F_m \in \mathcal{Z}_m$ and $\bigcap Z_n = \bigcap F_n \subset \bigcap A_n = \emptyset$. This brings us to the contradictory conclusion that $m(\bigcap Z_n) = 0$, i.e. $\bigcap Z_n \notin \mathcal{Z}_m$.

Conversely, suppose that $m \in \mathcal{M}_o$ is countably additive. To show that m ∈ υT we show that $z(M_m) = \{Z \in \mathcal{Z} \mid m(Z) = 1\}$ is stable under countable intersections. To this end, let (Z_n) be a countable family of sets from $z(M_m)$. Thus, for each $n \in \underset{\sim}{N}$, $m(Z_n) = 1$ and, since a countable intersection of zero sets is a zero set by (1.2-1) (e),

$$m(C \bigcap_{n \in \underset{\sim}{N}} Z_n)) = m(\bigcup_{n \in \underset{\sim}{N}} CZ_n) \le \sum_{n \in \underset{\sim}{N}} m(CZ_n) = 0 . \quad \nabla$$

To view Part (b) of the above theorem in a somewhat different light, note that the correspondence

$$M_o = \beta T \to Z_u$$
$$m \to z(M_m)$$

(see Theorem 1.3-1) between βT and the z-ultrafilters on T pairs the elements of υT with the <u>δ-z-ultrafilters</u>, those z-ultrafilters stable under the formation of countable intersections: In the theorem it was established that $z(M_m) = \{z(x) \mid m(z(x)) = 1\}$. Thus, by Part (b) of the theorem, there is a 1-1 correspondence between the measures (i.e. countably additive members) of M_o and the δ-z-ultrafilters.

In proving that υT consists of the measures in M_o we made no mention of how an $x \in C(T,\underset{\sim}{R})$ is extended up to a continuous function on υT. We did see, however, how to continuously extend bounded functions to βT: For $x \in C_b(T,\underset{\sim}{R})$, take

$$\hat{x} : \beta T = M_o \to \underset{\sim}{R}$$
$$m \to \int x dm$$

Thus it is not unreasonable to suspect that this same method might be used to extend the functions of $C(T,\underset{\sim}{R})$ up to elements of $C(\upsilon T,\underset{\sim}{R})$. Indeed if $x \in C(T,\underset{\sim}{R})$, then $x^{-1}([-n,n])$ is an increasing sequence of zero sets converging to T. Hence for any 0-1 measure m, $m(x^{-1}([-n,n])) = 1$ for some n. Thus

$$\left|\int x dm\right| = \left|\int_{x^{-1}([-n,n])} x dm\right| < \infty$$

as $|x|$ is bounded by n on $x^{-1}([-n,n])$. To see that the real-valued function

$$\hat{x} : \upsilon T \to \underset{\sim}{R}$$
$$m \to \int x dm$$

is continuous on υT, let (m_μ) be a net from υT converging to $m \in \upsilon T$ in the vague topology. As $x^{-1}((-n,n))$ is the complement of a zero set Z_n (see (1.2-3)) and $(x^{-1}((-n,n)))$ increases to T, there is an index N such that $m(Z_N) = 0$. Thus, by (1.7-1), $\limsup_\mu m_\mu(Z_N) \leq m(Z_N) = 0$ and $m_\mu(Z_N) \to 0$. Consequently there must be an index μ_o such that $m_\mu(Z_N) = 0$ for all $\mu \geq \mu_o$. Setting $x_N = \min(x,N)$, it follows that

$$\int x\,dm_\mu - \int x\,dm = \int x_N\,dm_\mu - \int x_N\,dm$$

for each $\mu \geq \mu_o$. Hence $\int x\,dm_\mu \to \int x\,dm$ and $\hat{x}$ is seen to be continuous.

At this juncture we examine Part (b) of Theorem 1.7-1 above for the class of discrete spaces. Our purpose is to provide a foundation for Shirota's result presented in the next section. There it is essentially shown that a completely regular Hausdorff space endowed with a complete compatible uniform structure is replete if and only if each closed discrete subspace is replete.

If T is discrete then it is evident that $\mathcal{Z}$ and $\mathcal{A}_z$ coincide with the collection $\mathcal{P}(T)$ of all subsets of T. Thus a discrete space T is replete if and only if each (countably additive) 0-1 measure on $\mathcal{P}(T)$ is concentrated at a point of T.

Definition 1.7-1. ULAM CARDINALS. A (countably additive) 0-1 measure defined on the collection of all subsets of a set T which is not concentrated at a point of T is called an Ulam measure. Since the existence of such a measure is clearly a property of the equipotence class of T rather than just T, the cardinal numbers $|T|$ for which Ulam measures exist are called Ulam (measurable) cardinals.

In this terminology Part (b) of Theorem 1.7-1 yields the result that: A discrete space is replete if and only if it is not of Ulam cardinal.

A complete uniform space $(T, \mathcal{U}_w)$ remains complete when equipped with a finer uniformity $\mathcal{U}_f$ provided $\mathcal{U}_f$ possesses a fundamental system of entourages which are closed in $\mathcal{T}_w \times \mathcal{T}_w$ (Bourbaki 1966a, p. 185, Prop. 7). Because of this and the facts that products and closed subspaces of replete spaces are replete, we have:

(1.7-2) NON-ULAM CARDINALITY IS HEREDITARY AND PRODUCTIVE. (a) If T is not of Ulam cardinality then neither is any subspace of T. (b) If (T_μ) is a family of sets none of which is of Ulam cardinal, then πT_μ is not of Ulam cardinality either.

Furthermore $|\underset{\sim}{N}|$ is not an Ulam cardinal. Indeed if m is a 0-1 measure on $(\underset{\sim}{N})$ then $1=m(\underset{\sim}{N})=\Sigma_{n\in N}\, m(\{n\})$ so that an integer n exists for which $m(\{n\})=1$. Hence $m=m_n$, i.e. m is concentrated at n. This fact together with (1.7-2) yields the conclusion that every cardinal less than or equal to a cardinal of the form

$$2^{2^{\cdot^{\cdot^{\cdot^{2^{|\underset{\sim}{N}|}}}}}}$$

is not an Ulam cardinal, a fact which indicates that if Ulam cardinals exist at all, they must be very large.

1.8 Complete Implies Replete

The fact that broad classes of spaces are replete was mentioned in Section 1.5. In Theorem 1.5-3, for example, it was shown that each Lindelöf space is replete. In the principal result of this section another broad class of such spaces is established: each space whose cardinality is not Ulam and whose topology is given by a complete uniform structure. For the sake of that demonstration it is helpful to single out the following notion of "discreteness."

Definition 1.8-1. d-DISCRETE. Let d be a pseudometric defined on the set T. A family (T_μ) of subsets of T is d-discrete of gauge $p > 0$ if $d(T_\mu,T_\lambda) = \inf\{d(s,t) \mid (s,t) \in T_\mu \times T_\lambda\} \geqq p$ for all distinct pairs of indices λ,μ. A set $S \subset T$ is d-discrete if $(\{s\})_{s\in S}$ is d-discrete.

The basic properties of d-discreteness follow.

(1.8-1) d-DISCRETENESS. Let d be a continuous pseudometric on the completely regular Hausdorff space T. Then

(a) Any d-closed set $A \subset T$ is a zero set (viz. the map $t \to d(A,t)$);

(b) the union of any d-discrete family (T_μ) of d-closed (i.e. closed with respect to the topology induced by d) sets is d-closed;

(c) any d-discrete set S is closed;

(d) any d-discrete set S is discrete.

Proof (a) Clear.

(b) Suppose that t is a d-adherence point of $\cup T_\mu$. For each $\varepsilon \leqq p$ the set $\{s \in T \mid d(s,t) < \varepsilon\}$ meets just one of the sets T_μ; it follows that $t \in cl_d T_\mu = T_\mu$.

(c) By (b) the set $\hat{S} = \bigcup_{s\in S} cl_d\{s\}$ is d-closed. If t is an adherence point of S it is also a d-adherence point of $\hat{S}$ so that $t \in cl_d\{s\}$ for some $s \in S$. If we assume that $t \neq s$ and if V_t is a neighborhood of t in the Hausdorff space T excluding s, then the neighborhood (in T), $V_t \cap \{r \in T \mid d(r,t) < p\}$ fails to meet S. Thus $t = s \in S$.

(d) If $s \in S$ where S is d-discrete, the neighborhood $\{t \mid d(s,t) < p\}$ of s meets S only in the point s. ∇

Theorem 1.8-1 (Shirota) COMPLETE IMPLIES REPLETE. A completely regular Hausdorff space T is replete if and only if there exists a complete compatible uniformity on T and no closed discrete subspace of T is of Ulam cardinality. Thus any complete Hausdorff uniform space which is not of Ulam cardinality is replete.

Proof From our earliest result on repleteness, Theorem 1.5-1, and the comment right after it, we know that the initial uniformity $\mathcal{C} = \mathcal{C}(T,\underset{\sim}{R})$ induced by $C(T,\underset{\sim}{R})$ on T is a compatible complete uniformity for replete T. As to the cardinality of closed discrete subspaces of T, (1) any closed subspace of a replete space is replete by Theorem 1.5-2, and (2) for discrete spaces, repleteness is equivalent to not being of Ulam cardinality by the discussion before and after Definition 1.7-1. So much for necessity.

To prove sufficiency, i.e. that the conditions imply $\upsilon T \subset T$, we choose $p \in \upsilon T$ and consider $z(M_p)$. Assuming $\mathcal{U}$ to be a complete compatible uniformity on T, we show $z(M_p)$ to be a $\mathcal{U}$-Cauchy filterbase. As such it converges to a point of T. Generally, however, the filterbase $z(M_p)$ converges to p. Since βT is Hausdorff, it follows that $p \in T$.

We must show that $z(M_p)$ contains sets that are U-small for any $U \in \mathcal{U}$. Since $\mathcal{U}$ is generated by the family of all $\mathcal{U}$-uniformly continuous pseudometrics on T, there exists a pseudometric d on T and a $p > 0$ such that $d(s,t) < 2p \Rightarrow (s,t) \in U$. By the well-ordering principle there is a well-ordering $\leqq$ on T. For each $n \in \underset{\sim}{N}$ we now define by transfinite induction a d-discrete family $(Z_{ns})_{s \in T}$ of U-small zero sets as follows:

$$Z_{ns} = \{ t \in T \mid d(s,t) \leqq p-p/n \quad \text{and} \quad d(Z_{nr},t) \geqq p/n \quad \text{for all} \quad r < s \}.$$

If u is the first element of T then $Z_{nu} \neq \emptyset$ for each $n \in \underset{\sim}{N}$; any other Z_{ns} may be empty.

The next stage of the proof consists of showing that some nonempty Z_{ns} is a (U-small) member of $z(M_p)$. To do this we introduce the sets

$$Z_n = \bigcup_{s \in T} Z_{ns} \qquad (n \in \underset{\sim}{N})$$

and show that some $Z_n \in z(M_p)$. We first contend that $\bigcup_n Z_n = T$.

For $t \in T$ let s be the first element of the set $\{ r \in T \mid d(r,t) < p\}$. Now choose $n \in \underset{\sim}{N}$ such that $d(s,t) \leqq p-p/n$. If $r < s$ then for any $q \in Z_{nr}$

$$d(q,t) \geqq d(r,t) - d(r,q) \geqq p - (p-p/n) = p/n$$

so

$$d(Z_{nr},t) \geqq p/n .$$

It follows that $t \in Z_{ns}$ and consequently that $T = \bigcup_n Z_n$. Next we show that some Z_n meets each member of $z(M_p)$. If not, then for each $n \in \mathbb{N}$ there is an $F_n \in z(M_p)$ such that $F_n \cap Z_n = \emptyset$. This leads to the following contradictory statements:

$$\bigcap_n F_n \in z(M_p)$$

but

$$\bigcap_n F_n = (\bigcap_n F_n) \cap T = (\bigcap_n F_n) \cap (\bigcup_m Z_m) = \emptyset$$

Since $z(M_p)$ is a z-ultrafilter it follows that the zero set Z_n belongs to $z(M_p)$.

To show that $Z_{ns} \in z(M_p)$ for this n and some $s \in T$, we introduce an auxiliary set $S \subset T$ to which the hypothesis may be applied. First single out the nonempty members Z_{nt} of Z_n; let the family of such sets denoted by $(Z_{n\mu})_{\mu \in \bar{M}}$. By the Axiom of Choice one element s_μ may be selected from each Z_μ and we may form the set $S = \{s_\mu | \mu \in M\}$. Since $(Z_{n\mu})_{\mu \in M}$ is d-discrete, S is a closed discrete subspace of T.

The family $(Z_{n\mu})_{\mu \in M}$ consists of disjoint sets so the map

$$f : \mathcal{P}(S) \to \mathcal{Z}$$

$$A \to \bigcup_{s_\mu \in A} Z_{ns_\mu}$$

from the power set $\mathcal{P}(S)$ of S into $\mathcal{Z}$, the zero sets of T, is injective. Since $z(M_p)$ is a z-ultrafilter, it now follows that

$$f^{-1}(z(M_p)) = \{A \subset S | f(A) \in z(M_p)\}$$

is an ultrafilter on S. Since $p \in \upsilon T$, $z(M_p)$ is stable under the formation of countable intersections, and it follows that $f^{-1}(z(M_p))$ too is stable under countable intersections. As such — since $|S|$ is not Ulam — the 0-1 measure associated with $f^{-1}(z(M_p))$ must be concentrated at some $s_\mu \in S$. In other words,

$$\{s_\mu\} \in f^{-1}(z(M_p)) .$$

Therefore

$$Z_{n\mu} = f(\{s_\mu\}) \in z(M_p) . \quad \nabla$$

Exercises 1

1.1 Complete Regularity and Algebraic Generality

For all topological spaces S there is a completely regular Hausdorff space T and a continuous surjection $f : S \to T$ such that the map $x \to x \cdot f$ is an isomorphism of $C(T,\underset{\sim}{R})$ onto $C(S,\underset{\sim}{R})$. To see this, first prove (a) and (b) below.

(a) If T is a Hausdorff space whose topology is the initial topology determined by some subfamily of $\underset{\sim}{R}^T$, then T is completely regular.

(b) If A is a subfamily of $C(T,\underset{\sim}{R})$ which induces T's topology, then a map f from a topological space S into T is continuous if and only if $x \cdot f$ is continuous for each $x \in A$.

Now define $s \sim t$ in S to mean that $x(s) = x(t)$ for all $x \in C(T,\underset{\sim}{R})$. Let T be the collection of all equivalence classes determined by this equivalence relation, and define $f : S \to T$ to be the canonical map sending $s \in S$ into the equivalence class $\bar{s}$ determined by s. For each $x \in C(S,\underset{\sim}{R})$ consider the map

$$\begin{aligned} x' : T &\to \underset{\sim}{R} \\ t &\to x(t) \end{aligned}$$

Let T carry the initial topology determined by the family A of all such maps x'. The continuity of f now follows from (b), the complete regularity of T from (a). Last, show that $A = C(T,\underset{\sim}{R})$.

1.2 Properties of βT

Let T be a completely regular Hausdorff space and βT be its Stone-Cech compactification. Then:

(a) If $T \subset S \subset \beta T$ then $\beta T = \beta S$.

(b) If S is a clopen (i.e. closed and open) subset of T then $cl_\beta S$ and $cl_\beta(T-S)$ are complementary open subsets of βT.

(c) An isolated point of T is isolated in βT.

(d) T is open in βT if and only if T is locally compact.

1.3 The Stone-Cech Compactification of $\underset{\sim}{N}$

With $\underset{\sim}{N}$ denoting the discrete space of natural numbers and $\beta\underset{\sim}{N}$ its Stone-Cech compactification we have:

(a) $|\beta\underset{\sim}{N}| = 2^{2^{|\underset{\sim}{N}|}}$, $|\ |$ denoting cardinality.

(b) If $M \subset \underset{\sim}{N}$ then M is open in $\beta\underset{\sim}{N}$.

(c) If E denotes the even positive integers, then βE, $\beta(\underset{\sim}{N}-E)$ and $\beta\underset{\sim}{N}$ are all homeomorphic.

(d) There is a homeomorphism $h : \beta \underset{\sim}{N} \rightarrow \beta \underset{\sim}{N}$ such that $h(\underset{\sim}{N}) = \underset{\sim}{N}$, h is its own inverse and $h(p) \neq p$ for each $p \in \beta\underset{\sim}{N} - \underset{\sim}{N}$.

1.4 Zero Sets

If I is any ideal of $C(T,\underset{\sim}{R})$ then $z(I) = \{z(x) \mid x \in I\}$ is a filterbase. For any point $p \in \beta T$ and $M_p = \{x \in C(T,\underset{\sim}{R}) \mid p \in cl_{\beta} z(x)\}$ as in the Gelfand-Kolmogorov theorem (Th. 1.4-1), show that $z(M_p) \rightarrow p$.

1.5 Rings of Integer-Valued Continuous Functions (Pierce 1961)

Here T is any topological space, $\underset{\sim}{Z}$ the discrete ring of integers, and $C(T,\underset{\sim}{Z})$ and $C_b(T,\underset{\sim}{Z})$ the rings of continuous and bounded continuous maps from T into $\underset{\sim}{Z}$ respectively. Some of the basic properties of $C(T,\underset{\sim}{Z})$ and $C_b(T,\underset{\sim}{Z})$ are set forth below.

(a) Adjoints. Let S be a topological space and $f : T \rightarrow S$ a continuous map. The map

$$f' : C(S,\underset{\sim}{Z}) \rightarrow C(T,\underset{\sim}{Z})$$
$$x \rightarrow x \cdot f$$

is the adjoint of f. The map f' is (i) a ring homomorphism, (ii) maps $C_b(S,\underset{\sim}{Z})$ into $C_b(T,\underset{\sim}{Z})$, and (iii) is injective if and only if S contains no nonempty clopen set disjoint from f(T).

(b) Analog of βT. There is a continuous map f of T into a compact totally disconnected (the component of any point is the point itself) Hausdorff space δT such that f' is an isomorphism of $C_b(\beta T,\underset{\sim}{Z})$ onto $C_b(T,\underset{\sim}{Z})$; moreover, up to homeomorphism, δT is the only compact totally disconnected space for which $C_b(T,\underset{\sim}{Z})$ is isomorphic to $C_b(\beta T,\underset{\sim}{Z})$. [Hint: Given a subset S of $C(T,\underset{\sim}{Z})$, let P_S denote the topological product of the discrete spaces $x(T)$, $x \in S$. Let f_S be the map sending $t \in T$ into $(x(t))_{x \in S} \in P_S$ and note that f_S is continuous. Let $S = C_b(T,\underset{\sim}{Z})$ and define δT to be the closure of $f_S(T)$ in P_S. Since $x(T)$ is finite for each $x \in S$, P_S must be compact. As for the uniqueness, let $B(T)$ and $B(\delta T)$ denote the Boolean algebras of clopen subsets of T and δT respectively. Since the isomorphism f' maps idempotents into idempotents, the Boolean algebras $B(T)$ and $B(\delta T)$ are isomorphic. By a result of Stone (1937), since δT is compact and totally disconnected, δT must be homeomorphic to the Boolean space of $B(\delta T)$.]

(c) T in δT. The natural injection of T into δT is 1-1 if and only if the clopen subsets of T separate points. It is bicontinuous if and only if T is a 0-dimensional T_1 space.

(d) βT Versus δT. If T is a completely regular Hausdorff space then δT is homeomorphic to βT if and only if each pair of subsets of T which are separated by $C(T,\underset{\sim}{R})$ are separated by a clopen subset of T.

1.6 Stone's Theorem: Rings A for Which $C_b(T,A)/M \simeq A$

If T is a completely regular Hausdorff space and M a maximal ideal in $C_b(T,\underset{\sim}{R})$, then $C_b(T,\underset{\sim}{R})/M$ is isomorphic to $\underset{\sim}{R}$ as follows readily from the results of Section 1.4. Due to its appearance in Stone 1937 (Theorem 76), we refer to this result as Stone's Theorem. If $\underset{\sim}{R}$ is replaced by the complex numbers or the quaternions, Stone's theorem still holds, as remarked in Section 1.4, but what if $\underset{\sim}{R}$ is replaced by a topological division ring A? I.e. after introducing a notion of boundedness in A, does Stone's theorem still hold in $C_b(T,A)$? The answer is no and necessary and sufficient conditions on A (essentially local compactness) for Stone's theorem to hold are developed in (b) below.

(a) Boundedness. Analogous to the way boundedness is defined in topological vector spaces, a subset S of a topological ring A is bounded if for any neighborhood W of 0 in A there are neighborhoods U and V of 0 such that $US \subset W$ and $SV \subset W$. If A is a metric ring, metric boundedness implies this notion of boundedness. If A is a valued field, the two notions coincide. The closure of a bounded set is bounded, as is any convergent sequence bounded. If $x_n \to 0$ and (y_n) is bounded, then $\lim_n x_n y_n = \lim_n y_n x_n = 0$. The collection of all continuous maps from a topological space T into A whose range is a bounded subset of A is denoted by $C_b(T,A)$. Division rings A where subsets H which are bounded away from 0 (i.e. for some neighborhood V of 0, $H \cap V = \emptyset$) must be such that H^{-1} is bounded are said to be of type V.

(b) Stone Rings. (Kowalsky 1955, Theorems 4 and 5) A topological division ring K is a Stone ring if, given any topological space T, $C_b(T,K)/M$ is isomorphic to K for any maximal ideal M of $C_b(T,K)$. A Stone ring of type V must be locally compact, and any locally compact topological division ring is a Stone ring. Other approaches to this result appear in Goldhaber and Wolk 1954 and Correl and Henriksen 1956; among other things the latter paper makes some corrections to Goldhaber and Wolk's results. The question of which rings are Stone rings was first raised in Kaplansky 1947, page 183; this paper also deals with other results about rings of ring-valued functions.

(c) Functions with Relatively Compact Range. (Correl and Henriksen 1956) Let T be a completely regular Hausdorff space, let K be a totally disconnected topological division ring, and let $C_c(T,K)$ denote the collection of all functions in $C(T,K)$ with relatively compact range in K. Then, for any maximal ideal M of $C_c(T,K)$, $C_c(T,K)/M$ is isomorphic to K.

(d) Integer-Valued Continuous Functions. (Pierce 1961, page 381, Cor. 3.2.4) Let T be any topological space, $\underset{\sim}{Z}$ the discrete space of integers and $C(T,\underset{\sim}{Z})$ the ring of continuous maps from T into $\underset{\sim}{Z}$. For any maximal ideal M in $C(T,\underset{\sim}{Z})$, $C(T,\underset{\sim}{Z})$ is isomorphic to the integers modulo p for some prime p.

1.7 A Construction of βT and υT Using Extensions (Chandler 1972)

The spaces βT and υT are C_b- and C-extensions of the completely regular Hausdorff space T (i.e. T is dense in βT and υT and functions from $C_b(T,\underset{\sim}{R})$ and $C(T,\underset{\sim}{R})$ can be continuously extended to βT and υT respectively). Moreover, as will be seen from what follows, βT and υT are the "largest" C_b- and C-extensions of T in the sense that each is a quotient of the disjoint union of all the C_b- or C-extensions of T.

Since each C_b- or C-extension of T has cardinality smaller than or equal to $2^{2^{|T|}}$, each such space can be viewed as a subspace of the power set $P^2(T) = P(P(T))$ with an appropriate Hausdorff topology. We obtain βT and υT by taking the disjoint union of all C_b- and C-extensions, respectively, of T from $P^2(T)$ and then (to insure that the resulting spaces will be Hausdorff, among other things) identifying points for which all the extensions agree: specifically, for C-extensions S_α of T, if Y is the union of all such S_α, s, t are elements of Y, define $s \sim t$ if $x^\alpha(s) = x^\beta(t)$ for each x in $C(T,\underset{\sim}{R})$ where x^α and x^β are the appropriate continuous extensions of x; then take υT to be the set of all such equivalence classes. For βT use the same equivalence relation on the class of all C_b-extensions. To each $x \in C(T,\underset{\sim}{R})$ there corresponds a unique $x^\upsilon : \upsilon T \to \underset{\sim}{R}$ defined by $x^\upsilon(\bar{t}_\alpha) = x^\alpha(t_\alpha)$. Let υT carry the initial topology determined by all the x^υ; βT is topologized similarly.

(a) βT and υT are C_b- and C-extensions of T respectively.

(b) βT and υT are completely regular Hausdorff spaces.

(c) βT is compact. (It suffices to show that each maximal ideal of $C_b(T,\underset{\sim}{R})$ is fixed at a point of βT; cf. Theorem 1.4-1.)

(d) υT is replete. (Assume the existence of a proper C-extension of T.)

1.8 Compactifications and Semicontinuous Functions

(Nielsen and Sloyer 1970)

As shown in Section 1.4 the Stone-Cech compactification of a completely regular Hausdorff space T is given by the collection of all maximal ideals of $C(T,\underset{\sim}{R})$. For T_1-spaces T a similar sort of thing can be done — i.e. a compactification μT can be obtained — using semicontinuous functions, rather than continuous ones. A lower semicontinuous function $x : T \to \underset{\sim}{R}$ is one for which $x^{-1}(a,\infty)$ is open for each $a \in \underset{\sim}{R}$. The collection L(T) of nonnegative lower semicontinuous functions with pointwise operations is not algebra however. Additively and multiplicatively it is a semigroup with identity; it is closed under multiplication by nonnegative scalars.

(a) Ideals and Filters. A subset I of L(T) is an ideal if $I + I \subset I$, $L(T)I \subset I$, and for each $x \in I$ there is an open set U such that $k_U x = x$ where k_U denotes the characteristic function of U. Letting c(T) denote the closed subsets of T, filters from c(T) are called c-filters. Under these conventions the map $x \to z(x)$ maps ideals of L(T) into c-filters.

(b) Fixed Ideals. An ideal I in L(T) is fixed if $\cap z(I) \neq \emptyset$; otherwise I is free. The fixed maximal ideals of L(T) are precisely the sets $I_t = \{x \in L(T) \mid x(t) = 0\}$. If T is compact, then every ideal is fixed.

(c) The Stone Topology. A base of open sets for the Stone topology $\mathcal{T}$ on the space $\mathcal{M}(T)$ of maximal ideals of L(T) is given by sets of the form $U(x) = \{M \in \mathcal{M}(T) \mid x \notin M\}$ as x runs through L(T). $(\mathcal{M}(T),\mathcal{T})$ is homeomorphic to T if T is compact; consequently, for compact T_1-spaces S and T, L(T) is isomorphic to L(S) if and only if S and T are homeomorphic. (The map $t \to I_t$ takes T injectively — since T is T_1 — into $\mathcal{M}(T)$.)

(d) Compactification. The map μ sending t into I_t embeds T homeomorphically as a dense subset of $\mathcal{M}(T)$. $\mathcal{M}(T)$ moreover is compact in its Stone topology.

1.9 Generalized Stone-Cech Compactification (Bachman, Beckenstein, Narici, and Warner 1975; Narici, Beckenstein, and Bachman 1971)

Let S be a Hausdorff space containing at least two points. For any topological space T, let $C_c(T,S)$ denote the set of all continuous maps of T into S with relatively compact range. (If S is a locally compact field, $C_c(T,S)$ is just the set of all bounded continuous functions.) A Hausdorff space T is S-completely regular if the initial topology determined by $C_c(T,S)$ on T is T's original topology. Thus in this terminology the completely regular Hausdorff spaces are precisely the $\underset{\sim}{R}$-completely regular spaces. The Hausdorff space T is <u>ultraregular</u> (or zero-dimensional) if there is a neighborhood base at each point consisting of clopen sets. (Any ultrametric space — i.e. any metric space in which $d(a,b) \leq \max(d(a,c),d(c,b))$ holds instead of the triangle inequality — is certainly ultraregular.)

(a) For any ultraregular Hausdorff space S containing at least two points, T is ultraregular if and only if T is S-completely regular.

(b) If T is S-completely regular then the initial uniformity $\mathcal{C}_c(T,S)$ determined by $C_c(T,S)$ on T is compatible with T's original topology. (The uniform spaces into which the functions $x \in C_c(T,S)$ map T are the compact spaces $\overline{x(T)}$.) The $\mathcal{C}_c(T,S)$-completion $\beta_S T$ of T is a compactification of T and each $x \in C_c(T,S)$ may be continuously extended to $\beta_S T$, the range of the extension being $\overline{x(T)}$. Moreover each continuous function from T into a compact S-completely regular space W has a continuous extension to $\beta_S T$.

(c) T is S-completely regular if and only if T is homeomorphic to a relatively compact subset of the cartesian product S^A for some index set A. Consequently cartesian products of S-completely regular spaces are S-completely regular.

(d) Let $\mathcal{G}$ be the algebra of clopen subsets of the ultraregular space T and let μ be a finitely additive measure mapping $\mathcal{G}$ into $\{0,1\}$, addition being performed modulo 2 in $\{0,1\}$. In addition assume that if $S,S' \in \mathcal{G}$, $S \supset S'$ and $\mu(S) = 0$, then $\mu(S') = 0$. (This last property would automatically be fulfilled if addition were not performed modulo 2 — if, say, $\{0,1\}$ were a subset of a field whose characteristic was not 2.) Let $M_o(T)$ (or just M_o) denote the collection of all such 0-1 "measures" endowed with the topology generated by the sets $V(\mu_o,S_1,\ldots,S_n) = \{\mu \in M_o \mid \mu_o(S_j) = \mu(S_j),\ j=1,\ldots,n\}$. $M_o(T)$ is also a compactification of T and, for T and F ultraregular, $\beta_F T$ is homeomorphic to $M_o(T)$.

(e) If T is an ultraregular space and F an ultraregular topological field, $C_c(T,F)$ is an F-algebra with respect to the usual pointwise operations. <u>Characters</u> of F-algebras are nontrivial homomorphisms of the algebra into F. Let $H_c(T,F)$ denote the characters of $C_c(T,F)$ endowed with the weakest topology with respect to which each of the maps $h \to h(x)$, $h \in H_c(T,F)$, $x \in C_c(T,F)$, is continuous. Show that $H_c(T,F)$ is homeomorphic to $\beta_F T$. (Cf. (1.6-2).)

(f) A <u>clopen partition</u> of a space T is a pairwise disjoint cover of consisting of clopen sets. Letting $\mathcal{U}$ denote the collection of all finite clopen partitions of the ultraregular space T, the sets $\bigcup(U_i \times U_i)$, $(U_i) \in \mathcal{U}$, form a base of entourages for a compatible uniformity for T. The completion $\beta_o T$ of T with respect to this uniform structure is a compactification of T which coincides (is homeomorphic) with $\beta_S T$ whenever S is ultraregular. Thus it is superfluous to write $\beta_S T$ in the class of ultraregular spaces S.

(g) A Hausdorff space T is <u>ultranormal</u> if, given any two disjoint closed subsets C and D, there is a clopen set U containing C and disjoint from D. If T is ultranormal, S is any Hausdorff space containing at least two points, then T is S-completely regular and $\mathcal{C}_c(T',S) = \mathcal{C}_c(T,\underset{\sim}{R})$. Consequently, for ultranormal T, $\beta T = \beta_S T$ for any Hausdorff space S containing at least two points.

1.10 <u>The Banaschewski Compactification</u> (Banaschewski 1955).

Let T and F be ultraregular spaces. For each cover $\mathcal{U}$ of T let $W(\mathcal{U}) = \bigcup_{U \in \mathcal{U}} U \times U$. The sets $\{W(\mathcal{U}) \mid \mathcal{U}$ is a finite open cover of T$\}$ form a fundamental system of entourages for a compatible uniformity $\mathcal{B}(T)$ on T. $\mathcal{B}(T)$ is precompact since each finite open cover of T consists of $W(\mathcal{U})$-small sets, so the $\mathcal{B}(T)$-completion $\beta_b T$ of T is a compactification of T. We refer to $\beta_b T$ as the <u>Banaschewski compactification of T</u>. If T is ultranormal then $\mathcal{C}_c(T,\underset{\sim}{R}) = \mathcal{B}(T)$ (where $\mathcal{C}_c(T,\underset{\sim}{R})$ is the initial uniformity determined by the class $C_c(T,\underset{\sim}{R})$ of continuous maps from T into R having relatively compact range in $\underset{\sim}{R}$) so $\beta T = \beta_b T$. (In an ultranormal space every open cover is refined by a finite open cover.)

For any ultraregular space F, $\mathcal{C}_c(T,F) = \mathcal{B}(T)$ so, with $\beta_F T$ as in Exercise 1.9, $\beta_F T = \beta_b T$ for any ultraregular space F containing at least two points. Moreover, as noted in Exercise 1.9 (f), this is another way to see that the "F" in $\beta_F T$ is superfluous in the class of ultraregular spaces F.

Another way to get $\beta_b T$ is as the $\mathcal{C}_c(T,\underset{\sim}{Z})$-completion of T where $\underset{\sim}{Z}$ denotes the discrete space of integers (Pierce 1961). Still another way is as the "E-compactification of T" (Exercise 1.11) where E is a two-point set.

1.11 <u>E-Complete Regularity and E-Compactness</u> (Mrowka 1956 and Engelking and Mrowka 1958)

Let T and E be Hausdorff space, let $C(T,E^n)$ denote the class of all continuous maps from T into the n-fold cartesian product E^n, and let $(T,E) = \bigcup_{n \in \underset{\sim}{N}} C(T,E^n)$. The space T is <u>E-completely regular</u> if for any closed subset A of T and point $t \notin A$ there is some $x \in (T,E)$ such that $x(t) \notin \text{cl}\, x(A)$.

(a) E-Complete Regularity. (Mrowka 1956) E-complete regularity is hereditary and productive; T is E completely regular if and only if T can be embedded in some cartesian product E^B for some index set B.

(b) E-Compactness. (Engelking and Mrowka 1958) An E-completely regular space T is E-compact if there is no E-completely regular space S which contains T as a dense subspace and is such that each $x \in C(T,E)$ may be continuously extended to some $x \in C(S,E)$. T is E-compact if and only if T is homeomorphic to a closed subspace of a cartesian product E^B for some index set B. Thus closed subspaces and cartesian products of E-compact spaces are E-compact.

(i) With the interval [0,1] carrying its usual topology, a space T is [0,1]-compact if and only if T is compact.

(ii) T is $\underset{\sim}{R}$-compact if and only if T is replete.

(iii) T is $\{0,1\}$-compact if and only if T is 0-dimensional (in the sense that there is a base of clopen sets for the topology on T) and compact.

(iv) With $\underset{\sim}{N}$ carrying the discrete topology, T is $\underset{\sim}{N}$-compact if and only if T is 0-dimensional and replete.

(c) E-Compactification. (Engelking and Mrowka 1958, Theorem 4) If T is E-completely regular then there is an E-completely regular extension $\beta_E T$ of T such that (i) $\beta_E T$ is E-compact and contains T as a dense subset and (ii) each $x \in C(T,E)$ possesses an extension to some $x \in C(\beta_E T,E)$.

If S is any E-compact space, then each $x \in C(T,S)$ possesses an extension $x \in C(\beta_E T,S)$.

The extension $\beta_E T$ is uniquely determined by (i) and (ii) in the sense that if X is any extension of T satisfying (i) and (ii) then there is a homeomorphism h of $\beta_E T$ onto X such that $h(t) = t$ for each $t \in T$.

1.12 Repleteness and the Repletion

Throughout this exercise T denotes a completely regular Hausdorff space.

(a) Properties of υT. (i) If X is replete then each continuous function $x: T \to X$ has a continuous extension to υT. (ii) If a sequence (Z_n) of zero sets of T has empty intersection then so does $(\text{cl}_\upsilon Z_n)$ in υT. (iii) If (Z_n) is a sequence of zero sets in T then $\text{cl}_\upsilon(\bigcap_n Z_n) = \bigcap_n \text{cl}_\upsilon(Z_n)$.

If X is a replete space containing a dense homeomorphic image of T and satisfying one of the properties given above, then there is a homeomorphism h taking X onto T which reduces to the identity on T.

(b) Intersections of Replete Spaces. An arbitrary intersection of replete subspaces of a given space is replete.

(c) Continuous Inverse Images of Replete Spaces. If f is a continuous function taking the replete space T into the completely regular Hausdorff space X and F is a replete subspace of X then $f^{-1}(F)$ is replete. (Note that not every open subset of a replete space is replete as evidenced by $[0,\Omega)$ in $[0,\Omega]$ with the order topology.) Furthermore if every subspace of a replete space is replete, then all the one-point subsets of the space are G_δ-sets (cf. (g)).

(d) Unions of Replete Spaces. In any completely regular space, the union of a replete subspace with a compact subspace is replete. On the other hand, it is not necessarily the case that the union of two replete subspaces is replete (Mrowka 1958).

(e) Ultrafilters and Replete Spaces. T is replete if and only if each ultrafilter $\mathcal{F}$ on T for which $\mathcal{F} \cap \mathcal{Z}$ ($\mathcal{Z}$ denotes the zero sets of T) has the countable intersection property converges to a point of T.

(f) Continuous Functions with Compact Support on a Replete Space. Let $C_K(T,\underset{\sim}{R})$ denote the collection of all continuous functions on T which vanish outside some compact subset of T, the subset depending on the particular function. If T is replete then $C_K(T,\underset{\sim}{R}) = \bigcap_{p \in \beta T - T} M_p$ where M_p is as in the Gelfand-Kolmogorov theorem (Th. 1.4-1).

(g) When are all Subspaces Replete? Each subspace of T is replete if and only if (i) given a completely regular Hausdorff space X for which a continuous map $f:X \to T$ exists such that $f^{-1}(\{t\})$ is compact for each t in T then X must be replete or (ii) each 1-1 continuous image of T is replete or (iii) $T-\{t\}$ is replete for each t in T.

1.13 A Generalized Repletion (Bachman, Beckenstein, Narici and Warner 1975)

Let F denote a complete Hausdorff space containing at least two points, and let T be an F-completely regular topological space as defined in the preceding exercise. Let $\mathcal{C}(T,F)$ denote the initial uniformity determined by the continuous functions C(T,F) on T. The completion $\upsilon_F T$ of T with respect to $\mathcal{C}(T,F)$ is called the F-repletion of T and has a number of properties in common with the repletion υT of a completely regular (or $\underset{\sim}{R}$-completely regular in this context) Hausdorff space T. T is <u>F-replete</u> if $T = \upsilon_F T$. For example, $\upsilon_F T$ is the largest subspace of $\beta_F T$ (see preceding exercise) to which each $x \in C(T,F)$ may be continuously extended and this property characterizes T up to homeomorphism in the following sense: If T is a dense subset of an F-completely regular space S such that each $x \in C(T,F)$ may be continuously extended to S, then there is a unique homeomorphism h mapping $\upsilon_F T$ onto $\upsilon_F S$ such that $h(t) = t$ for all $t \in T$. In fact $\upsilon_F T$ also admits the following characterization.

(a) According to part (b) of the preceding exercise, each $x \in C(T,F)$ has a unique continuous extension to $\hat{x} \in C(\beta_F T, \beta_F F)$. In this notation

$$\upsilon_F T = \left\{ t \in \beta_F T \,\middle|\, \hat{x}(t) \in F \quad \text{for all} \quad x \in C(T,F) \right\}.$$

(b) If F is F-completely regular, then F is F-replete. Thus $\underset{\sim}{R}$ is $\underset{\sim}{R}$-replete or, simply, replete.

(c) If T is F-completely regular then T is F-replete if and only if T is homeomorphic to a closed subset of the cartesian product F^B for some index set B. Consequently cartesian products and closed subspaces of F-replete spaces are F-replete.

(d) If F is replete then every F-replete space T is replete.

(e) In analogy with the characterization of υT given in Theorem 1.5-1 (b), referring to the space $\beta_o T$ defined in part (f) of Exercise 1.9 , we single out the subspace $\upsilon_o T$ of $\beta_o T$ when T is ultraregular consisting of those points $t \in \beta_o T$ such that for each sequence (V_n) of neighborhoods of t in $\beta_o T$, $(\bigcap V_n) \bigcap T \neq \emptyset$. If T is ultranormal then $\upsilon_o T = \upsilon_F T$. Otherwise we have (f) and (g) below.

(f) If T is ultraregular and F is replete and ultranormal then $\upsilon_o T = \upsilon_F T$.

(g) If the idempotent entourages (V such that $V \circ V = V$) form a base for F's uniformity — as happens for example when F is ultrametrizable — and F is not compact, then $\upsilon_F T \subset \upsilon_o T$.

(h) If T is an ultraregular Lindelöf space then $T = \upsilon_o T$ (cf. the analogous result for υT in Th. 1.5-3). If the idempotent entourages form a base for F's uniformity and F is not compact, then every ultraregular Lindelöf space T is F-replete, i.e. $T = \upsilon_F T$.

(i) With reference to part (d) of the preceding exercise, for T ultraregular and F an ultraregular topological field, $\upsilon_o T$ is homeomorphic to the subcollection $M_\sigma(T)$ of countably additive members of $M_o(T)$.

(j) Let F denote a complete ultraregular topological field and let T be ultraregular, and let $\hat{x}$ denote the continuous extension of $x \in C(T,F)$ to $\upsilon_F T$. For each $p \in \upsilon_F T$, let $\hat{p}$ denote the map

$$\hat{p} : C(T,F) \to F$$
$$x \to \hat{x}(p)$$

The map $p \to \hat{p}$ is a homeomorphism from $\upsilon_F T$ onto the space $H(T,F)$ of all characters (nontrivial homomorphisms into F) of $C(T,F)$ endowed with the weakest topology with respect to which the maps

$$H(T,F) \to F$$
$$h \to h(x)$$

are continuous for each $x \in C(T,F)$. (Cf. (1.6-2).)

(k) If F is a complete nonarchimedean valued field with nontrivial valuation, then F is replete if and only if F has non-Ulam cardinality.

1.14 Pseudocompactness

In any topological space T, the following conditions are equivalent:

(a) T is pseudocompact.

(b) The continuous image of T, in any topological space S, is pseudocompact.

(c) The continuous image of T in any T_5 space (sets A and B such that neither meets the other's closure may be separated by open sets) is countably compact.

(d) The continuous image of T in any metric space is compact.

(e) Every real-valued continuous function x on T attains the values inf x(T) and sup x(T).

(f) Every countable collection of nonempty zero sets in T has nonempty intersection.

(g) Every locally finite collection $\mathcal{U}$ of complements of zero sets of T — i.e. such that each point of T has a neighborhood which meets only finitely many members of $\mathcal{U}$ — is finite.

In the class of weakly normal Hausdorff spaces — Hausdorff spaces in which closed sets disjoint from countable closed sets may be separated by open sets — pseudocompactness and countable compactness coincide. In the class of completely regular Hausdorff spaces, pseudocompactness is equivalent to the uniform closure of each ideal of $C(T,\underset{\sim}{R})$ again being an ideal. (Nanzetta and Plank 1972)

The product of pseudocompact spaces need not be pseudocompact. In fact (Comfort 1967) if (T_n) is a countable family of completely regular spaces, ΠT_n is not necessarily pseudocompact even tho the product of any finite number of them is. Under additional hypotheses such as sequential compactness for example on the factors, pseudocompactness of product can be guaranteed.

Clearly not every subspace of a pseudocompact space need be pseudocompact (consider (0,1) in [0,1] for example). If T is a completely regular pseudocompact space, then the closure of every open subset of T is pseudocompact. Cf. also Stephensen 1968, Glicksberg 1959, and Bagley, Connell and McKnight 1958.

1.15 Products

Let S and T be infinite completely regular Hausdorff spaces.

(a) Stone-Cech Compactifications. $\beta(S\times T) = \beta S\times\beta T$ if and only if $S\times T$ is pseudocompact (first proved in Glicksberg 1959, reproved in Frolik 1960).

(b) Repletions. So far it is known (Comfort 1968) that the following conditions suffice for $\upsilon(S\times T)$ to $= \upsilon S\times\upsilon T$.

(i) $S\times T$ is not of Ulam cardinal, T is a k-space (each subset of T which has a closed intersection with each compact set must itself be closed (Section 2.3)), υS is locally compact.

(ii) $S\times T$ is not of Ulam cardinal, T is a k-space and S is pseudocompact.

(iii) $S\times T$ is not of Ulam cardinal and $\upsilon S\times T$ and $\upsilon S\times\upsilon T$ are k-spaces.

(iv) Each compact subset of T and each pseudocompact subset of υS is not of Ulam cardinal, T is a k-space and υS is locally compact.

1.16 Connectedness and the Stone-Cech Compactification

Let T be a completely regular Hausdorff space.

(a) An open subset U of βT is connected if and only if $U \cap T$ is connected. (Easy to show $U \cap T$ to be disconnected if U is. If $U \cap T$ is disconnected there are nonempty disjoint open subsets V and W of T such that $U \cap T = V \cup W$. Now $cl_\beta(U\cap T) = cl_\beta V \cup cl_\beta W \supset U$ and if $cl_\beta V \cap cl_\beta W = \emptyset$ then $U = (cl_\beta V\cap U) \cup (cl_\beta W\cap U)$ is disconnected and the proof is complete. If $p \in cl_\beta V \cap cl_\beta W$, choose $x \in C(\beta T,\underset{\sim}{R})$ such that $x(p) = 0$ and $x(\beta T-U) = 1$. Define y on T by taking $x(t) = y(t)$ except for $t \in W$ and $y(t) < 1/2$; let $y(t) = 1/2$ otherwise. The continuous extension y^β of $y \in C_b(T,\underset{\sim}{R})$ to βT coincides with x on $cl_\beta V$ so $y^\beta(p) = 0$, yet $y^\beta \geq 1/2$ on $cl_\beta W$. This contradiction shows U to be disconnected.)

(b) At each $t \in T$, βT is locally connected at t if and only if T is locally connected at t. (Use (a) and the fact that a family (V_μ) of open neighborhoods of $t \in T$ in βT is a base of open neighborhoods at t in βT if and only if $(V_\mu \cap T)$ is a base of open neighborhoods of t in T.)

The next result serves as a lemma to (d).

(c) υT is not locally connected at any point not in υT. (For $p \in \beta T - \upsilon T$ there exists $x \in C(T,\underset{\sim}{R})$ such that $\hat{x}(p) = \infty$ where $\hat{x}$ denotes the continuous extension of x to $C(\beta T, \underset{\sim}{R} \cup \{\infty\})$ mentioned in (1.5-1). For $i = 0,1,2,3$, let Z_i be the set of all $t \in T$ such that $n \leq x(t) \leq n+1$ for some integer $n \equiv i \pmod 4$. The sets Z_i cover T so p is in one of their closures in βT, $p \in cl_\beta Z_i$ say. Now $p \notin cl_\beta Z_3$ since Z_1 and Z_3 are completely separated, so there must be a neighborhood U of p disjoint from Z_3. If βT is locally connected at p, then U contains a connected open neighborhood U' of p. By (a), $U' \cap T$ is connected, so there must be some integer n such that $4n \leq x(t) \leq 4n+3$ for all $t \in U' \cap T$, contradicting the fact that $\hat{x}(p) = \infty$.)

(d) βT is not locally connected unless T is locally connected and pseudocompact. (Recall that T is pseudocompact if and only if $\beta T = \upsilon T$ ((1.5-4)).)

Rather than give hints to the remaining parts, we simply list the results and refer to Henriksen and Isbell 1957 for proofs.

(e) Let uT denote the completion of T with respect to the finest compatible uniform structure on T. Clearly $uT \subset \upsilon T$. T is locally connected if and only if uT is locally connected.

(f) βT is locally connected if and only if T is locally connected and pseudocompact.

(g) βT is locally connected if and only if every (completely regular Hausdorff) space S containing T as a dense subspace is locally connected.

TWO

Topological Vector Spaces of Continuous Functions

IN CHAPTER ONE exclusive attention was devoted to the algebraic properties of the algebra $C(T,\underset{\sim}{R})$ of real-valued continuous functions on a completely regular Hausdorff space T, with special focus directed at how these algebraic properties were the progeny of topological properties of T. Here we switch to F-valued functions on T, $\underset{\sim}{F}=\underset{\sim}{R}$ or $\underset{\sim}{C}$, and endow $C(T,\underset{\sim}{F})$ with the compact-open topology, the weakest topology for $C(T,\underset{\sim}{F})$ with respect to which each of the seminorms $p_K(x)=\sup|x(K)|$, K a compact subset of T, is continuous. $C(T,\underset{\sim}{F})$ so topologized is denoted by $C(T,\underset{\sim}{F},c)$ and is a locally convex Hausdorff topological vector space; some brief discussion of its most fundamental properties is given in Chapter 0. The point of this chapter is the development of the interaction between topological properties of T and topological vector space properties of $C(T,\underset{\sim}{F},c)$.

Since this is the <u>compact</u>-open topology, it is to be expected that the structure and abundance of the compact subsets of T will play a central role in determining certain properties of the topological vector space $C(T,\underset{\sim}{F},c)$. This is very much the case in characterizing completeness and metrizability of $C(T,\underset{\sim}{F},c)$, as shown in Secs. 1.1 and 1.2 where necessary and sufficient conditions on T for $C(T,\underset{\sim}{F},c)$ to be metrizable and complete, respectively, are obtained. Tho the structure of the compact subsets of T figures less prominently in subsequent results, they still play a significant role.

A way of representing the elements of the continuous dual $C(T,\underset{\sim}{F},c)'$ of $C(T,\underset{\sim}{F},c)$ as integrals with respect to certain Borel measures is obtained in Sec. 2.4. The representation is used in developing the idea of support of a continuous linear functional on $C(T,\underset{\sim}{F},c)$ — a compact subset K of T related to the continuous linear functional x' in such a way that if the continuous function x vanishes on K, x' must vanish on x — which is used to develop necessary and sufficient conditions on T for $C(T,\underset{\sim}{F},c)$ to be barreled.

Among the other results obtained are necessary and sufficient conditions on T for $C(T,\underset{\sim}{F},c)$ to be infrabarreled, bornological, and separable. These characterizations enable us to give an example (Example 2.6-1) of a barreled space which is not bornological. In Sec. 2.8 such things as when $C(T,\underset{\sim}{F},c)$ is Montel or reflexive are considered. Both of these in particular have a very simple answer: T must be discrete.

2.1 Metrizability of $C(T,\underset{\sim}{F},c)$ and Hemicompactness

Metrizability and completeness of $C(T,\underset{\sim}{F},c)$, the algebra of continuous functions mapping the completely regular Hausdorff space T into $\underset{\sim}{F}$ ($=\underset{\sim}{R}$ or $\underset{\sim}{C}$) with compact-open topology, naturally enough have to do with the abundance and structure of the compact sets in T. If T is compact, for example, $C(T,\underset{\sim}{F},c)$ is complete and metrizable: it is a Banach algebra in this case. More generally completeness and metrizability of $C(T,\underset{\sim}{F},c)$ are equivalent to T being a k_R-space and T being hemicompact respectively.

Definition 2.1-1. HEMICOMPACTNESS. A topological space T is hemicompact if there is a countable family (K_n) of compact subsets of T such that each compact subset of T is contained in some K_n.

Since one-point sets are always compact, the sets (K_n) must cover T if T is hemicompact.

Hemicompactness is clearly stronger than σ-compactness which merely requires that the space be a countable union of compact sets. In the presence of local compactness however, σ-compactness is enough to guarantee hemicompactness. Indeed suppose that $T=\bigcup_n K_n$, where each K_n is compact. By covering K_1 with relatively compact open sets and using the compactness of K_1, a relatively compact open set U_1 is seen to exist which contains K_1. Similarly a relatively compact U_2 exists containing $\text{cl } U_1 \cup K_2$. Thus, by induction, a countable family (U_n) of relatively compact open sets exists such that $\text{cl } U_{n-1} \subset U_n$ for each n and $T=\bigcup_n U_n$. Now if K is an arbitrary compact set, there is a finite collection $U_1,\ldots,U_k$ which covers K and it follows that $K \subset \bigcup_{i=1}^k \text{cl } U_i = \text{cl } U_k$. Since $T=\bigcup_n \text{cl } U_n$, T is hemicompact. Thus a fortiori any locally compact second countable space is hemicompact.

Lest it be thought that local compactness is unduly strong for the purpose of forcing a σ-compact space to be hemicompact, we note that in a first countable Hausdorff space, hemicompactness implies local compactness as we now show. To this end let (K_n) be an increasing cover of compact subsets of T such that each compact subset of T is a subset of some K_n and let (V_n) be a decreasing base of open neighborhoods of a point $t \in T$. If no $\text{cl } V_n$ is compact, then $V_n \not\subset K_n$ for any n. For $t_n \in V_n - K_n$, then $t_n \to t$ and $\{t_n\} \cup \{t\}$ is compact. Thus $\{t_n\} \cup \{t\} \subset K_m$ for some m which contradicts the way in which t_m was chosen. Therefore T is locally compact.

The space $\underset{\sim}{Q}$ of rational numbers with Euclidean topology is clearly not locally compact. As $\underset{\sim}{Q}$ is metrizable, it is first countable; thus, as a result of what we have just shown, it is not hemicompact. Hence $\underset{\sim}{Q}$ is a σ-compact space which is not hemicompact. Some of these conclusions are listed in (2.1-1) below.

<u>(2.1-1) LOCAL COMPACTNESS AND HEMICOMPACTNESS</u> (a) A locally compact σ-compact space is hemicompact. (b) A hemicompact first countable Hausdorff space is locally compact.

<u>Theorem 2.1-1. METRIZABILITY OF $C(T,\underset{\sim}{F},c)$.</u> $C(T,\underset{\sim}{F},c)$ is metrizable iff the completely regular Hausdorff space T is hemicompact.

<u>Proof</u> Suppose that T is hemicompact and let (K_n) be a family of compact subsets of T such that each compact subset K of T is a subset of one of them. For $p_K(x)=\sup|x(K)|$, $x \in C(T,\underset{\sim}{F},)$, then $p_K(x) \le p_{K_n}(x)$ for some n. It follows that the compact-open topology is generated by the countable family (p_{K_n}) of seminorms and is therefore metrizable.

Conversely if $C(T,\underset{\sim}{F},c)$ is metrizable, it must have a countable neighborhood base at 0 of the form $a_nV_{p_{K_n}} = a_n\{x \in C(T,\underset{\sim}{F}) \mid p_{K_n}(x) < 1\}$ where (a_n) is a sequence of positive numbers and the K_n are compact subsets of T. Thus for each compact set K of T there is an integer n such that $a_nV_{p_{K_n}} \subset V_{p_K}$. Now suppose that $K \not\subset K_n$, i.e. that there is a $t \in K-K_n$. Because T is a completely regular Hausdorff space, there is an $x \in C(T,\underset{\sim}{F})$ such that $x(t)=1$ and $x(K_n)=\{0\}$. Hence $x \in a_nV_{p_{K_n}}$ while $x \notin V_{p_K}$. This contradiction implies that $K \subset K_n$ and T is seen to be hemicompact.∇

2.2 <u>Completeness of $C(T,\underset{\sim}{F},c)$ and k_R-Spaces</u>

In this section we show that $C(T,\underset{\sim}{F},c)$ is complete iff T is a "k_R-space." Other facts about k_R-spaces in general, as well as the related notion of "k-space", are discussed in the next section.

For $C(T,\underset{\sim}{F},c)$ to be complete, each Cauchy filterbase must converge. A filterbase $\mathcal{B}$ of functions is Cauchy in $C(T,\underset{\sim}{F},c)$ iff $\mathcal{B}|_K=\{B|_K \mid B \in \mathcal{B}\}$, where $B|_K=\{x|_K \mid x \in B\}$, is a Cauchy filterbase in $C(K,\underset{\sim}{F},c)$ for each compact subset K of T. By the completeness of $C(K,\underset{\sim}{F},c)$ then $\mathcal{B}$ is seen to converge to a continuous function on each compact subset of T. Certainly then if T is a topological space with the property that an $\underset{\sim}{F}$-valued function is continuous whenever its restriction to each compact subset is continuous, then $\mathcal{B}$ will converge to a function continuous on all of T and $C(T,\underset{\sim}{F},c)$ will be complete. Completely regular Hausdorff spaces on which continuity on compact sets implies continuity are called k_R-spaces so that a sufficient condition for $C(T,\underset{\sim}{F},c)$ to be complete is that T be a k_R-space. As it happens this condition is also necessary, a fact which we prove in our next result.

For $E \subset T$, p_E denotes the seminorm defined at each $x \in C(T,\underset{\sim}{F})$ by $p_E(x)=\sup|x(E)|$; $V_{p_E}=\{x \in C(T,\underset{\sim}{F}) \mid p_E(x) < 1\}$ and $\overline{V}_{p_E}=\{x \in C(T,\underset{\sim}{F}) \mid p_E(x) \leq 1\}$.

<u>Theorem 2.2-1. COMPLETENESS AND k_R-SPACES</u>. Let T be a completely regular Hausdorff space. Then the following statements are equivalent:

(a) $C(T,\underset{\sim}{F},c)$ is complete.

(b) $C(T,\underset{\sim}{F},c)$ is quasi-complete (i.e. every closed and bounded subset of $C(T,\underset{\sim}{F},c)$ is complete).

(c) T is a k_R-space.

<u>Proof</u>. The implication (c) → (a) has already been established; (a) → (b) is obvious.

<u>(b) → (c)</u>. Suppose that $C(T,\underset{\sim}{F},c)$ is quasi-complete and let x be a bounded $\underset{\sim}{F}$-valued function on T with continuous restriction to each compact subset. It remains to be shown that x is continuous.

Since βT is compact and Hausdorff - hence normal - $x|_K$ can be extended to a continuous $\underset{\sim}{F}$-valued function y_K on βT such that $p_{\beta T}(y_K)= p_K(x|_K)$ for each compact subset K of T. Letting x_K denote the restriction of y_K to T, we obtain a net (x_K) in $p_T(x)\overline{V}_{p_T}$ indexed by the family of compact subsets K of T, directed by taking $K \leq K'$ iff $K \subset K'$. It is easy to see that (x_K) is Cauchy in $C(T,\underset{\sim}{F},c)$ and pointwise convergent to x. Thus by the completeness of the closed and bounded set $p_T(x)\overline{V}_{p_T}$, $x \in C(T,\underset{\sim}{F})$. Next suppose that x is an unbounded real-valued function on T with continuous restriction to each compact set K. For each integer $n \geq 0$ we define

$$x_n(t) = \begin{cases} x(t) \text{ if } |x(t)| \leq n \\ n \text{ if } x(t) > n \\ -n \text{ if } x(t) < -n \end{cases}$$

The restriction of x_n to K is certainly continuous for each compact K so, by the previous argument, $x_n \in C(T,\underset{\sim}{F})$. Clearly (x_n) converges pointwise to x on T. Moreover if K is compact then, as x is continuous, and therefore bounded, on K, $x_n|_K=x|_K$ for all sufficiently large n. Thus $x_n \to x$ uniformly on each compact set and (x_n) is Cauchy in $C(T,\underset{\sim}{F},c)$. Letting B denote the closure of $\{x_n \mid n \in \underset{\sim}{N}\}$ in $C(T,\underset{\sim}{F},c)$, it follows by the quasi-completeness of $C(T,\underset{\sim}{F},c)$ that B is complete; thus (x_n) has a limit y in $C(T,\underset{\sim}{F},c)$. It follows that $x_n \to y$ pointwise and $x=y \in B \subset C(T,\underset{\sim}{F})$. In the event that x is an unbounded complex-valued function on T with continuous restriction to each compact set, the above argument may be applied to the real and imaginary parts of x. ∇

2.3 k-Spaces, k_R-Spaces and Pseudofinite Spaces

What kinds of spaces are k_R-spaces? If sequential continuity implies continuity, for example, then the space is a k_R-space, for then continuity on compact sets yields continuity on the compact set $\{t_n\} \cup \{t\}$ for any $t_n \to t$. Thus any first countable space (completely regular and Hausdorff) is a k_R-space.

Another class of k_R-spaces is the **k-spaces, spaces in which a set is open if its intersection with each compact subset is open in the compact set.** Indeed if T is a k-space and x is an $\underset{\sim}{F}$-valued function which is continuous on each compact subset K of T, then for each open subset G of $\underset{\sim}{F}$, $x^{-1}(G) \cap K = x|_K^{-1}(G)$ so that $x^{-1}(G)$ is open in T and x is continuous on T. Thus each k-space is a k_R-space. Familiar classes of spaces that are also k-spaces include the locally compact spaces and the first countable spaces (see Exercise 2.2(b)). It is not the case however that every completely regular Hausdorff k_R-space is a k-space.

Example 2.3-1. A k_R-SPACE WHICH IS NOT A k-SPACE. Let A be an uncountable set, let W denote the nonnegative integers with discrete topology, let W^A carry the product topology, and let

$$S = \{t \in W^A \mid t(a) = 0 \text{ for all but countably many } a \in A\}$$

with its subspace topology. W^A is a k_R-space but not a k-space. In proving that W^A is a k_R-space we make use of the facts that S is a k_R-space and that its repletion υS is W^A, both of which follow from (a) below.*

(a) If x is a sequentially continuous real-valued function on S, i.e. $x(t_n) \to x(t)$ whenever $t_n \to t$ in S, then x is continuous. Moreover associated with each such sequentially continuous x there is a countable subset C of A and a continuous real-valued function x_C on W^C such that $x = x_C \cdot pr_C$ where pr_C is the projection

$$pr_C : S \to W^C$$
$$t \to t|_C$$

Once the existence of x_C has been established the continuity of $x = x_C \cdot pr_C$ follows from the continuity of x_C and pr_C. Furthermore, as

---

* It is also true that S is a k-space (Exercise 2.2(g)) so that S is an example of a k-space whose repletion is not a k-space.

will be seen, the representation $x=x_C \cdot pr_C$ is a consequence of the existence of a countable set $C \subset A$ with the property that for each $t \in S$, $x(t)=x(t_C)$ where $t_C=tk_C$, k_C denoting the characteristic function of C. Indeed if such a set C exists, then for each $s \in W^C$ we define $x_C(s)$ to be $x(t)$ where t is chosen to be any element of S such that $t|_C=s$. The sequential continuity of x_C follows from the sequential continuity of x and thus continuity of x_C follows from the second countability of W^C. Hence it remains to demonstrate the existence of a countable set $C \subset A$ such that $x(t)=x(t_C)$ for each $t \in S$.

Prior to doing this however we digress somewhat in establishing three preliminary technicalities, the last one of which is used in the proof of the existence of C.

(i) If $(A_b)_{b \in B}$ is an uncountable collection of countable subsets of A such that any one a from A belongs to at most a countable number of the A_b, then there is an uncountable subset $B_o \subset B$ such that $A_b \cap A_c = \emptyset$ whenever b and c are distinct elements from B_o.

Proof (i): Let $b \in B$ and let U_b be the collection of all $c \in B$ for which there is a finite sequence $b=b_o, b_1, \ldots, b_n=c$ in B such that $A_{b_{i-1}} \cap A_{b_i} \neq \emptyset$ for $1 \leqq i \leqq n$. As A_b is countable and for each $a \in A_b$ there are only a countable number of c such that $a \in A_b \cap A_c$ there are only a countable number of such sequences for n=1. It follows by induction that the cardinality of the class of sequences $b=b_o, \ldots, b_n=c$ is countable for any positive integer n so that U_b is countable. $U_b \cap U_c = \emptyset$ or $U_b=U_c$ for each pair b,c from B and we can produce an acceptable B_o by choosing exactly one element from each of the distinct U_b's.

(ii) If $(A_b)_{b \in B}$ is an uncountable collection of finite subsets of A, each containing at most k elements, then there exists a finite subset $Z \subset A$ and an uncountable subset $B_o \subset B$ such that $A_b \cap A_c = Z$ for distinct $b,c \in B_o$.

Proof (ii): Let m be the largest integer such that a set $Z \subset A$ having m elements and an uncountable subset $B' \subset B$ exist such that $Z \subset A_b$ for all $b \in B'$ (m may be 0). Then for fixed $a \in A$, $a \in A_b - Z$ for at most a countable number of elements b from B' for otherwise a could be adjoined to Z. Thus by (i) there is an uncountable set $B_o \subset B'$ such that $(A_b - Z) \cap (A_c - Z) = \emptyset$ (or, equivalently, $A_b \cap A_c = Z$) for each pair of distinct elements $b,c \in B_o$.

Let k be a positive integer and S^k be the collection of all $t \in W^A$ that vanish at all but at most k values of A.

(iii) If $(t_b)_{b\in B}$ is an uncountable family of functions from S^k then there is a sequence (b_n) of distinct elements from B such that $\lim t_{b_n}=t$. Furthermore there is some finite set $Z\subset A$ and $b\in B$ such that $(t_b)_Z=t$ (for any $t\in S$ recall that $t_Z=tk_Z$ where k_Z is the characteristic function of Z).

Proof (iii): Let $A_b=\{a\in A\,|\,t_b(a)\neq 0\}$. Each A_b contains at most k elements so by the previous result there is an uncountable set $B_o\subset B$ and a finite set $Z\subset A$ such that $A_b\cap A_c=Z$ for all distinct elements $b, c\in B_o$. If $Z=\emptyset$ choose any sequence (b_n) of distinct elements from B_o and any $b\in B_o$. Now each a belongs to at most one A_{b_n} so that $t_{b_n}(a)\to 0=(t_b)_Z(a)$. If $Z\neq\emptyset$ then consider the family of restrictions $(t_b|Z)_{b\in B_o}$. Since each t_b can assume only integer values at each of the finite number of a's in Z, there is an uncountable collection of t_b's that agree on Z. From this collection choose some t_b and a sequence of distinct elements b_n. Then, since $t_{b_n}=t_b$ on Z, it follows that $t_{b_n}(a)\to t_b(a)$ for each $a\in Z$; $t_{b_n}(a)\to 0=(t_b)_Z(a)$ whenever $a\notin Z$. This establishes the result.

Given a sequentially continuous map $x:S\to\underset{\sim}{R}$ we are now ready to prove that a countable set $C\subset A$ exists such that $x(t)=x(t_C)$ for each $t\in S$. For each positive integer k let A_k be the set of all $a\in A$ for which there is a $t_a\in S^k$ such that $x(t_a)\neq x(t_a')$ where $t_a'=(t_a)_{A-\{a\}}$. We contend that each A_k is countable. Suppose that this is not the case for some k. Denoting the diagonal of $\underset{\sim}{R}\times\underset{\sim}{R}$ by D, it follows that for each $a\in A_k$, $(x(t_a),x(t_a'))\in\underset{\sim}{R}\times\underset{\sim}{R}-D$. But $\underset{\sim}{R}\times\underset{\sim}{R}-D=\bigcup_{m\in\underset{\sim}{N}}F_m$ where each F_m is closed in $\underset{\sim}{R}\times\underset{\sim}{R}$. Thus at least one of the F_m's, say F_{m_o}, contains an uncountable number of the pairs $(x(t_a),x(t_a'))$, i.e. there is an uncountable set $B\subset A_k$ such that $(x(t_a),x(t_a'))\in F_{m_o}$ for each $a\in B$. By the previous result there is a sequence of distinct elements $b_m\in B$, $b\in B$, and a finite subset $Z\subset A$ such that $t_{b_m}\to(t_b)_Z$. Since the b_m's are distinct, for each $a\in A$ there is a point in the sequence beyond which $b_m\neq a$ so that $t_{b_m}'(a)=t_{b_m}(a)\to(t_b)_Z(a)$. Thus both sequences (t_{b_m}) and (t_{b_m}') converge to the function $(t_b)_Z$ which belongs to S^k. It follows that $\lim(x(t_{b_m}),x(t_{b_m}'))=(x((t_b)_Z),x((t_b)_Z))$ is contained in the closed set F_{m_o} and $F_{m_o}\cap D=\emptyset$; we have reached a contradiction - a contradiction which arose from the assumption that A_k was uncountable. Hence $C=\bigcup_k A_k$ is countable and we claim that this is the desired set. First we verify that $x(t)=x(t_C)$ for $t\in S^k$. If $t\in S^k$ let $a_1,\ldots,a_r$ be the elements of A not in C at which t does not vanish. Then, as each $a_i\notin A_k$ and

$$t_C = ((t_{A-\{a_1\}})_{A-\{a_2\}}\cdots)_{A-\{a_r\}},$$

it follows that

$$x(t) = x(t_{A-\{a_1\}}) = x((t_{A-\{a_1\}})_{A-\{a_2\}}) = \cdots = x(t_C).$$

Next suppose that $t \in S$ and let $t_k \to t$ where $t_k \in S^k$. Clearly $\lim (t_k)_C = t_C$ so that

$$x(t) = \lim x(t_k) = \lim x((t_k)_C) = x(\lim (t_k)_C) = x(t_C)$$

and this completes the proof of (a).

(b) S is a k_R-space and $\upsilon S = W^A$.

That S is a k_R-space follows from (a) once we observe that any real-valued function on S which is continuous on each compact subset of S must be sequentially continuous.

The product W^A is homeomorphic to the closed subspace of $\underset{\sim}{R}^A$ consisting of all functions assuming nonnegative integer values and is thereby a replete space (Theorem 1.5-2). Since S is dense in W^A, it follows by (1.6-3) that $\upsilon S = W^A$ once it has been noted that each continuous real-valued function on S is actually a restriction of a continuous function on W^A. This fact follows from the representation given in (a) once it has been observed that pr_C may be viewed as a continuous map on all of W^A rather than just S.

(c) W^A is a k_R-space.

To show that W^A is a k_R-space it suffices to show that $C(W^A, \underset{\sim}{R}, c)$ is complete (Theorem 2.2-1). Hence suppose that $(x_i)_{i \in I}$ is a Cauchy net in $C(W^A, \underset{\sim}{R}, c)$; we must show that it converges uniformly on each compact subset to a function which is continuous on W^A. As compact subsets of S are also compact in the product, the net $(x_i|_S)_{i \in I}$ is Cauchy in $C(S, \underset{\sim}{R}, c)$. Because S is a k_R-space, so that $C(S, \underset{\sim}{R}, c)$ is complete, and W^A is the repletion of S, there is an $x \in C(W^A, \underset{\sim}{R}, c)$ such that $x_i|_S \to x|_S$ in $C(S, \underset{\sim}{R}, c)$. We claim that $x_i \to x$ in $C(W^A, \underset{\sim}{R}, c)$. It is enough to show that $x_i \to x$ pointwise on W^A since a net of functions which is uniformly Cauchy and pointwise convergent on a set must be uniformly convergent on that set. Let $t \in W^A$. If $\mathcal{F}$ denotes the collection of all finite subsets of A directed by set inclusion, then $(t_F)_{F \in \mathcal{F}}$ is a net from S convergent to t in W^A. Furthermore each element of this net and its limit lie in the compact set $K = \prod_{a \in A} \{0, t(a)\} \subset W^A$. Now consider

$$|x_i(t) - x(t)| \leq |x_i(t) - x_j(t)| + |x_j(t) - x_j(t_F)|$$

$$+ |x_j(t_F) - x(t_F)| + |x(t_F) - x(t)|.$$

Since $(x_i)_{i\in I}$ is uniformly Cauchy on each compact subset of W^A, there is an index i_o such that $|x_i(t)-x_j(t)| < \epsilon/4$ whenever $i, j \geq i_o$. As $(x_i)_{i\in I}$ is uniformly Cauchy on K it is uniformly Cauchy on $\{t_F | F\in\mathcal{F}\}$. Moreover $x_i \to x$ pointwise on S and therefore on $\{t_F | F\in\mathcal{F}\}$; thus $(x_i)_{i\in I}$ converges uniformly to x on $\{t_F | F\in\mathcal{F}\}$. Consequently there is an index j_o such that $|x_j(t_F)-x(t_F)| < \epsilon/4$ whenever $j \geq j_o$ and any $F\in\mathcal{F}$. Fixing $j \geq i_o, j_o$, we may choose F such that $|x_j(t)-x_j(t_F)|$ and $|x(t_F)-x(t)|$ are less than $\epsilon/4$. Hence $|x_i(t)-x(t)| < \epsilon$ whenever $i \geq i_o$ and we have shown that $C(W^A,\underset{\sim}{R},c)$ is complete.

(d) W^A is not a k-space.

We shall exhibit a non-closed set $Q\subset W^A$ such that $Q\cap K$ is closed in K for each compact set K thereby establishing the fact that W is not a k-space. Let Q' be the collection of all $t \in W^A$ such that for some $n \in W$, $t(a)=n$ for all but at most n values of a and 0 otherwise. To begin with we claim that Q' is closed. Indeed if the net $(t_i)_{i\in I}$ from Q' converges to t then for each $a \in A$, $t_i(a) \to t(a)$. Thus each $t(a)\in W$. If $t=0$ then $t \in Q'$. If $t \neq 0$ then there is some $b \in A$ such that $t(b)=n\neq 0$. Let $B\subset A$ be the collection of elements at which t is 0 (B may be empty). If we assume that B contains a subset B' having n+1 elements then, since $t_i \to t$, there is an i such that $t_i=t$ on $B'\cup\{b\}$. But this implies the contradictory facts that $t_i(b)=t(b)=n$ and $t_i(a)=0$ for at least n+1 values of a. Thus B has cardinality less than or equal to n. Now suppose that $a\neq b$ and $a \notin B$. Then for some i, $t_i=t$ on $\{a,b\}$. Since $t_i(a)=t_i(b)=n$, it follows that $t(a)=t(b)=n$ and $t \in Q'$.

Next we prove that 0 is a limit point of Q' so that $Q=Q'-\{0\}$ is not closed. The collection $\mathcal{F}$ of all finite subsets of A is directed by set inclusion. For each $F \in\mathcal{F}$ let t^F be that element of Q' which is 0 on F and m on A-F, where m is the number of elements in F. Clearly $t^F \in Q'$ for all $F\neq\emptyset$ and $t^F \to 0$.

It remains to show that $Q\cap K$ is closed in K for each compact set $K\subset W^A$. Since each K is contained in a compact set of the form $\prod_{a\in A} L_a$, where each L_a is a finite subset of W, and $Q'=Q\cup\{0\}$ is closed in W^A, it suffices to prove that 0 is not an element of the closure of $Q \cap \prod_{a\in A} L_a$ in $\prod_{a\in A} L_a$. To do this, first choose an integer n such that $\max_{k\in L_a} k=n$ for an infinite number of indices a and let F be a finite subset of A with more than n elements. If we assume that 0 is a member of the closure of $Q \cap \prod_{a\in A} L_a$ in $\prod_{a\in A} L_a$ then it follows that an element $t\in Q\cap \prod L_a$ exists such that t=0 on F.

Now since $t \in Q$, t is not identically 0 and therefore F', the set of indices at which t vanishes, is finite and contains F. This leads directly to the fact that there is an $a \in A-F'$ for which $\max_{k \in L_a} k=n$. Now, since $t \in Q$ and vanishes on a set containing more than n elements, $t(a) > n$ which contradicts the fact that $t \in \Pi L_a$ and it follows that 0 is not a limit point of $Q \cap \Pi L_a$.∇

Even if T is not a k-space, the compact subsets of T "generate" a topology $\mathcal{J}_k$ which renders T a k-space. Indeed the collection of all subsets $U \subset T$ that meet each compact subset K of T in an open subset of K is a topology called the <u>k-extension topology</u> of T. It is the strongest topology on T for which all the injection maps $i_K: K \to T$, $t \to t$, K compact, are continuous. Clearly the k-extension topology is at least as fine as the original topology. This, combined with the fact that each $i_K: K \to (T, \mathcal{J}_k)$ is continuous, implies that the original topology and the k-extension topology have the same compact sets. Thus T is a k-space in the k-extension topology. In addition we make note that both the original and k-extension topologies induce the same subspace topology on each compact subset of T.

With the notion of the k-extension topology at our disposal we are in a position to describe the circumstances under which a k_R-space is a k-space.

<u>(2.3-1) WHEN IS A k_R-SPACE A k-SPACE?</u> Let T_k denote the space T equipped with the k-extension topology. A k_R-space T is a k-space iff T_k is completely regular.

<u>Proof</u>. Since the k-extension topology coincides with the original topology when T is a k-space, $T_k=T$ is completely regular. Conversely suppose that T_k is completely regular. As the k-extension topology is at least as fine as the original topology, $C(T,\underset{\sim}{F}) \subset C(T_k,\underset{\sim}{F})$. On the other hand, if $x \in C(T_k,\underset{\sim}{F})$, then x is continuous on each compact subset of T_k. Since T and T_k have the same compact sets, induce identical subspace topologies on the compact subsets, and T is a k_R-space, then $x \in C(T,\underset{\sim}{F})$ and $C(T,\underset{\sim}{F})=C(T_k,\underset{\sim}{F})$. Now, since T and T_k are completely regular, then $T=T_k$ and it follows that T is a k-space.∇

Though it is not generally true that a completely regular Hausdorff space T is a k-space when $C(T,\underset{\sim}{F},c)$ is complete, there are categories of spaces for which it is true - two such are the hemicompact spaces and the pseuofinite spaces (defined below).

<u>Definition 2.3-1. PSEUDOFINITE SPACES</u>. A completely regular Hausdorff space T is <u>pseudofinite</u> if each compact subset $K \subset T$ is finite.

Certainly every discrete space is pseudofinite. Every P-space (i.e. those completely regular Hausdorff spaces T for which each prime ideal in $C(T,\underset{\sim}{R})$ is maximal) is pseudofinite and there is a plentiful supply of non-discrete P-spaces (Gillman and Henriksen, 1954). Most of the remainder of this section is devoted to characterizing completeness of $C(T,\underset{\sim}{R},c)$ for pseudofinite spaces and hemicompact spaces. First we characterize pseudofinite spaces.

<u>(2.3-2) CHARACTERIZATIONS OF PSEUDOFINITE</u> The following are equivalent for a completely regular Hausdorff space T.

(a) T is pseudofinite.

(b) The compact-open and point-open topologies coincide on $C(T,\underset{\sim}{F})$.

(c) T_k is discrete.

<u>Proof</u>. Since the compact subsets of T are finite, the implication (a) → (b) is clear. To obtain the converse, i.e. the implication (b) → (a), let K be a compact subset of T. By (b) there is a finite set $F \subset T$ such that $\overline{V}_{p_F} \subset \overline{V}_{p_K}$ where

$$\overline{V}_{p_K} = \{x \in C(T,\underset{\sim}{F}) \,|\, p_K(x) = \sup |x(K)| \leq 1\}$$

etc.. Now suppose that $K \not\subset F$, i.e. that there is a point $t \in K-F$. Since T is completely regular there is an $x \in C(T,\underset{\sim}{F})$ such that $x(t)=2$ while $x(F)=\{0\}$ so that $x \in \overline{V}_{p_F}$ and $x \notin \overline{V}_{p_K}$. Thus $K \subset F$ and K is finite.

As T_k (as in (2.3-1)) and T have the same compact sets, the implication (c) → (a) is clear. On the other hand if T is pseudofinite, each compact (hence finite) subset of T is discrete. Therefore any subset U of T meets each compact set K in a relatively open set and T_k must be discrete.∇

It is now clear by (a) → (c) of (2.3-2) and (2.3-1) that any pseudofinite k_R-space is a k-space. However, as we shall presently see, the only pseudofinite k_R-spaces are the discrete ones.

<u>(2.3-3) PSEUDOFINITE k_R-SPACES ARE DISCRETE</u> If T is pseudofinite, then the following statements are equivalent.

(a) $C(T,\underset{\sim}{F},c)$ is complete (i.e. T is a k_R-space, by Theorem 2.2-1).

(b) T_k is discrete.

(c) T is a k-space.

<u>Proof.</u> Since T is pseudofinite, by (2.3-2), T_k is discrete. Thus the equivalence of (b) and (c) is clear. To see that the implication (a) → (b)

holds, it is enough to note that - by (2.3-2)(b) and the fact that $C(T,\underset{\sim}{F})$ is dense in the product space $T^{\underset{\sim}{F}}$ - $C(T,\underset{\sim}{F})=T^{\underset{\sim}{F}}$. Finally (c) → (a) follows by Theorem 2.2-1 and the fact that k-spaces are k_R-spaces. ▽

As we mentioned earlier the class of hemicompact spaces is also a class of spaces in which k_R-spaces are k-spaces.

(2.3-4) HEMICOMPACT k_R-SPACES Let the completely regular Hausdorff space T be hemicompact. Then $C(T,\underset{\sim}{F},c)$ is complete (or, equivalently, T is a k_R-space) iff T is a k-space.

Proof. The implication $k \to k_R$ is obvious. Conversely, suppose that T is a k_R-space. Since T is a k-space whenever T_k is completely regular by (2.3-1), it suffices to prove that if T is hemicompact, then T_k is normal. Furthermore T_k must be hemicompact if T is, because T_k has the same compact sets as T. Thus T_k is Lindelöf so it is only necessary to show that T_k is regular. To this end let $t \in T_k$ and U be an open neighborhood of t. As T is hemicompact, T may be written as the union of an increasing sequence of compact sets K_n having the property that each compact subset of T is contained in some K_n. We may assume without loss of generality that $t \in K_1$.

Our first contention is that there exists an increasing sequence (W_n) of closed neighborhoods of t in K_n such that (i) $W_n \subset U \cap K_n$ for each n, (ii) $W_n \cap K_m = W_m$ for all $m \leq n$, and (iii) $(\mathrm{int}_n W_n) \cap K_m = \mathrm{int}_m W_m$ for all $m \leq n$ where int_n denotes the interior taken in K_n. Since K_1 is a compact Hausdorff space and therefore regular there is a closed neighborhood W of t in K_1 contained in $U \cap K_1$. Let us assume that closed neighborhoods $W_1, \ldots, W_{n-1}$ of t exist satisfying the three properties stated above. As W_{n-1} is a closed subset of $U \cap K_n$ in the normal space K_n there is an open subset V_n of K_n such that $W_{n-1} \subset V_n \subset \mathrm{cl}\, V_n \subset U \cap K_n$. Since W_{n-1} is a neighborhood in K_{n-1}, an open set $U_n \subset K_n$ exists which meets K_{n-1} in the set $\mathrm{int}_{n-1} W_{n-1}$. Hence $\mathrm{cl}\, U_n \cap K_{n-1} = \mathrm{cl}(\mathrm{int}_{n-1} W_{n-1}) \subset W_{n-1}$ and it follows that the closed neighborhood of t in K_n, $W_n = (\mathrm{cl}\, U_n \cap \mathrm{cl}\, V_n) \cup W_{n-1}$ meets K_{n-1} in the set W_{n-1} and is a subset of $U \cap K_n$. Furthermore the relations $\mathrm{int}_n W_n \supset V_n \cap U_n$, $K_{n-1} \cap U_n = \mathrm{int}_{n-1} W_{n-1}$, and $V_n \cap K_n \supset \mathrm{int}_{n-1} W_{n-1}$ imply that $\mathrm{int}_n W_n \cap K_{n-1} \supset \mathrm{int}_{n-1} W_{n-1}$. The reverse inclusion is obvious and the contention has been established.

Next we claim that $W = \bigcup_{n \in \underset{\sim}{N}} W_n$ is a closed neighborhood of t in T_k contained in U, thereby proving that T_k is regular. It is clear that $W \subset U$. That W is closed in the hemicompact k-space T_k follows immediately from the relation $W \cap K_n = W_n$ for each n. Certainly $W \supset \bigcup_{n \in \underset{\sim}{N}} \mathrm{int}_n W_n$. Since $\bigcup_n \mathrm{int}_n W_n \cap K_m = \mathrm{int}_m W_m$ for each m, $\bigcup_n \mathrm{int}_n W_n$ is an open subset of W in T_k

containing t. Thus W is a neighborhood of t and the proof is complete. ∇

Thus it is natural to inquire when $C(T,\underset{\sim}{F},c)$ is fully complete. In the event that the completely regular Hausdorff space T is hemicompact or pseudofinite, $C(T,\underset{\sim}{F},c)$ is fully complete iff T is a k-space. Indeed if T is a hemicompact k-space, $C(T,\underset{\sim}{F},c)$ is a Frechet space by Theorem 2.1-1 and (2.3-4). As any Frechet space is fully complete (Husain, 1965, 4.1, Prop. 3) the desired result follows. The converse is trivial. As for the case of pseudofinite spaces, we saw in the proof of (2.3-3) that $C(T,\underset{\sim}{F})=\underset{\sim}{F}^T$ and the compact-open and product topologies agree whenever T is a pseudofinite k-space. This, combined with the fact that an arbitrary product of real lines is fully complete in the product topology (Husain, 1965, 5.5, Prop. 14), leads to the conclusion that $C(T,\underset{\sim}{F},c)$ is fully complete.

Unfortunately it is not generally the case the $C(T,\underset{\sim}{F},c)$ is fully complete when T is a k-space, and necessary and sufficient conditions on T for $C(T,\underset{\sim}{F},c)$ to be fully complete are not known. Ptak (1953) has shown that T is a k-space whenever $C(T,\underset{\sim}{F},c)$ is fully complete and has given a counter example for the converse (see Exercise 2.3(b)).

Some of the interrelationships between k-spaces and k_R-spaces, etc., are summarized in Table 1 below. T denotes a completely regular Hausdorff space and K, with or without subscript, a compact subset of T.

2.4 <u>The Continuous Dual of $C(T,\underset{\sim}{F},c)$ and the Support</u>

A matter of considerable importance in the theory of locally convex spaces X is characterizing the continuous dual X', the space of continuous linear functionals on X. The considerations of this section have primarily to do with the continuous dual $C(T,\underset{\sim}{F},c)'$ of $C(T,\underset{\sim}{F},c)$ where T is a completely regular Hausdorff space. $C(T,\underset{\sim}{F},c)'$ is characterized here in two ways, one involving the evaluation maps $T^* = \{t^* | t\in T\}$ $(t^*(x)=x(t)$ for $x\in C(T,\underset{\sim}{F}))$ and the other, more important for our purposes, represents the elements of $C(T,\underset{\sim}{F},c)'$ as integrals with respect to certain set functions defined on the Borel subsets of T. This result has predecessors as far back as 1909 when Riesz first characterized $C([0,1],\underset{\sim}{R})'$ in terms of Riemann-Stieltjes integrals. Subsequent generalizations were made which culminated in the well-known characterization of $C(T,\underset{\sim}{F},c)'$ for compact Hausdorff spaces T in terms of integrals with respect to regular set functions defined on the Borel subsets of T (see Theorem 2.4-1). As the elements of $C(T,\underset{\sim}{F})$ are all bounded when T is compact, it was natural for others to consider the space $C_b(T,\underset{\sim}{R})$ (with sup norm) and integral representations of the elements of

property	definition	implied by	equivalent to
hemi-compact	(Def. 2.1-1) A countable family (K_n) exists such that each K K_n for some n.	loc. compact + σ-compact	$C(T,\underset{\sim}{F},c)$ metrizable
k_R-space	(Sec. 2.1) For $x:T \to \underset{\sim}{R}$ if $x\|_K$ is continuous.	seq. continuity → continuity or k-space	$C(T,\underset{\sim}{F},c)$ complete
k-space	(Sec. 2.3) For each $K, G \cap K$ open in K → G open in T.	k_R+pseudo-finite or k_R+ hemicompact	

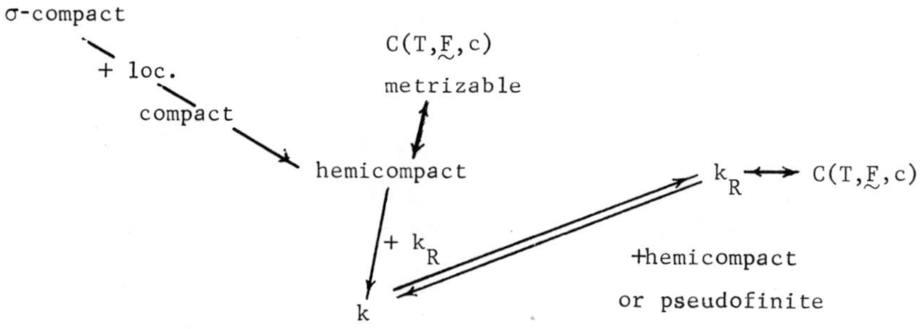

Table 1

$C_b(T,\underset{\sim}{R})'$ were obtained for T satisfying various normality conditions (see, for example, Dunford and Schwartz, 1958 and Alexandrov (1940, 1941, 1943)). By abandoning the space of regular set functions on the Borel subsets of a topological space T for the larger class of regular set functions on the Baire subsets of T it was possible to establish integral representations of the elements of $C(T,\underset{\sim}{F},c)'$. We follow this approach in obtaining a representation of $C_b(T,\underset{\sim}{F})'$ (where $C_b(T,\underset{\sim}{F})$ carries the sup norm) for completely regular Hausdorff spaces T, but then take an alternate route for $C(T,\underset{\sim}{F},c)'$- one which utilizes knowledge of the dual of $C(T,\underset{\sim}{F},c)$ for compact T. In so doing we obtain integral representations in terms of set functions on the Borel subsets. These representations are then used to prove the existence of the "support" of an element $x'\in C(T,\underset{\sim}{F},c)'$, a minimal compact subset S of T with the property that if $y\in C(T,\underset{\sim}{F})$ vanishes on S, then x' must vanish on y. The notion of support is put to use in the next section where necessary and sufficient conditions on T are obtained for $C(T,\underset{\sim}{F},c)$ to be barreled and bornological respectively.

The following result is useful later on.

<u>(2.4-1) T* AND THE DUAL OF $C(T,\underset{\sim}{F},c)$</u> Let $\mathcal{K}$ denote the family of compact subsets of the completely regular Hausdorff space T, $K^* = \{t^* | t\in K\}$, and K_{bc} the balanced convex hull of K. Then $C(T,F,c)'=[\bigcup_{K\in\mathcal{K}} cl_{\sigma(X^*,X)}(K^*)_{bc}]=H$ where $X=C(T,\underset{\sim}{F},c)$, X^* the algebraic dual of X of <u>all</u> linear functionals on X, and the square brackets denote linear span. If T is compact, then $C(T,F,c)'=[cl_{\sigma(X^*,X)}(T^*)_{bc}]$.

<u>Proof</u> Since $t^*\in H$ for each $t \in T$, then (X,H) is a dual pair. By the Mackey-Arens theorem we need only show that the compact-open topology is a topology of uniform convergence on a collection of balanced convex $\sigma(H,X)$ compact subsets of H to prove that H is the dual of X. To do this we show first that each set $E=cl_{\sigma(X^*,X)}(K^*)_{bc}$ is absolutely convex and $\sigma(H,X)$-compact. And to do this it is sufficient to show that E is $\sigma(X^*,X)$-bounded for it is already $\sigma(X^*,X)$-closed, and the closure of an absolutely convex set is absolutely convex; $\sigma(X^*,X)$-boundedness, in turn, is shown by showing that for each $x \in X$, $\sup|\langle x,E\rangle| < \infty$. To this end let $t_1,\dots,t_n\in K$ and $\mu_1,\dots,\mu_n\in\underset{\sim}{F}$ be such that $\Sigma|\mu_i| \le 1$. Then for any $x \in X$

$$|\Sigma\mu_i t_i^*(x)| = |\Sigma\mu_i x(t_i)| \le (\Sigma|\mu_i|)p_K(x) \le p_K(x)$$

where $p_K(x)=\sup|x(K)|$. It follows that E is bounded and that for each $f \in cl_{\sigma(X^*,X)}(K^*)_{bc}$, $|f(x)| \leq p_K(x)$.

The next and final step is to show that

$$(cl_{\sigma(X^*,X)}(K^*)_{bc})^{o} = \bar{V}_{p_K} = \{x \in X \,|\, p_K(x) \leq 1\}.$$

Since $|f(x)| \leq p_K(x)$ for each $x \in X$ and each $f \in cl_{\sigma(X^*,X)}(K^*)_{bc}$, it is clear that

$$\bar{V}_{p_K} \subset (cl_{\sigma(X^*,X)}(K^*)_{bc})^{o}.$$

To obtain the reverse inclusion, consider any $x \in (cl_{\sigma(X^*,X)}(K^*)_{bc})^{o}$ and any $t \in K$. Then $|x(t)| \leq 1$ and it follows that $p_K(x)=\sup|x(K)| \leq 1$.∇

In Sec. 1.7 we had occasion to consider the class of finitely additive signed measures (differences of finitely additive nonnegative regular real-valued set functions) defined on the algebra of sets generated by the zero sets of a completely regular Hausdorff space. We have need of some of the measure-theoretic notions set forth there but somewhat more general versions of them must be obtained first. The algebras generated by the zero sets $\mathcal{Z}$ and the closed sets $\mathcal{C}$ of T are denoted by $\mathcal{A}_Z$ and $\mathcal{A}_c$ respectively while the σ-algebras they generate are denoted by $\mathcal{B}_a$ and $\mathcal{B}$, called the Baire and Borel sets respectively. The total variation $|\mu|$ of a finitely additive set function μ defined on any of these algebras is the set function defined at each set A in the algebra to be the sup of the sums $\Sigma|\mu(E_i)|$ taken over all finite pairwise disjoint collections (E_i) of subsets of A taken from the algebra. In the event that μ is a nonnegative finitely additive set function defined on an <u>algebra</u> of sets, it is bounded. Thus a difference of two such nonnegative finitely additive set functions is also a bounded set function. Conversely, a bounded real-valued finitely additive set function μ defined on an algebra of sets has a finite-valued total variation $|\mu|$ [Dunford and Schwartz, 1958, p. 97] and can be decomposed [ibid., p. 98, Th. 8] into a difference of nonnegative additive set functions μ^+ and μ^- such that $\mu=\mu^+ + \mu^-$. Moreover a complex-valued additive set function μ defined on an algebra of sets is bounded iff $|\mu|$ is finite-valued and, in this case, may be written in the form $\mu=\mu_r+i\mu_i$ where μ_r and μ_i are bounded real-valued additive set functions.

A finitely additive set function μ defined on $\mathcal{A}_Z$ or $\mathcal{B}_a$ is <u>regular</u> if for each A on which μ is defined and $\epsilon > 0$ there are $Z, Z' \in \mathcal{Z}$ such that $Z \subset A \subset CZ'$ and $|\mu|(CZ'-Z) < \epsilon$. Similarly if μ is defined on $\mathcal{A}_c$ or $\mathcal{B}$ μ is regular whenever B is in the domain of definition of μ and $\epsilon > 0$ there are closed and open sets C and U such that $C \subset A \subset U$ and $|\mu|(U-C) < \epsilon$.

Our first goal is to characterize $C_b(T,\underset{\sim}{F})'$ and, as might be expected, we begin with the real-valued case. As it happens the elements of $C_b(T,\underset{\sim}{R})'$ can be decomposed into a difference of "positive" components. More precisely, if $x' \in C_b(T,\underset{\sim}{R})'$ then $x' = x_p' - x_n'$ where x_p' and x_n' are positive linear functionals, i.e. $x_p'(x)$ and $x_n'(x)$ are ≥ 0 whenever $x \geq 0$. Prior to establishing such a decomposition we verify that positive linear functionals are always continuous.

<u>(2.4-2) POSITIVE → CONTINUOUS</u> Any positive linear functional defined on $C_b(T,\underset{\sim}{F})$ where T is a completely regular Hausdorff space is continuous.

<u>Proof</u> To see that a positive linear functional h defined on $C_b(T,\underset{\sim}{F})$ is continuous when $C_b(T,\underset{\sim}{F})$ carries the sup norm topology, let $x \in C_b(T,\underset{\sim}{F})$. Clearly then

$$-p_T(x)1 \leq x \leq p_T(x)1$$

where $1(t)=1$ for each $t \in T$. Thus, since h is positive,

$$-p_T(x)h(1) \leq h(x) \leq p_T(x)h(1)$$

and it follows that $|h(x)| \leq |h(1)|p_T(x)$. If x is complex-valued, it follows that $|h(x)| \leq 2h(1)p_T(x)$ so that h is continuous.∇

<u>(2.4-3) $C_b(T,\underset{\sim}{R})'$ AND POSITIVE LINEAR FUNCTIONALS</u> If T is a completely regular Hausdorff space and $C_b(T,\underset{\sim}{R})$ carries the topology induced by the sup norm, then corresponding to each $x' \in C_b(T,\underset{\sim}{R})'$ there are positive linear functionals x_p' and x_n' such that $x' = x_p' - x_n'$.

<u>Proof</u> First we define x_p' on the nonnegative elements of $C_b(T,\underset{\sim}{R})$: if $x \geq 0$ then

$$x_p'(x) = \sup\{x'(y) \mid y \in C_b(T,\underset{\sim}{R}),\ 0 \leq y \leq x\}.$$

Clearly $x_p'(ax) = ax_p'(x)$ for $a \geq 0$ and $x \geq 0$. Next we claim that $x_p'(x+y) = x_p'(x) + x_p'(y)$ whenever $x, y \geq 0$. To see this suppose that $0 \leq w \leq x+y$. Clearly* $0 \leq w \wedge x \leq x$ and $0 \leq w-(w \wedge x) \leq y$. Thus

$$x'(w) = x'(w \wedge x) + x'(w-(w \wedge x)) \leq x_p'(x) + x_p'(y),$$

and, taking the supremum over all such w, we obtain

---

* The <u>meet</u> $x \wedge y$ and <u>join</u> $x \vee y$ of real-valued functions x and y defined on a set T are the functions $t \to \min(x(t),y(t))$ and $t \to \max(x(t),y(t))$.

$$x_p'(x+y) \leq x_p'(x) + x_p'(y).$$

On the other hand, if $0 \leq v \leq x$ and $0 \leq w \leq y$ then $0 \leq v + w \leq x + y$ and

$$x'(v) + x'(w) = x'(v+w) \leq x_p'(x+y)$$

from which it follows that $x_p'(x) + x_p'(y) \leq x_p'(x+y)$.

We can now extend x_p' to all of $C_b(T,\underset{\sim}{R})$ as follows: if $x \in C_b(T,\underset{\sim}{R})$ write $x=x^+-x^-$ for some $x^+, x^- \geq 0$ and $x^+, x^- \in C_b(T,\underset{\sim}{R})$ (e.g. $x=x \vee 0-(-x) \vee 0$). Then define $x_p'(x) = x_p'(x^+) - x_p'(x^-)$. To see that x_p' is well-defined, suppose that $x = u-y = v-w$ where $u,y,v,w \geq 0$. Then $u+w = y+v$ and since $u,y,v,w,u+w, y+v \geq 0$,

$$x_p'(u)+x_p'(w) = x_p'(u+w) = x_p'(y+v) = x_p'(y) + x_p'(v).$$

Hence $x_p'(u)-x_p'(y) = x_p'(v)-x_p'(w)$. Next we show that x_p' is linear on $C_b(T,\underset{\sim}{R})$. Let $x,y \in C_b(T,\underset{\sim}{R})$ and consider

$$x_p'(x)+x_p'(y)=[x_p'(x \vee 0)-x_p'((-x) \vee 0)]+[x_p'(y \vee 0)-x_p'((-y) \vee 0)]$$

$$=x_p'(x \vee 0+y \vee 0)-x_p'((-x) \vee 0+(-y) \vee 0).$$

Now $x+y = (x \vee 0 + y \vee 0) - ((-x) \vee 0 + (-y) \vee 0)$ and $(x \vee 0) + (y \vee 0)$, $(-x) \vee 0 + (-y) \vee 0 \geq 0$; thus

$$x_p'(x) + x_p'(y) = x_p'(x+y).$$

To establish the homogeneity of x_p' we note first that $x_p'(ax) = ax_p'(x)$ for $a \geq 0$ and $x \in C_b(T,\underset{\sim}{R})$ follows directly from the fact that it is valid whenever $x \geq 0$. On the other hand for $x \in C_b(T,\underset{\sim}{R})$ we have

$$x_p'(-x) = x_p'((-x)^+ - (-x)^-)$$

$$= x_p'(x^- - x^+)=x_p'(x^-)+x_p'(-x^+)=x_p'(x^-)-x_p'(x^+)=-x_p'(x).$$

Hence x_p' is a positive linear functional. We complete the proof by observing that $x_n' = x_p' - x'$ is a positive linear functional. ∇

Now how are positive linear functionals represented?

<u>(2.4-4) POSITIVE LINEAR FUNCTIONALS ARE "GENERATED" BY REGULAR ADDITIVE NONNEGATIVE SET FUNCTIONS</u> If T is a normal (completely regular) Hausdorff space and h is a positive linear functional on $C_b(T,\underset{\sim}{R})$, then there exists a finite nonnegative regular additive set function μ defined on the algebra $\mathcal{A}_c$ ($\mathcal{A}_Z$) generated by the closed subsets $\mathcal{C}$ (zero sets $\mathcal{Z}$) of T such that $h = \int d\mu$ ($\int d\mu$ is taken here in the same sense as in Sec. 1.7).

Proof Both situations - where T is normal and where T is completely regular - are dealt with similarly. In fact if the terms "zero set" and "complement of zero set" are substituted for "closed" and "open" respectively in the argument given below for the case where T is normal, we obtain the proof for completely regular T.

We begin by defining a real-valued set function on the class of all subsets of T which we prove to be subadditive. We subsequently show that this set function when restricted to the appropriate algebra is, in fact, regular and additive and it is finally established that h and the regular additive set function μ are related by the formula given above.

Suppose that T is a normal Hausdorff space.

(a) Definition of μ on $\mathcal{P}(T)$. If U is an open subset of T we define $\mu(U)$ to be the supremum of the values $h(x)$ where $x \in C_b(T,\underset{\sim}{R})$ and $0 \leqq x \leqq k_U$, k_U denoting the characteristic function of U. If A is an arbitrary subset of T, $\mu(A)$ is defined to be the infimum over all open $U \supset A$ of the values $\mu(U)$.

Clearly μ is a nonnegative monotone set function and $\mu(\emptyset)=0$. Furthermore if U is open and $0 \leqq x \leqq k_U$, then $0 \leqq x \leqq 1$ and $0 \leqq h(x) \leqq h(1)$ so that $\mu(U) < \infty$. Hence it follows that μ is finite-valued on all of the power set $\mathcal{P}(T)$.

(b) μ is subadditive. To see that μ is subadditive, first consider any two open sets U and V and let $x \in C_b(T,\underset{\sim}{R})$ be such that $0 \leqq x \leqq k_{U \cup V}$ and $h(x) > \mu(U \cup V) - \epsilon$ where $0 < \epsilon < 1/2$. Next consider the closed set $H = x^{-1}([\epsilon,1]) \cap CU \subset V$ and choose $u, v \in C_b(T,\underset{\sim}{R})$ such that $u(H)=\{0\}$, $u(CV)=\{1\}$, $v(CV)=\{0\}$, and $v(u^{-1}(0))=\{1\}$. Since $v(t)=1$ whenever $u(t)=0$, we may set $y=xv^2/(u^2+v^2)$ and obtain a continuous function on T such that $0 \leqq y \leqq x$, $y=x$ on H, and $y(CV)=\{0\}$. Hence $y \leqq k_V$. As $y \leqq x$, the function $w'=x-y$ satisfies $0 \leqq w' \leqq 1$. Furthermore we claim that the closed set $F=w'^{-1}([2\epsilon,1]) \subset U$ for if it is not then, for some $t \in F$, we have $t \in CU$. But $w'(t) \geqq 2\epsilon$ certainly implies that $x(t) \geqq \epsilon$ so that $t \in H$. This however is a contradiction for we know that $w'(H)=\{0\}$. Thus $F \subset U$ and we may choose $r \in C_b(T,\underset{\sim}{R})$ such that $0 \leqq r \leqq 1$ and $r(F)=\{1\}$ while $r(CU)=\{0\}$ so that $w=rw' \leqq k_U$. Recalling that $y \leqq k_V$, it follows that $h(y)+h(w) \leqq \mu(V)+\mu(U)$ so that the desired conclusion - $\mu(U \cup V) \leqq \mu(U)+\mu(V)$ - follows once we show that $x \leqq w+y+2\epsilon$ for then

$$\begin{aligned}\mu(U \cup V) < h(x) + \epsilon &\leqq h(w) + h(y) + (2h(1) + 1)\epsilon \\ &\leqq \mu(U) + \mu(V) + 2(h(1) + 1)\epsilon\end{aligned}$$

for each $0 < \epsilon < 1/2$. To see that $x \leq w+y+2\epsilon$ first note that on F, where $w=w'$, we have $x=w+y$. On the other hand if $t \notin F$, then $x(t)-y(t)=w'(t)<2\epsilon$ and

$$x(t) < y(t) + 2\epsilon \leq w(t) + y(t) + 2\epsilon$$

Having just established subadditivity for open sets, let A and B be arbitrary subsets of T and $U \supset A$, $V \supset B$ be open subsets such that $\mu(U) < \mu(A)+\epsilon$ and $\mu(V) < \mu(B)+\epsilon$ where ϵ is some pre-assigned positive number. Then $U \cup V \supset A \cup B$ so that

$$\mu(A \cup B) \leq \mu(U \cup V) \leq \mu(U) + \mu(V) < \mu(A) + \mu(B) + \epsilon$$

and subadditivity of μ on $\mathcal{P}(T)$ follows.

Thus μ is "outer regular" on the class of all subsets of T. By restricting μ to $\mathcal{A}_c$ - the smallest algebra of sets containing $\mathcal{C}$ - we wish to obtain a regular additive set function.

<u>(c) $\mu|_{\mathcal{A}_c}$ is regular and additive</u>. To prove this consider the class $\mathcal{A}$ of all subsets of T on which μ is "inner regular", i.e.

$$\mathcal{A} = \{A \subset T \,|\, \mu(A) = \sup\{\mu(F) \,|\, A \supset F,\ F \in \mathcal{C}\}\}$$

We show that μ is a (perforce nonnegative and regular) additive set function on (the algebra) $\mathcal{A}$ and that $\mathcal{A} \supset \mathcal{A}_c$.

First we show μ to be additive on the closed subsets of T. To this end suppose that F and H are disjoint closed subsets of T and, corresponding to the positive number ϵ, let W be an open set containing $F \cup H$ such that $\mu(W) < \mu(F \cup H)+\epsilon$. Since T is normal, disjoint open sets $U \supset F$ and $V \supset H$ exist. Choose $u, v \in C_b(T,\underset{\sim}{R})$ such that $0 \leq u, v \leq 1$, $0 \leq u \leq k_{U \cap W}$, $0 \leq v \leq k_{V \cap W}$, $h(u) > \mu(U \cap W) - \epsilon/2$, and $h(v) > \mu(V \cap W) - \epsilon/2$. Then

$$\begin{aligned}\mu(F) + \mu(H) &\leq \mu(U \cap W) + \mu(V \cap W) \\ &< (h(u) + \epsilon/2 + h(v) + \epsilon/2 \\ &= h(u + v) + \epsilon \\ &\leq \mu((U \cap W) \cup (V \cap W)) + \epsilon \\ &\leq \mu(W) + \epsilon \\ &< \mu(F \cup H) + 2\epsilon\end{aligned}$$

Consequently $\mu(F)+\mu(H) \leq \mu(F \cup H)$. Equality follows from the subadditivity of μ. Next suppose that the disjoint sets A, B $\in \mathcal{A}$ so that closed sets F and H exist such that $A \supset F$, $B \supset H$, $\mu(F) > \mu(A) - \epsilon/2$, and $\mu(H) > \mu(B) - \epsilon/2$. Thus

$$\mu(A) + \mu(B) < \mu(F) + \mu(H) + \epsilon = \mu(F\cup H) + \epsilon \leqq \mu(A\cup B) + \epsilon$$

and additivity on $\mathcal{A}$ follows once we establish the fact that $A\cup B \in \mathcal{A}$. Since μ is subadditive on $\mathcal{P}(T)$ we can say that

$$\mu(A\cup B) \leqq \mu(A) + \mu(B) < \mu(F\cup H) + \epsilon .$$

Hence $\sup\{\mu(E) \mid E \in \mathcal{C},\ E\subset A\cup B\} = \mu(A\cup B)$ and $A\cup B \in \mathcal{A}$. There is just one more thing to prove before we can say that $\mathcal{A}$ is an algebra containing $\mathcal{C}$ - namely $CA\in\mathcal{A}$ whenever $A \in \mathcal{A}$. A prerequisite for doing this is the knowledge that each open $U \in \mathcal{A}$. Indeed if $\mu(U)=0$ then, as $\emptyset\subset U$ and $\mu(\emptyset)=0$, $\sup\{\mu(F) \mid F\subset U, F\in\mathcal{C}\} \geqq \mu(U)$. The reverse inequality follows from the definition of μ and $U \in\mathcal{A}$. If $\mu(U) > 0$ and $1/2 > \epsilon > 0$ then there is some $x \in C_b(T,\underset{\sim}{R})$ such that $0 \leqq x \leqq k_U$ and $h(x) > \mu(U) - \epsilon$. Clearly $V=x^{-1}([2\epsilon,1])\subset F=x^{-1}([\epsilon,1])\subset U$. Now $x-2\epsilon \leqq 0$ on CV so the continuous function $y=\max(x-2\epsilon,0) \leqq k_V$ and therefore

$$h(x) - 2\epsilon h(1) = h(x - 2\epsilon) \leqq h(y) \leqq \mu(V).$$

It follows that $\mu(F) \geqq \mu(V) \geqq h(x)-2\epsilon h(1) > \mu(U)-(2h(1)+1)\epsilon$, whence we conclude that $U \in\mathcal{A}$. Now suppose that $A \in \mathcal{A}$. We wish to show that $CA \in \mathcal{A}$. To this end let F be closed and U be open such that $F\subset A\subset U$, $\mu(F) > \mu(A)-\epsilon/2$ and $\mu(A) > \mu(U)-\epsilon/2$. Hence $\mu(U)-\mu(F) < \epsilon$. As μ is additive on $\mathcal{A}$, $U-F$ is open and belongs to $\mathcal{A}$, and $U=F\cup(U-F)$, it follows that $\mu(U-F)=\mu(U)-\mu(F)<\epsilon$. But

$$\mu(CA) - \mu(CU) \leqq \mu(CF) - \mu(CU) = \mu(CF-CU) = \mu(U-F) < \epsilon$$

and, since CU is closed, $\mu(CA)=\sup\{\mu(E) \mid E\subset CA,\ E \text{ closed}\}$. Thus $CA \in\mathcal{A}$ and $\mathcal{A}$ is an algebra containing the closed sets $\mathcal{C}$. In view of our previous remarks then μ is a regular additive nonnegative set function on $\mathcal{A}_c$.

It only remains for us to prove that $h(x)=\int xd\mu$ for each $x\in C_b(T,\underset{\sim}{R})$; in fact we need only establish the inequality $h(x) \leq \int xd\mu$ for each nonzero $x\in C_b(T,\underset{\sim}{R})$ for the inequality for $x=0$ is trivial and the reverse inequality follows from this one with the substitution of $-x$ for x. We obtain these inequalities by constructing a simple integrable function F with the properties that $\int Fd\mu$ be less than or equal to $\int xd\mu + \epsilon\mu(T)$ and $h(x)$ be less than or equal to $\int Fd\mu$ plus a function of ϵ ($\epsilon > 0$) which tends to zero as $\epsilon \to 0^+$.

(d) $h(x) \leq \int xd\mu$ for $x\in C_b(T,\underset{\sim}{R})$: Let x be a nonzero element of $C_b(T,\underset{\sim}{R})$ and consider $0 < \epsilon \leq \|x\|$. First we partition T into a finite number of disjoint sets $B_k \in \mathcal{A}_c$ on each of which the oscillation of x is less than $\epsilon/2$.

To do this realize that $x(T) \subset [-||x||, ||x|| + 1)$ and choose n large enough so that $(2||x|| + 1)/n < \epsilon/2$. Then partition $[-||x||, ||x|| + 1)$ into the n subintervals of the form

$$I_k = [-||x|| + (k-1)(2||x|| + 1)/n, -||x|| + k(2||x|| + 1)/n]$$

and set $B_k = x^{-1}(I_k)$ for $k=1,\ldots,n$. It follows that the B_k's are pairwise disjoint elements of $\mathcal{A}_c$ ($B_k \in \mathcal{A}_Z \subset \mathcal{A}_c$ for each k), $T = \cup B_k$, and x has an oscillation less than $\epsilon/2$ on each B_k. Let c_k be the right endpoint of I_k plus $\epsilon/2$ and $F = \Sigma c_k k_{B_k}$. Then $x(t) < c_k = F(t) < x(t) + \epsilon$ for each $t \in B_k$ and $k=1,\ldots,n$. Hence

$$\int F d\mu \leq \int x d\mu + \epsilon\mu(T).$$

In the remainder of the proof we show that h(x) is less than or equal to $\int F d\mu$ plus a function ϵ with limit equal to 0 as $\epsilon \to 0^+$. Since $x < c_k$ on B_k, $W_k = x^{-1}(-\infty, c_k) \supset B_k$. Recalling that each $B_k \in \mathcal{A}_c$ and that μ is regular on $\mathcal{A}_c$, we see that an open set $U_k \supset B_k$ exists such that $\mu(U_k) < \mu(B_k) + \epsilon/n$ for $1 \leq k \leq n$. Setting $V_k = W_k \cap U_k$ we obtain $x < c_k$ on V_k and $\mu(V_k) < \mu(B_k) + \epsilon/n$ for each k. Thus

$$\Sigma c_k \mu(V_k) < \Sigma c_1 \mu(B_k) + (\Sigma c_k)\epsilon/n < \int F d\mu + (||x|| + 1 + \epsilon)\epsilon.$$

Now there is a continuous partition of unity* $\{x_1,\ldots,x_n\}$ subordinate to $\{V_1,\ldots V_n\}$ so that $0 \leq x_k \leq k_{V_k}$. (In the case where T is completely regular and each V_i is the complement of a zero set, we may select $0 \leq y_i \leq 1$ such that $V_i = C(y_i^{-1}(0))$ by (1.2-1)(d) so that the functions $x_i = y_i/\Sigma y_j$ $(i=1,\ldots,n)$ sum to 1 on T and $0 < x_i \leq k_{V_i}$ for each i.) It follows that $h(x_k) \leq \mu(V_k)$ for each k. Furthermore $0 \leq xx_k \leq c_k x_k$ for each k; thus

---

*Let T be a topological space and $x: T \to \underset{\sim}{R}$ be nonnegative and continuous. The support of x, supp(x), is $cl_T\{t \in T \mid x(t) \neq 0\}$. A <u>(continuous) partition of unity subordinate to a cover</u> (A_i) of T is a family F of continuous nonnegative-valued functions (x_i) on T such that for each i, $0 \leq x_i \leq 1$, $\Sigma_i x_i(t) = 1$ for each $t \in T$, and $supp(x_i) \subset A_i$ for each i. It is not difficult to show that if T is a normal Hausdorff space and $\{U_1,\ldots,U_n\}$ an open cover of T, then there is a partition of unity subordinate to $\{U_1,\ldots,U_n\}$.

$$h(x) = h(\Sigma xx_k) = \Sigma\, h(xx_k) \le \Sigma\, h(c_k x_k)$$

$$= \Sigma\, c_k h(x_k) \le \Sigma\, c_k \mu(V_k).$$

Combining the above inequalities, we obtain

$$h(x) < \int x d\mu + \epsilon(\mu(T) + \|x\| + 1 + \epsilon)$$

for each $\epsilon > 0$ and the proof is complete. ▽

It is clear now that each element of $C_b(T,\underset{\sim}{R})'$ is representable in terms of an integral with respect to a bounded regular additive real-valued set function on $\mathcal{Q}_Z$ if T is completely regular and Hausdorff or a bounded regular additive real-valued set function on $\mathcal{Q}_c$ if T is a normal Hausdorff space. The result is equally clear for the complex case.

Certainly if μ is a given bounded real- or complex-valued regular additive set function then

$$\left|\int x d\mu\right| \le \int |x| d|\mu| \le \|x\|\ |\mu|(T)$$

so that a linear correspondence $\mu \to x'$ is established between the bounded $\underset{\sim}{F}$-valued regular additive set functions and $C_b(T,\underset{\sim}{F})'$.

It follows from the inequality above that $\|x'\| \le |\mu|(T)$. To obtain the reverse inequality, let $\epsilon > 0$ be given and let $A_1,\ldots,A_n \in \mathcal{Q}_c$ (or $\mathcal{Q}_Z$) be pairwise disjoint sets such that $\Sigma_i |\mu(A_i)| > |\mu|(T)-\epsilon$. As μ is regular, closed sets (zero sets) C_i and open sets (complements of zero sets) U_i exist such that $C_i \subset A_i \subset U_i$ and $|\mu|(U_i - C_i) < \epsilon/n$, for $i=1,\ldots,n$. Now $C_1,\ldots,C_n$ are pairwise disjoint so that pairwise disjoint open sets (complements of zero sets) V_i exist such that $C_i \subset V_i$: This is clear if T is normal. When T is completely regular and μ is defined on $\mathcal{Q}_Z$ we demonstrate the existence of the V_i for the case $n=2$; the general case follows with the aid of the observation that a finite union of zero sets is a zero set [(1.2-1)(c)]. By Theorem 1.2-2, $cl_\beta C_1$ and $cl_\beta C_2$ are disjoint so that there is a continuous function x^β, $0 \le x^\beta \le 1$, defined on βT such that $x^\beta(cl_\beta C_1)=\{0\}$ while $x^\beta(cl_\beta C_2)=\{1\}$. Thus, denoting the restriction of x^β to T by x, it follows by (1.2-3) that $V_1=x^{-1}(-\infty,1/4)$ and $V_2=x^{-1}(3/4,\infty)$ serve the desired purpose.

Now let $W_i=U_i \cap V_i$; W_i is open when T is normal and is the complement of a zero set when T is completely regular. As $|\mu|$ is monotone, $|\mu|(W_i - C_i) < \epsilon/n$ for $i=1,\ldots,n$. By the normality of T (or the normality of βT when μ is defined on $\mathcal{Q}_Z$) continuous functions $0 \le x_i \le 1$ defined on T exist such that $x_i(C_i)=\{1\}$ and $x_i(CW_i)=\{0\}$. Letting $a_i \in \underset{\sim}{F}$ be such that

$a_i\mu(A_i)=|\mu(A_i)|$ and setting $x=\Sigma_i a_i x_i$, it follows that

$$\left|\int x d\mu - \Sigma_i |\mu(A_i)|\right| = \left|\Sigma_i a_i \int_{W_i} x_i d\mu - |\mu(A_i)|\right|$$

$$= \left|\Sigma_i [a_i \int_{C_i} x_i d\mu - |\mu(A_i)|] + \Sigma_i a_i \int_{W_i - C_i} x_i d\mu\right|$$

$$= \left|\Sigma_i a_i[\mu(C_i) - \mu(A_i)] + \Sigma_i a_i \int_{W_i - C_i} x_i d\mu\right|$$

$$\leq \Sigma_i |\mu|(A_i - C_i) + \Sigma_i |\mu|(W_i - C_i) < 2\epsilon.$$

Since the W's are pairwise disjoint, $||x|| \leq 1$ and so $||x'|| > |\mu|(T) - 3\epsilon$. Thus $||x'||=|\mu|(T)$ and we may state:

(2.4-5) CONTINUOUS LINEAR FUNCTIONALS ON $C_b(T,\underset{\sim}{F})$ AND REGULAR ADDITIVE SET FUNCTIONS If T is completely regular (normal) and Hausdorff, then $C_b(T,\underset{\sim}{F})'$ is isometrically isomorphic to the linear space of all bounded finitely additive regular $\underset{\sim}{F}$-valued set functions defined on $\mathcal{A}_Z(\mathcal{A}_c)$ equipped with the total variation norm. (When $\underset{\sim}{F}=\underset{\sim}{R}$ these are the elements of $\mathcal{M}$, the finitely additive measures of Sec. 1.7.)

Certainly both results of (2.4-5) apply to the compact Hausdorff situation. Indeed in this case we may say even more. As a consequence of our next result the set functions may be considered to be countably additive set functions defined on $\mathcal{B}_a$ or $\mathcal{B}$ as the case may be.

(2.4-6) (Alexandrov) A BOUNDED REGULAR ADDITIVE SET FUNCTION IS COUNTABLY ADDITIVE If μ is a bounded regular, real- or complex-valued, additive set function defined on $\mathcal{A}_c$, the algebra generated by the closed subsets $\mathcal{C}$ of a compact Hausdorff space T, then μ is countably additive on $\mathcal{A}_c$, i.e. $\mu(\bigcup A_n)=\Sigma\, \mu(A_n)$ whenever (A_n) is a countable family of pairwise disjoint sets from $\mathcal{A}_c$ with union in $\mathcal{A}_c$. Moreover μ has a unique regular countably additive extension to the σ-algebra $\mathcal{B}$ of Borel subsets of T, i.e. the σ-algebra generated by $\mathcal{C}$.

Remark As in previous results, e.g. (2.4-4), the theorem remains valid if the closed sets $\mathcal{C}$ are replaced by the zero sets $\mathcal{Z}$, $\mathcal{A}_c$ by $\mathcal{A}_Z$ and $\mathcal{B}$ by $\mathcal{B}_a$ the σ-algebra generated by $\mathcal{Z}$ (sometimes called the Baire sets).

Proof First we show that the total variation $|\mu|$ is countably additive on $\mathcal{A}_c$. Let $\epsilon > 0$ be given and let (A_n) be as in the statement of the theorem. Choose $E \in \mathcal{C}$ such that $E \subset A$ and $|\mu|(A-E) < \epsilon$ and for each $n \geq 1$ let U_n be an open subset of T containing A_n such that $|\mu|(U_n - A_n) < \epsilon/2^n$. As E is compact, there are $U_1,\ldots,U_n$ that cover E so that

$$\Sigma_k |\mu|(A_k) \geq \Sigma_k |\mu|(U_k) - \epsilon \geq \Sigma_{k=1}^{n} |\mu|(U_k) - \epsilon .$$

Since $|\mu|$ is a nonnegative finitely additive set function, it is monotone and finitely subadditive and it follows that

$$\Sigma_{k\in\underset{\sim}{N}} |\mu|(A_k) \geq \Sigma_{k=1}^{n} |\mu|(U_k) - \epsilon \geq |\mu|(\cup_{k=1}U_k) - \epsilon$$

$$\geq |\mu|(E) - \epsilon \geq |\mu|(A) - 2\epsilon.$$

Thus $\Sigma_k |\mu| (A_k) \geq |\mu|(A)$. On the other hand

$$\Sigma_{k=1}^{n} |\mu| (A_k) = |\mu| (\cup_{k=1}^{n} A_k) \leq |\mu|(A)$$

for any n so $|\mu|$ is seen to be countably additive.

As $|\mu|$ is a finite-valued set function on an algebra, it is bounded, so that $|\mu|(A) = \Sigma_k |\mu|(A_k) < \infty$ and

$$|\mu|(\cup_{k\geq n} A_k) = \Sigma_{k\geq n} |\mu| (A_k) \to 0$$

as $n \to \infty$. Thus

$$|\mu(A) - \Sigma_{k=1}^{n-1} \mu(A_k)| = |\mu(\cup_{k\geq n} A_k)| \leq |\mu| (\cup_{k\geq n} A_k) \overset{n}{\to} 0$$

and μ is countably additive on $\mathcal{A}_c$.

Since a bounded real-valued additive set function is regular iff its positive and negative components with respect to its Jordan decomposition are regular and a complex-valued additive set function is regular iff its real and imaginary parts are regular, we may reduce the problem of existence of an extension of μ to $\mathcal{B}$ to the case where μ is a regular nonnegative additive set function on $\mathcal{A}_c$. Given such a set function μ first we extend it to an outer measure (i.e. a nonnegative monotone countably subadditive set function which takes $\emptyset$ into 0) μ^* on $\mathcal{P}(T)$. To this end let

$$\mu^*(S) = \inf\{\Sigma_{k\in\underset{\sim}{N}} \mu(A_k) | A_k \in \mathcal{A}_c, S \subset \cup_k A_k\}$$

for each $S \subset T$. Clearly μ^* takes $\emptyset$ into 0 and is monotone. To see that it is also countably subadditive, suppose that $S = \cup_k S_k$ and $\epsilon > 0$ is given. For each $k \geq 1$ choose a sequence (A_{kn}) from $\mathcal{A}_c$ such that $S_k \subset \cup_{n\in\underset{\sim}{N}} A_{kn}$ and $\mu^*(S_k) > \Sigma_{n\in\underset{\sim}{N}} \mu(A_{kn}) - \epsilon/2^k$. Now $S \subset \cup_k \cup_n A_{kn}$ so that

$$\mu^*(S) \leq \Sigma_k \Sigma_n \mu(A_{kn}).$$

On the other hand

$$\Sigma_k \mu^*(S_k) > \Sigma_k \Sigma_n \mu(A_{kn}) - \epsilon$$

and it follows that $\mu^*(S) \leq \Sigma_k \mu^*(S_k)$.

To obtain a class of sets on which μ^* is countably additive, we single out the family of μ^*-sets: A subset $R \subset T$ is a μ^*-set if (this condition sometimes being referred to as the Caratheodory condition)

$$\mu^*(S) = \mu^*(S \cap R) + \mu^*(S \cap CR)$$

for each $S \subset T$. First we claim that the class of μ^*-sets is an algebra on which μ^* is additive. Clearly T is a μ^*-set, as is the complement of any μ^*-set. It only remains to show that the intersection of two μ^*-sets is again a μ^*-set. Let X and Y be μ^*-sets so that for each $S \subset T$

(1) $\mu^*(S) = \mu^*(S \cap Y) + \mu^*(S \cap CY)$

(2) $\mu^*(S \cap Y) = \mu^*((S \cap Y) \cap X) + \mu^*((S \cap Y) \cap CX)$

and

(3) $\mu^*(S \cap C(X \cap Y)) = \mu^*(S \cap C(X \cap Y) \cap Y) + \mu^*(S \cap C(X \cap Y) \cap CY)$

$$= \mu^*(S \cap CX \cap Y) + \mu^*(S \cap CY).$$

Now by using Eqs. (2) and (3) to substitute first for $\mu^*(S \cap Y)$ in Eq. (1) and then for $\mu^*(S \cap CX \cap Y)$ in the resulting equation we obtain

$$\mu^*(S) = \mu^*(S \cap (X \cap Y)) + \mu^*(S \cap C(X \cap Y)) \quad (S \subset T)$$

so that $X \cap Y$ is a μ^*-set. To see that μ^* is additive on the μ^*-sets, first observe that for disjoint μ^*-sets X and Y

$$\mu^*(S \cap (X \cup Y)) = \mu^*(S \cap (X \cup Y) \cap X) + \mu^*(S \cap (X \cup Y) \cap CX)$$

$$= \mu^*(S \cap X) + \mu^*(S \cap Y).$$

Thus, by induction,

(4) $\mu^*(S \cap \bigcup_{k=1}^{n} X_k) = \Sigma_{k=1}^{n} \mu^*(S \cap X_k)$

for any finite class of disjoint μ^*-sets $X_1, \ldots, X_n$ and any $S \subset T$. Finite additivity follows by substituting T for S in (4).

What about countable unions of μ^*-sets? Are they μ^*-sets? Let (X_k) be a sequence of μ^*-sets with union X; then for each $n \geq 1$, $\bigcup_{k=1}^{n} X_k$ is a μ^*-set and for each $S \subset T$

$$\mu^*(S) = \mu^*(S \cap \cup_{k=1}^{n} X_k) + \mu^*(S \cap C(\cup_{k=1}^{n} X_k)).$$

Then, by (4),

$$\mu^*(S) = \Sigma_{k=1}^{n} \mu^*(S \cap X_k) + \mu^*(S \cap C(\cup_{k=1}^{n} X_k))$$

for each $n \in \underset{\sim}{N}$. By the monotonicity of μ^*

$$\mu^*(S) \geq \Sigma_{k=1}^{n} \mu^*(S \cap X_k) + \mu^*(S \cap CX)$$

and, letting $n \to \infty$, it follows that

$$\mu^*(S) \geq \Sigma_{k \in \underset{\sim}{N}} \mu^*(S \cap X_k) + \mu^*(S \cap CX).$$

As μ^* is countably subadditive,

$$\mu^*(S) \geq \Sigma_k \mu^*(S \cap X_k) + \mu^*(S \cap CX) \geq \mu^*(S \cap X) + \mu^*(S \cap CX) \geq \mu^*(S)$$

so that two things follow: X is a μ^*-set and

$$\mu^*(S \cap X) = \Sigma_{k \in \underset{\sim}{N}} \mu^*(S \cap X_k)$$

for each $S \subset T$. By the substitution of T for S in this equation, μ^* is seen to be countably additive on the μ^*-sets.

Next we show that μ^* and μ agree on $\mathcal{A}_c$ and each element of $\mathcal{A}_c$ is a μ^*-set so that μ^* is a countably additive nonnegative extension of μ to the Borel sets $\mathcal{B}$. If $A \in \mathcal{A}_c$ then $A \subset A$ and, by the definition of μ^*, $\mu^*(A) \leq \mu(A)$. On the other hand, if (A_n) is a countable family of sets from $\mathcal{A}_c$ and $A \subset \cup_n A_n$ then the sets B_n where $B_1 = A_1$ and $B_n = A_n - \cup_{k=1}^{n-1} A_k$ for $n \geq 2$, are pairwise disjoint elements of $\mathcal{A}_c$ with union equal to $\cup A_n$. Hence by the monotonicity and countable additivity of μ on $\mathcal{A}_c$

$$\mu(A) = \mu(A \cap (\cup_n B_n)) = \mu(\cup_n (A \cap B_n)) = \Sigma_n \mu(A \cap B_n)$$

$$\leq \Sigma \mu(B_n) \leq \Sigma \mu(A_n).$$

Thus $\mu(A) \leq \mu^*(A)$ and $\mu^*|\mathcal{A}_c = \mu$. To see that each $A \in \mathcal{A}_c$ is a μ^*-set, let $S \subset T$, $\epsilon > 0$, and choose a sequence (A_n) from $\mathcal{A}_c$ such that $S \subset \cup_n A_n$ and

$$\epsilon + \mu^*(S) > \Sigma_n \mu(A_n).$$

Hence, by the definition of μ^* and its subadditivity

$$\epsilon + \mu^*(S) > \Sigma_n \mu(A_n) = \Sigma_n(\mu(A_n \cap A) + \mu(A_n \cap CA))$$

$$\geq \mu^*(S \cap A) + \mu^*(S \cap CA) \geq \mu^*(S).$$

As ϵ is arbitrary, it follows that A is a μ^*-set.

Having demonstrated the existence of a nonnegative countably additive extension of μ to $\mathcal{B}$ we now prove its uniqueness. Let λ be another such extension, $B \in \mathcal{B}$, and (A_n) be any sequence from $\mathcal{a}_c$ such that $B \subset \bigcup_{n \in \mathbb{N}} A_n$. Since λ is countably additive and nonnegative, it is monotone and countably subadditive so that

$$\lambda(B) \leq \lambda(\bigcup_n A_n) \leq \Sigma_n \lambda(A_n) = \Sigma_n \mu(A_n).$$

Thus, by the definition of μ^*, $\lambda(B) \leq \mu^*(B)$. As $CB \in \mathcal{B}$ we also have $\lambda(CB) \leq \mu^*(CB)$. Consequently

$$\mu(T) = \lambda(T) = \lambda(B) + \lambda(CB) \leq \mu^*(B) + \mu^*(CB) = \mu^*(T) = \mu(T).$$

Hence $\lambda(B)+\lambda(CB)=\mu^*(B)+\mu^*(CB)$ so $\lambda(B)=\mu^*(B)$ for each $B \in \mathcal{B}$.

Finally then it only remains to show that μ^* is regular on $\mathcal{B}$. Let $B \in \mathcal{B}$ and choose a sequence (A_n) from $\mathcal{a}_c$ such that

$$\mu^*(B) \geq \Sigma_n \mu(A_n) - \epsilon .$$

As μ is regular on $\mathcal{a}_c$ there exist open sets $U_n \supset A_n$ such that $\mu(A_n) > \mu(U_n) - \epsilon/2^n$ for each n. Thus

$$\mu^*(B) \geq \Sigma_n \mu(U_n) - 2\epsilon \geq \mu(\bigcup_n U_n) - 2\epsilon$$

where the last inequality follows by the countable subadditivity of μ on $\mathcal{a}_c$. As $\bigcup_{n \in \mathbb{N}} U_n$ is open, μ^* is seen to be regular on $\mathcal{B}$. ∇

As was mentioned earlier there is a characterization of $C(T,\underset{\sim}{F},c)'$ for completely regular Hausdorff spaces T in terms of set functions defined on the Baire sets. For an outline of this result, see Exercise 2.4. At this time we present a characterization of $C(T,\underset{\sim}{F},c)'$ in terms of "Borel set functions" with "compact support." A set function μ defined on the Borel subsets of a completely regular Hausdorff space T has <u>compact support</u> if there is a compact set $K \subset T$ with the property that $|\mu|(T-K)=0$.

<u>Theorem 2.4-1 A MEASURE THEORETIC CHARACTERIZATION OF $C(T,\underset{\sim}{F},c)'$</u> If T is a completely regular Hausdorff space then the continuous dual $C(T,\underset{\sim}{F},c)'$ of $C(T,\underset{\sim}{F},c)$ is algebraically isomorphic to the space of countably additive regular $\underset{\sim}{F}$-valued set functions with compact support on the Borel subsets $\mathcal{B}$ of T via the map $\mu \to \int \cdot d\mu$.

Proof Certainly the collection of all countably additive regular $\underset{\sim}{F}$-valued set functions with compact support is a vector space over $\underset{\sim}{F}$ and the mapping $\mu \to \int \cdot d\mu$ is a linear map into the algebraic dual (i.e. the collection of all linear functionals) of $C(T,\underset{\sim}{F})$. To see that $\int \cdot d\mu$ is a continuous linear functional, let K be a "supporting" compact subset for μ, i.e. $|\mu|(T-K)=0$. Then for each $x \in C(T,\underset{\sim}{F})$

$$|\int x d\mu| = |\int_K x d\mu| \leq \int_K |x| d|\mu| \leq |\mu|(K) p_K(x)$$

and $\int \cdot d\mu$ is continuous on $C(T,\underset{\sim}{F},c)$. As for 1-1-ness, suppose that $\mu \neq 0$ so that $|\mu|(K) > 0$ where K is a supporting compact subset of T. Let $B_1,\ldots,B_n \in \mathcal{B}$ be a pairwise disjoint decomposition of K with the property that $\Sigma|\mu(B_i)| > (3/4)|\mu|(K)$. Since μ is regular, there are closed sets C_i and open sets U_i such that $C_i \subset B_i \subset U_i$ $(i=1,\ldots,n)$ and $|\mu|(U_i - C_i) < (1/4n)|\mu|(K)$ for each i, $1 \leq i \leq n$. Utilizing the fact that C_i and $CU_i \cap K$ are compact, it follows by the complete regularity of T that there are continuous functions x_i on T, $0 \leq x_i \leq 1$, such that $x_i(CU_i \cap K)=\{0\}$ while $x_i(C_i)=\{1\}$. (One could find such a continuous function defined on βT and restrict it to T.) Choose $a_i = |\mu(C_i)|/\mu(C_i)$ provided $\mu(C_i) \neq 0$ and 0 otherwise, and let $x=\Sigma_i a_i x_i$. Now consider

$$\int x d\mu = \int_K x d\mu = \Sigma_i a_i \int_K x_i d\mu = \Sigma_i a_i \int_{U_i} x_i d\mu$$

$$= \Sigma_i a_i (\int_{U_i - C_i} x_i d\mu + \int_{C_i} x_i d\mu) = \Sigma_i a_i \int_{U_i - C_i} x_i d\mu + \Sigma\mu(C_i).$$

But $|\int_{U_i - C_i} x_i d\mu| \leq |\mu|(U_i - C_i) < |\mu|(K)/4n$ for each i so it follows that $|\Sigma a_i \int_{U_i - C_i} x_i d\mu| < |\mu|(K)/4$. Furthermore

$$\Sigma|\mu(C_i)| = \Sigma|\mu(B_i)| - \Sigma(|\mu(B_i)| - |\mu(C_i)|)$$

$$> (3/4)\ |\mu|(K) - \Sigma\ |\mu|(B_i - C_i)$$

$$> (3/4)\ |\mu|(K) - (1/4)|\mu|(K) = (1/2)|\mu|(K).$$

Hence $|\int x d\mu| > (1/4)|\mu|(K) > 0$ and the mapping is seen to be 1-1.

Finally we contend that the mapping is onto. If $x' \in C(T,\underset{\sim}{F},c)'$, we shall show first that there is a compact set $K \subset T$ and a continuous linear functional f_K defined on $C(K,\underset{\sim}{F},c)$ such that $x'(x)=f_K(x|_K)$ for each $x \in C(T,\underset{\sim}{F})$.

Since x' is continuous there is a neighborhood of the origin $V_{p_K} = \{x \in C(T,\underset{\sim}{F}) \mid p_K(x) < 1\}$ on which x' is bounded. But the ideal $p_K^{-1}(0) \subset V_{p_K}$ so that $x'(p_K^{-1}(0))=\{0\}$. Thus if $y \in C(K,\underset{\sim}{F})$ we may continuously extend it up to y_e on T and then define $f_K(y)=x'(y_e)$. This definition is unambiguous since if y_e and y_e' are any two extensions of y to T, $y_e'-y_e \in p_K^{-1}(0)$ so that $x'(y_e')=x'(y_e)$. As $x'(V_{p_K})=f_K(S_1(0))$ (where $S_1(0)$ is the open unit ball in $C(K,\underset{\sim}{F},c)$) f_K is continuous.

Focusing attention on f_K we see by (2.4-5,6) that there is a countably additive regular $\underset{\sim}{F}$-valued set function μ_K defined on the Borel subsets $\mathcal{B}_K$ of K such that $f_K(\cdot)=\int \cdot \, d\mu_K$. It only remains to show that a countably additive regular $\underset{\sim}{F}$-valued set function μ can be defined on the Borel subsets $\mathcal{B}$ of T by the equation $\mu(B)=\mu_K(B\cap K)$ $(B \in \mathcal{B})$ for then it follows that μ has compact support and

$$x'(x) = f_K(x|_K) = \int_K x|_K d\mu_K = \int x d\mu$$

for each $x \in C(T,\underset{\sim}{F})$. First it is necessary to show that $B\cap K \in \mathcal{B}_K$ for each $B \in \mathcal{B}$. Let $\mathcal{D}$ denote the collection of all subsets of T of the form $E \cup (B-K)$ where $E \in \mathcal{B}_K$ and $B \in \mathcal{B}$. It is straightforward to verify that $\mathcal{D}$ is a σ-algebra. Furthermore if C is closed in T then $C\cap K \in \mathcal{B}_K$ and $C = (C\cap K) \cup (C-K) \in \mathcal{D}$. Hence, as $\mathcal{D}$ contains all the closed subsets of T, $\mathcal{D} \supset \mathcal{B}$. Thus $\mathcal{D} \cap K=\{D\cap K \mid D \in \mathcal{D}\} \supset \mathcal{B}\cap K=\{B\cap K \mid B \in \mathcal{B}\}$. But $\mathcal{D}\cap K= \mathcal{B}_K$ so that $\mathcal{B}_K$ contains $\mathcal{B}\cap K$ and the set function μ has a meaningful definition. Obviously μ is countably additive, but what about regularity? Let $B \in \mathcal{B}$. As $B\cap K \in \mathcal{B}_K$ given any $\epsilon > 0$ there is a closed set $C \subset K$ and an open subset U of K such that $C \subset B\cap K \subset U$ and $|\mu_K|(U-C) < \epsilon$. Let $\tilde{U}$ be any open subset of T such that $\tilde{U} \supset CK$ and $\tilde{U}\cap K=U$. If $B_1,\ldots,B_k$ are pairwise disjoint Borel subsets of T with union contained in $\tilde{U}-C$ then $\cup(B_i\cap K) \subset U-C$. Thus

$$\Sigma|\mu(B_i)| = \Sigma\ |\mu_K(B_i\cap K)| < \epsilon$$

and, by taking the supremum over all such sums, it follows that $|\mu|(\tilde{U}-C) \leq \epsilon$. Since $C \subset B \subset \tilde{U}$, μ is regular and the proof is complete. ∇

Prior to presenting the result which both defines and establishes the existence of "support of a linear functional" it is necessary to introduce the notion of support for set functions.

<u>(2.4-7) THE SUPPORT OF A SET FUNCTION</u> If μ is a nontrivial regular countably additive set function on $\mathcal{B}$, the Borel subsets of the compact Hausdorff space T, then there exists a unique compact set $K \subset T$ called the <u>support of μ</u>

such that

$$\int x d\mu = \int_K x d\mu$$

for each $x \in C(T,\underset{\sim}{F})$ and no proper compact subset of K has this property. K is denoted by supp(μ).

Proof Let $K=T-\cup\mathcal{J}$ where $\mathcal{J}$ denotes the collection of all open subsets U of T such that $|\mu|(U)=0$. Since μ is regular then for each $k\in\underset{\sim}{N}$ there exists a compact set K_k and an open set V_k such that $K_k\subset K\subset V_k$, $|\mu|(V_k-K_k) < 1/k$. Now $V_k-K\subset V_k-K_k$ so $|\mu|(V_k-K) < 1/k$ and it follows that

$$|\mu|(V_k) = |\mu|(V_k-K) + |\mu|(K) \overset{k}{\rightarrow} |\mu|(K).$$

Consider the compact set $T-V_k$; since $K=T-\cup\mathcal{J}$, it is contained in $\cup\mathcal{J}$ and, therefore, a finite collection $U_1,\ldots,U_n\in\mathcal{J}$ exists such that $T-V_k\subset\cup_{i=1}^n U_i$. Hence $|\mu|(T-V_k)=0$ and consequently $|\mu|(T)=|\mu|(V_k)$ for each k. Clearly then $|\mu|(T)=|\mu|(K)$.

Next we note that

$$\left|\int_T x d\mu - \int_K x d\mu\right| \leq \int_{T-K} |x|\, d|\mu| \leq \sup|x(T)|\, |\mu|(T-K) = 0.$$

Thus $\int x d\mu=\int_K x d\mu$ for each $x\in C(T,\underset{\sim}{F},c)$. To see that K is minimal in the class of compact sets with respect to this property, suppose that a proper compact subset K' of K exists such that $\int_{K'} x d\mu=\int_T x d\mu$ for each $x\in C(T,\underset{\sim}{F},c)$. We claim that a contradiction can be reached if we show that $|\mu|(K-K')=0$. Indeed if this is so then

$$|\mu|(T-K') = |\mu|(T-K) + |\mu|(K-K') = 0.$$

But T-K' is open and T-K' contains $\cup\mathcal{J}$ properly.

To see that $|\mu|(K-K')=0$, let $E_1,\ldots,E_n\in\mathcal{B}$ be a pairwise disjoint decomposition of K-K'. By the regularity of μ, compact sets $C_k\subset E_k$ exist such that $|\mu|(E_k-C_k) < \epsilon/n^2 2$. Since T is normal, there are continuous functions x_{ij} $(1\leq i,j\leq n, i\neq j)$ such that $0\leq x_{ij}\leq 1$, $x_{ij}(C_j)=\{0\}$ and $x_{ij}(C_i)=\{1\}$. Setting $x_i=\Sigma_{j\neq i}x_{ij}/n-1$ we see that $0\leq x_i\leq 1$, $x_i(C_j)=\delta_{ij}$. Now let $x=\Sigma_i d_i x_i$ where $d_i=|\mu(C_i)|/\mu(C_i)$ if $\mu(C_i)\neq 0$ and 0 otherwise. Consider

$$0 = \int_{K-K'} x d\mu = \int_{\cup E_k} x d\mu = \Sigma\int_{E_k} x d\mu = \Sigma\int_{C_k} x d\mu + \Sigma\int_{E_k-C_k} x d\mu$$

$$= \Sigma\int_{C_k} d_k x_k d\mu + \Sigma\int_{E_k-C_k} x d\mu$$

$$= \Sigma|\mu(C_k)| + \Sigma\int_{E_k-C_k} x d\mu$$

Thus $\Sigma|\mu(C_k)| = \Sigma \int_{E_k-C_k} -x d\mu$ and it follows that

$$\Sigma|\mu(C_k)| \leq \Sigma|\mu|(E_k-C_k)n < (\epsilon/2n^2)n^2$$

Consider

$$\Sigma|\mu(E_k)|$$

$$\leq \Sigma|\mu(C_k)| + \Sigma|\mu(E_k-C_k)|$$

$$\leq \Sigma|\mu(C_k)| + \Sigma|\mu|(E_k-C_k) < \epsilon.$$

Since ϵ is arbitrary, $\Sigma|\mu(E_k)| = 0$. Finally, taking the supremum of all such sums, we obtain $|\mu|(K'-K)=0$.

As for uniqueness suppose that C is another such set. Consider the open set T-C. It suffices to show that $|\mu|(T-C)=0$ for it follows first that $T-C \in \mathcal{J}$. This implies that $C \subset K$ and, by the minimality of K, we see that C=K. To show that $|\mu|(T-C)=0$, we proceed in exactly the same way as we did to prove the minimality of K and we leave the remaining details to the reader. ∇

(2.4-8) SUPPORT OF A CONTINUOUS LINEAR FUNCTIONAL If T is a completely regular Hausdorff space and $x' \in C(T,\underset{\sim}{F},c)'$ then there exists a compact subset supp(x') of T, called the support of x', with the properties:

(a) whenever $x \in C(T,\underset{\sim}{F})$ vanishes on supp(x'), $x'(x)=0$;

(b) if E is a closed subset of T having property (a), i.e. $x'(x)=0$ whenever $x(E)=\{0\}$, then $E \supset \text{supp}(x')$.

Proof In the proof of Theorem 2.4-1 it was established that given x' there is a compact set $K \subset T$ and countably additive regular $\underset{\sim}{F}$-valued set functions μ and μ_K defined on the Borel sets $\mathcal{B}$ of T and $\mathcal{B}_K$ of K respectively such that $\mu_K(B \cap K)=\mu(B)$ for each $B \in \mathcal{B}$ and $x'(x)=\int_K x d\mu = \int x|_K d\mu_K$ for each $x \in C(T,\underset{\sim}{F})$. We claim that the support of μ_K, whose existence was demonstrated in (2.4-7), is actually the support of x'. As $x'(x)=\int_{\text{supp}(\mu_K)} x|_K d\mu_K = \int_{\text{supp}(\mu_K)} x d\mu$ for $x \in C(T,\underset{\sim}{F})$, $x'(x)=0$ whenever x vanishes on $\text{supp}(\mu_K)$ and $\text{supp}(\mu_K)$ is seen to satisfy (a). To prove (b), let E be closed in T with the property that $x'(x)=0$ whenever $x(E)=\{0\}$. It will follow that $E \supset \text{supp}(\mu_K)$ if we can show that $|\mu_K|(\text{supp}(\mu_K) \cap CE) = |\mu|(\text{supp}(\mu_K) \cap CE)=0$ for then the minimality of $\text{supp}(\mu_K)$ ((2.4-7)) implies that $\text{supp}(\mu_K)=\text{supp}(\mu_K) \cap E$. To this end suppose $\epsilon > 0$ has been given and let $B_1,\ldots,B_n \in \mathcal{B}$ be a pairwise disjoint decomposition of $\text{supp}(\mu_K) \cap CE$. By the regularity of μ there are pairwise disjoint compact sets $K_1,\ldots K_n$ such that $B_i \supset K_i$ and $|\mu|(B_i-K_i)<\epsilon/2n^2$

for i=1,...,n. Thus

$$\Sigma|\mu(B_i)| \leq \Sigma|\mu(K_i)| + \epsilon/2n.$$

Since K_i is compact and disjoint from the closed set $E \cup \bigcup_{j\neq i}K_i$, there is a continuous real-valued function x_i defined on the completely regular Hausforff space T such that $0 \leq x_i \leq 1$, $x_i(K_i)=\{1\}$, and $x_i(E \cup \bigcup_{j\neq i}K_j)=\{0\}$. Let $x=\Sigma d_i x_i$ where $d_i=|\mu(K_i)|/\mu(K_i)$ if $\mu(K_i)\neq 0$ and 0 otherwise. As x vanishes on E we have

$$0 = x'(x) = \int_{supp(\mu_K)} x d\mu = \int_{supp(\mu_K)\cap CE} x d\mu = \Sigma \int_{B_i} x d\mu$$

$$= \Sigma \int_{K_i} x d\mu + \Sigma \int_{B_i - K_i} x d\mu = \Sigma|\mu(K_i)| + \Sigma \int_{B_i - K_i} x d\mu .$$

It is clear that $0 \leq x(t) \leq n$ for each $t \in B_i - K_i$ and each i=1,...,n; hence $\Sigma|\mu(K_i)| < \epsilon/2$ and it follows that $\Sigma|\mu(B_i)| < \epsilon$. Since the disjoint decomposition (B_i) and ϵ are arbitrary, $|\mu|(supp(\mu_K)\cap CE)=0$ and the proof is complete. ▽

2.5 Barreledness of C(T,F,c) The notion of barreled space, introduced in Bourbaki (1953a), has its roots in a desire to determine a wider class of spaces for which a Banach-Steinhaus principle holds, taking as that principle the statement: If H is a family of linear maps of the Banach space X into the normed space Y such that $H(x)=\{h(x) | h \in H\}$ is a bounded subset of Y for each $x \in X$, then $\|H\|$ is bounded. (In function-theoretic language, H is uniformly bounded.) Thus, in the situation just mentioned, pointwise boundedness implies equicontinuity. The intrinsic properties of spaces yielding this kind of connection are principally what was sought. In view of this desire it is natural to define barreledness externally, via conditions on the space of continuous linear maps of such spaces: The locally convex space X is barreled if for any locally convex space Y, any pointwise bounded family H of continuous linear maps of X into Y is equicontinuous. The external description is equivalent to the internal description: A locally convex space X is barreled if every barrel in X is a neighborhood of 0.

Obviously Banach spaces are barreled; incomplete normed spaces, however, needn't be. Thus if T is compact and Hausdorff, C(T,F,c) is barreled, but there should certainly be completely regular Hausdorff spaces T for which C(T,F,c) is not barreled. Necessary and sufficient conditions on T for $C(T,\underset{\sim}{F},c)$ to be barreled were obtained independently by Nachbin (1954) and Shirota (1954) - $C(T,\underset{\sim}{F},c)$ is barreled iff for each closed non-compact subset E of T there exists $x\in(T,\underset{\sim}{F})$ which is unbounded on E - and this section is devoted to proving that result. A later ramification, due to Warner (1958), concerning necessary and sufficient conditions for $C(T,\underset{\sim}{F},c)$ to be infrabarreled is also presented: $C(T,\underset{\sim}{F},c)$ is infrabarreled iff for each closed non-compact subset E of T there is a nonnegative lower semi-continuous function y on T which is unbounded on E and bounded on each compact subset of T.

Theorem 2.5-1 BARRELEDNESS OF $C(T,\underset{\sim}{F},c)$ If T is a completely regular Hausdorff space then $C(T,\underset{\sim}{F},c)$ is barreled iff for each closed non-compact subset S of T there is some $x\in C(T,\underset{\sim}{F})$ which is unbounded on S. We refer to such spaces T as NS-spaces.

Proof To see that the condition is necessary, suppose that $C(T,\underset{\sim}{F},c)$ is barreled. Let S be a closed non-compact subset of T and consider $\overline{V}_{p_S} = \{x\in C(T,\underset{\sim}{F}) \mid \sup|x(S)| < 1\}$. The requirement that there be some $x\in C(T,\underset{\sim}{F})$ which is unbounded on S is clearly equivalent to $\overline{V}_{p_S}$ not being absorbent. Since $\overline{V}_{p_S}$ is almost a barrel, being closed and absolutely convex, to prove the necessity of the condition, it is enough to show that $\overline{V}_{p_S}$ is not a neighborhood of 0 in the barreled space $C(T,\underset{\sim}{F},c)$. Suppose that $\overline{V}_{p_S}$ is a neighborhood of 0 so that a compact set K and a positive number ϵ exist such that $\epsilon V_{p_K} \subset \overline{V}_{p_S}$. This, however, implies that $S\subset K$ for if $t\in S-K$ then there must be some $x\in C(T,\underset{\sim}{F})$ such that $x(t)=2$ while $x(K)=\{0\}$. Hence $x\in(\epsilon V_{p_K})$ but $x\notin\overline{V}_{p_S}$ which is contradictory. Thus $\epsilon V_{p_K}\subset\overline{V}_{p_S}$ implies that $S\subset K$ which means that S is compact, contrary to assumption. Therefore $\overline{V}_{p_S}$ can contain no basic neighborhood of 0. The necessity of the condition is now evident.

To establish the sufficiency, we shall prove indirectly that an arbitrary barrel V is a neighborhood of 0. First we use the barreledness of the Banach algebra (with sup norm) of uniformly bounded continuous $\underset{\sim}{F}$-valued functions on T to construct a closed subset K of T with the property that a scalar multiple of V_{p_K} is contained in V. Then, to show that K is compact, we assume that it is not and, by using the condition, we obtain a contradiction to the fact that V is absorbent.

To this end let Y be the subalgebra of $C(T,\underset{\sim}{F})$ consisting of all uniformly bounded continuous functions. Equipped with the uniform norm, Y is a Banach algebra, and, as such, it is barreled. Now $V\cap Y$ is certainly absolutely convex and absorbent. That it is closed in Y and therefore a barrel in Y follows from the facts that V is closed in $C(T,\underset{\sim}{F},c)$ and that the norm topology is finer than the induced compact-open topology of Y. Thus $V\cap Y$ is a Y-barrel so a $d > 0$ exists such that

$$d\bar{V}_{p_T} \subset V\cap Y \subset V.$$

Next we establish the important technicality that for a subset $S\subset T$ if all the elements of $C(T,\underset{\sim}{F})$ that vanish on S belong to the absolutely convex set V, then $aV_{p_S} \subset V$ for some $a > 0$, that is if the elements which vanish on S must belong to V, then so must all sufficiently small elements on S belong to V. First we consider the case $\underset{\sim}{F}=\underset{\sim}{R}$. Our choice for a is d/2.

Suppose that $p_S(x) \leq d/2$ and let $y=\max(x,d/2)+\min(x,-d/2)$. Since 2y vanishes on S, $2y\in V$. Moreover, it is easy to see that $p_T(2(x-y)) \leq d$ so $2(x-y)\in V$. Hence, by the absolute convexity of V, it follows that $x=(1/2)(2y)+(1/2)(2(x-y))\in V$ which is the desired conclusion.

Next let $\underset{\sim}{F}=\underset{\sim}{C}$. In this case $a=d/4$ will suffice. If $p_S(x) \leq d/4$ and x_r is the real part of x, then $p_S(x_r) \leq d/4$. It follows as in the real case (by choosing $y=\max(x_r,d/4)+\min(x_r,-d/4)$) that $2x_r\in V$. Similarly, letting x_i denote the imaginary part of x, $2x_i\in V$. Thus by the absolute convexity of V, $x=(1/2)(2x_r)+(i/2)(2x_i)\in V$.

In view of the result just established, it is now a matter of producing a compact set K which has the property that each of the elements vanishing on it belong to V. Furthermore $V=V^{oo}$, and for x to belong to V^{oo} it certainly suffices for each $x'\in V^o$ to vanish on x. We also know that supp(x'), the support of x' (as defined in (2.4-8)), has the property that $x'(x)=0$ whenever x vanishes on supp(x'). Thus the natural choice for K is the closure of $\bigcup_{x'\in V^o}$ supp(x'), since whenever x vanishes on it, $x'(x)=0$ for each $x'\in V^o$ and $x\in V^{oo}=V$.

It remains only to show that $K=cl_T(\bigcup_{x'\in V^o}\text{supp}(x'))$ is compact; it is in doing this that we use the hypothesis for the first and only time. Suppose that K is not compact. Then, by the condition, there is an element $y\in C(T,\underset{\sim}{F})$ which is unbounded on K. Thus the open sets $U_n=|y|^{-1}(n,\infty)$ constitute a decreasing sequence such that $\bigcap_n cl\ U_n=\emptyset$ and $U_n\cap K\neq\emptyset$ for each $n\in\underset{\sim}{N}$. Choose $x_n'\in V^o$ such that $U_n\cap \text{supp}(x_n')\neq\emptyset$. Since $CU_n \not\supset \text{supp}(x_n')$, it follows that a function $x_n\in C(T,\underset{\sim}{F})$ exists such that x_n vanishes on CU_n and $x_n'(x_n)\neq 0$. Moreover we may assume that $x_n'(x_n)=1$. Since the sets $(cl\ U_n)$ are a decreasing sequence and $\bigcap_{n\in\underset{\sim}{N}} cl\ U_n=\emptyset$, it follows that the family $(C(cl\ U_n))$ of complements of the U_n is increasing and $\bigcup_{n\in\underset{\sim}{N}} C(cl\ U_n)=T$. As

$x_m(C(\text{cl } U_n))=\{0\}$ for $m \geq n$ then, for any sequence (c_n) of complex numbers, $x=\Sigma_{n\in\mathbb{N}} c_n x_n$ reduces to a finite sum of continuous functions on each of the open covering sets $C(\text{cl } U_n)$ and it follows that $x\in C(T,\mathbb{F})$. By (2.4-8) each of the sets $\text{supp}(x_m')$ is compact so, for m fixed, $\text{cl } U_n \cap \text{supp}(x_m')=\emptyset$ for all but finitely many values of n and, by going to a subsequence of the U_n's if necessary, we may assume that $\text{cl } U_n \cap \text{supp}(x_m')=\emptyset$ for all $n > m$. Thus, for each $m\in\mathbb{N}$, c_m may be chosen so that

$$x_m'(x) = \Sigma c_n x_m'(x_n) = c_m + \Sigma_{n=1}^{m-1} c_n x_m'(x_n) = m.$$

To complete the proof we show that V cannot absorb x. Let a be an arbitrary positive number and suppose that $x\in aV$. Since $aV=(a^{-1}V^o)^o$ the absurd conclusion follows that $(a^{-1})x_m'(x)=(a^{-1})m \leq 1$ for each $m \geq 1$. ∇

The one and only time that the condition was used in proving its sufficiency for $C(T,\mathbb{F},c)$ to be barreled was in establishing the compactness of the set $\text{cl}(\bigcup_{x'\in V^o} \text{supp}(x'))$ for the arbitrary barrel V. Therefore it suffices to substitute the condition: $\text{cl}(\bigcup_{x'\in B} \text{supp}(x'))$ is compact for each weakly bounded set $B\subset C(T,\mathbb{F},c)'$ for the condition stated in the preceding theorem to obtain the conclusion that $C(T,\mathbb{F},c)$ is barreled. Conversely, if $C(T,\mathbb{F},c)$ is barreled, then the condition of Theorem 2.5-1 holds and it follows from the proof of the theorem that $\text{cl}(\bigcup_{x'\in V^o} \text{supp}(x'))$ is compact for each barrel V of $C(T,\mathbb{F},c)$. Since a set $B\subset C(T,\mathbb{F},c)'$ is weakly bounded iff it is contained in the polar of some barrel, $\text{cl}(\bigcup_{x'\in B} \text{supp}(x'))$ is compact for each weakly bounded B. Thus we have established another equivalent condition for $C(T,\mathbb{F},c)$ to be barreled and it is recorded in our next result.

If $B\subset C(T,\mathbb{F},c)'$ the set $\text{cl}(\bigcup_{x'\in B} \text{supp}(x'))$ is called the <u>support of B</u> and is denoted by supp(B).

<u>(2.5-1) SUPPORT AND BARRELEDNESS</u> Let T be a completely regular Hausdorff space. $C(T,\mathbb{F},c)$ is barreled iff supp(B) is compact for each weakly bounded set $B\subset C(T,\mathbb{F},c)'$.

As every barreled space carries its Mackey topology, $C(T,\mathbb{F},c)$ carries the Mackey topology whenever the support of each weakly bounded set $B\subset C(T,\mathbb{F},c)'$ is compact. Actually for $C(T,\mathbb{F},c)$ to carry its Mackey topology this condition need only hold for absolutely convex weakly compact B. We prove this and its converse next.

<u>(2.5-2) SUPPORT AND MACKEY TOPOLOGY</u> If T is a completely regular Hausdorff space then $C(T,\mathbb{F},c)$ carries Mackey topology iff supp(B) is compact for each weakly compact absolutely convex set $B\subset C(T,\mathbb{F},c)'$.

Proof Since a locally convex Hausdorff space carries its Mackey topology iff every weakly compact absolutely convex subset of its dual is equicontinuous, we shall show that the weakly compact absolutely convex set B is equicontinuous iff its support is compact. If B is equicontinuous there is a compact set $K \subset T$ and a positive number a such that $B \subset (aV_{p_K})^{o}$. We claim that $supp(B) \subset K$ and is therefore compact. Suppose that $x \in C(T,\underset{\sim}{F})$ vanishes on K. Then $p_K(mx) < a$ for each integer m and it follows that $|x'(mx)| \leq 1$ for each such m and any $x' \in B$. Thus $x'(x)=0$ and, by the minimality of $supp(x')$, $supp(x') \subset K$ for each $x' \in B$. Hence $supp(B) \subset K$.

Conversely, suppose that supp(B) is compact. Since B is weakly compact, the set $B(x)=\{x'(x) \mid x' \in B\}$ is bounded for each $x \in C(T,\underset{\sim}{F})$. Corresponding to each $x' \in B$ we define a linear functional x'' on $C(supp(B),\underset{\sim}{F})$ by the formula $x''(y)=x'(y')$ where $y \in C(supp(B),\underset{\sim}{F})$ and y' is any continuous extension of y up to T. That x'' is well-defined follows immediately from the observation that any two extensions agree on supp(B), therefore on supp(x'). To see that $x'' \in C(supp(B),\underset{\sim}{F},c)'$, it is necessary to show that x'' is bounded on $\{y \in C(supp(B),\underset{\sim}{F}) \mid \sup|y(supp(B)| \leq 1\}$. Thus it suffices to show that x' is bounded on $V_{p_{supp(B)}}=\{x \in C(T,\underset{\sim}{F}) \mid \sup|x(supp(B)| \leq 1\}$. To this end, suppose that x is real-valued and $\sup|x(supp(B))| \leq 1$. Let $w=\max(x,1)+\min(x,-1)$. Then w vanishes on supp(B), which contains supp(x'), so $x'(w)=0$. By the continuity of x' there is a compact set K such that x' is bounded on $\overline{V}_{p_K}$, i.e. there is a positive number M such that $x'(z) \leq M$ for all $z \in \overline{V}_{p_K}$. Since x-w is clearly bounded by 1 on all of T, it follows that $|x'(x-w)| \leq M$.

Having shown that each $x'' \in C(supp(B),\underset{\sim}{F},c)'$, we set $B''=\{x'' \mid x' \in B\}$. As $B''(y)=B(y')$ for each $y \in C(supp(B),\underset{\sim}{F})$ and each continuous extension y' of y to T, we see that B'' is a pointwise bounded collection of linear functionals on the Banach space $C(supp(B),\underset{\sim}{F},c)$. Hence, by the Banach-Steinhaus theorem, B'' is equicontinuous, i.e. for each $a > 0$ there is a positive d such that $\sup B''(y) < a$ whenever $\sup|y(supp(B))| < d$. By the definition of B'' it follows that $\sup|B(x)| < a$ whenever $\sup|x(supp(B))| < d$ and B is equicontinuous.▽

In proving that the weakly compact set B is equicontinuous whenever supp(B) is compact, the weak compactness of B is used only to guarantee that B(x) is bounded for each $x \in C(T,\underset{\sim}{F})$. Consequently the implication is also valid for weakly bounded B.

(2.5-3) SUPPORT AND EQUICONTINUITY A weakly bounded subset B of $C(T,\underset{\sim}{F},c)'$, T completely regular and Hausdorff, is equicontinuous whenever supp(B) is compact.

We now make use of (2.5-3) in establishing a necessary and sufficient condition for $C(T,\underset{\sim}{F},c)$ to be infrabarreled, a condition analogous to the one given in Theorem 2.5-1.

<u>(2.5-4) INFRABARRELEDNESS OF $C(T,\underset{\sim}{F},c)$</u> Let T be a completely regular Hausdorff space. $C(T,\underset{\sim}{F},c)$ is infrabarreled (i.e. every bornivorous barrel is a neighborhood of 0) iff for each closed noncompact subset S of T there is a nonnegative lower semicontinuous function y defined on T which is unbounded on S and bounded on each compact subset of T.

<u>Proof</u> Recall that a locally convex Hausdorff space is infrabarreled iff each strongly bounded subset of the continuous dual is equicontinuous (Horvath 1966, p. 217, Prop. 6).

Now, to prove necessity, let S^* be the homeomorphic image in the weakly topologized space $C(T,\underset{\sim}{F},c)'$ of the closed non-compact set S, i.e. the set of evaluation maps t^* as t runs through S. Since S^* is not weakly compact,

S^* is not equicontinuous; moreover since the space $C(T,\underset{\sim}{F},c)$ is infrabarreled, S^* is not strongly bounded (Horvath 1966, p. 217, Prop. 6). Thus a weakly bounded set A $C(T,\underset{\sim}{F},c)$ exists such that $A(S)=S^*(A)$ is unbounded. Consider the nonnegative function $t \to y(t)=\sup|A(t)|$. As A is weakly bounded, $t^*(A)=A(t)$ is bounded for each $t\in T$ and y is real-valued. Moreover each function $t \to |x(t)|$ is continuous on T so the map $t \to \sup_{x\in A}|x(t)|$ is lower semicontinuous (Dieudonné 1970, (12.7.6), p. 25). Finally $\sup|y(S)| = \sup|A(S)| = \infty$ while for each compact $K\subset T$, $\sup|y(K)| = p_K(A)<\infty$ since A is bounded and y is the desired function.

Conversely suppose that the condition holds. We must prove that each strongly bounded subset $H\subset C(T,\underset{\sim}{F},c)'$ is equicontinuous. Since H is weakly bounded it suffices by (2.5-3) to show that supp(H) is compact. Assuming that supp(H) is not compact, we use the condition to contradict the strong boundedness of H. Indeed if the closed set supp(H) is not compact there is a real-valued nonnegative lower semicontinuous function y defined on T and unbounded on supp(H). Thus $y^{-1}(n,\infty)$ meets supp(H) for each n. As $y^{-1}(n,\infty)$ is open and $\text{supp}(H)=\text{cl}(\bigcup_{x'\in H}\text{supp}(x'))$ then, for each $n\in\underset{\sim}{N}$, there is an $x_n'\in H$ whose support meets $y^{-1}(n,\infty)$. Next we construct a weakly bounded subset $\{y_n\}$ of $C(T,\underset{\sim}{F},c)$ on which H is unbounded, thereby establishing the desired contradiction. First note that $\text{supp}(x_n')\not\subset y^{-1}(-\infty,n]$ so that by (2.4-8)(b) a function $x_n\in C(T,\underset{\sim}{F})$ exists such that x_n vanishes on $y^{-1}(-\infty,n]$ and $x_n'(x_n)=1$. We define a weakly bounded sequence of functions $y_n=\Sigma_{k=1}a_k x_k$ such that $x_n'(y_n)=n$ for each n. To see that the set $\{y_n\}$ is bounded in $C(T,\underset{\sim}{F},c)$ - hence weakly bounded - let K be any compact subset of T. Since y is bounded on K, $y^{-1}(n,\infty)\cap k=\emptyset$ for all n larger than or equal to some m. Thus for each $t\in K\subset y^{-1}(-\infty,n]$, $x_n(t)=0$ for $n\geq m$ so that $y_n(K)=y_m(K)$ for all $n\geq m$. Hence with $p_K(w)=\sup_{t\in K}|w(t)|$, $\{p_K(y_n)|n\in\underset{\sim}{N}\}$ is a bounded set for each K and $\{y_n\}$ is bounded. As for the specific values of a_n, they are easily established by the equations $\Sigma_{k=1}^{n} a_k x_n'(x_k)=n$ $(n=1, 2,\ldots)$. This establishes the contradiction and completes the proof. ∇

<u>2.6 Bornologicity of $C(T,\underset{\sim}{F},c)$</u> A linear map between normed spaces is continuous iff it maps bounded sets into bounded sets. A wider class of spaces for which boundedness implies continuity is the bornological spaces, first defined by Mackey (1946). The external description of such spaces is that a locally convex space X is <u>bornological</u> if, for any locally convex space Y, a linear map $A:X\to Y$ which maps bounded sets into bounded sets must be continuous. An equivalent internal description is that a locally convex space X is bornological if each balanced convex bornivore in X is a neighborhood of 0.

There are incomplete normed spaces which are bornological (since they are metrizable) but not barreled.* In the converse direction, each complete bornological space must be barreled and, as will be shown in this section ((2.6-1)), if $C(T,\underset{\sim}{F},c)$ is bornological, it must be barreled. Quite early on Dieudonne (1953) raised the question of whether there are barreled spaces which are not bornological. Shortly afterwards the question was answered in the negative by Nachbin (1954) and Shirota (1954) who determined necessary and sufficient conditions on T for $C(T,\underset{\sim}{R},c)$ to be barreled (Theorem 2.5-1: $C(T,\underset{\sim}{F},c)$ is barreled iff T is an NS-space) and to be bornological (Theorem 2.6-1): The bornologicity of $C(T,\underset{\sim}{R},c)$ is equivalent to T being replete. Thus to exhibit a barreled space which is not bornological, it suffices to produce an NS-space which is not replete, and this is done in Example 2.6-1.

---

* The subspace X of the normed space ℓ_2 consisting of the sequences (μ_n) which are eventually 0, being dense in ℓ_2, has ℓ_2 as its continuous dual X'. Since X is metrizable, X is bornological. But there are weakly bounded sets in X' - $\{(\mu_n)\in\ell_2 \,|\, |\mu_n|\leq n,\ n\in\underset{\sim}{N}\}$ for example - which are not strongly (=in the norm here) bounded, hence not equicontinuous. Hence the Banach-Steinhaus theorem does not hold and X is not barreled.

Theorem 2.6-1 $C(T,\underset{\sim}{F},c)$ IS BORNOLOGICAL IFF T IS REPLETE For $C(T,\underset{\sim}{F},c)$ to be bornological it is necessary and sufficient that the completely regular Hausdorff space T be replete, i.e. $\upsilon T=T$.

Proof Necessity: Suppose that T is not replete, so that there is some $t_o \in \upsilon T-T$. It follows that the map t_o^*, $x \to x^{\upsilon}(t_o)$, x^{υ} denoting the unique continuous extension of $x \in C(T,\underset{\sim}{F})$ to υT, is a nontrivial $\underset{\sim}{F}$-valued homomorphism of $C(T,\underset{\sim}{F})$. Only those t^*'s from T can produce continuous homomorphisms of $C(T,\underset{\sim}{F},c)$, so t_o^* must be discontinuous.† To conclude that $C(T,\underset{\sim}{F},c)$ is not bornological, we need only show that t_o^* maps bounded sets into bounded sets. If t_o^* does not have this property then there is a bounded set $B \subset C(T,\underset{\sim}{F},c)$ and a sequence (x_n) of points of B such that $x_n^{\upsilon}(t_o)=t_o^*(x_n) \to \infty$. Now consider the open sets $V_n=\{t \in \beta T \mid x_n^{\upsilon}(t) > x_n^{\upsilon}(t_o) -1\}$. As the V_n's are neighborhoods of $t_o \in \upsilon T$, there is a point $s \in (\cap V_n) \cap T$ by Theorem 1.5-1(b). But $s \in \cap V_n$ implies that $s^*(x_n)=x_n^{\upsilon}(s) \to \infty$ which contradicts the fact that s^* (which must be continuous since $s \in T$) must take B into a bounded subset of F. Hence $t_o^*(B)$ is bounded and the necessity of the condition has been demonstrated.

Sufficiency: In proving that $C(T,\underset{\sim}{F},c)$ was barreled in Theorem 2.5-1, an arbitrary barrel V was shown to be a neighborhood of 0 in three steps. First the existence of a $d > 0$ was established such that $d\overline{V}_{p_T} \subset V$. It was proved next that for any subset S of T with the property that $x \in V$ whenever $x(S)=\{0\}$, there was an $a > 0$ such that $a\overline{V}_{p_S} \subset V$. It then only remained to produce a compact subset K of T with the above property. A similar approach is used here. For $C(T,\underset{\sim}{F},c)$ to be bornological, it is necessary and sufficient for each absolutely convex set that absorbs all bounded sets to be a neighborhood of 0. We shall prove that a stronger condition actually holds, i.e. that every absolutely convex set V that absorbs all bounded sets of a certain type (order segments) is a neighborhood of 0. First we define an order segment and prove it to be bounded. If x and y are real valued continuous functions on T such that $x \le y$, then the order segment $[x,y]$ consists of all $w \in C(T,\underset{\sim}{R})$ such that $x \le w \le y$. If K is compact then $p_K(w) \le \max\,(p_K(x), p_K(y))$ for each $w \in [x,y]$ so that $[x,y]$ is bounded.

---

†An apology is perhaps due the reader here for this result is proved in Example 4.10-2. Our desire to place Theorem 2.6-1 near to closer relatives motivated us to locate it here, rather than in Chap. 4.

Proceeding in the fashion outlined above we show that a $d > 0$ exists such that $d\bar{V}_{p_T} \subset V$. Since V absorbs all order segments there is a $b > 0$ such that $[-b1,b1] \subset V$. Now choose $d=b/2$ and suppose that $p_T(x) \leq d$. If $\underset{\sim}{F}=\underset{\sim}{R}$, it follows that $x\in[-b1,b1]\subset V$. In the event that $\underset{\sim}{F}=\underset{\sim}{C}$ we see that $2x_r$ and $2x_i$ belong to $[-b1,b1]$ where x_r and x_i are the real and imaginary parts of x respectively. Thus, by the absolute convexity of V, $x=(1/2)(2x_r)+(i/2)(2x_i)\in V$. Now, as was shown in the proof of the previous theorem, for any subset S of T with the property that x belongs to the absolutely convex set V whenever $x(S)=\{0\}$, there exists a positive number a such that $a\bar{V}_{p_S}\subset V$. Thus it is just a matter of producing a compact S with the above property.

To do this we begin by defining the notion of a support set of V. A closed subset K of βT is a <u>support set</u> of V if the continuous function $x\in V$ whenever $x^{\beta}(K)=0$ (here x^{β} denotes the unique extended real-valued extension of $x\in C(T,\underset{\sim}{R})$ to βT which exists by (1.5-1)). An example of such a set is βT itself. The intersection of all such support sets is called the <u>support of V</u> and is denoted by $K(V)$. After showing that $K(V)$ is a support set of V we shall complete the proof by showing that $K(V)\subset T$.

The fact that $K(V)$ is a support set will be established with the aid of two facts:

(1) A closed subset K of βT is a support set of V iff $x\in V$ whenever x^{β} vanishes on a neighborhood of K (i.e. a superset of K in βT whose interior contains K).

(2) Any finite intersection of support sets is a support set of V.

One half of (1) is trivial. To obtain the other half, suppose that the condition holds, i.e. that $x\in V$ whenever x^{β} vanishes on a neighborhood of K. To see that K is a support set of V, suppose that x^{β} vanishes on K. It remains to show that x^{β} vanishes on some neighborhood of K for then x will belong to V. First suppose that $\underset{\sim}{F}=\underset{\sim}{R}$ and let $G=\{t\in\beta T \mid x^{\beta}(t) < d/2$ for d such that $d\bar{V}_{p_T}\subset V\}$. G is clearly an open neighborhood of K. Next we define $y=\max(x,d/2)+\min(x,-d/2)$ and claim that $(2y)^{\beta}$ vanishes on G. By the uniqueness of the extension $w \to w^{\beta}$, it now follows that $y^{\beta}=\max(x^{\beta},(d/2)1)+\min(x^{\beta},(-d/2)1)$. Hence $(2y)^{\beta}$ must vanish on G and $2y\in V$. Furthermore $2(x-y)\in d\bar{V}_{p_T}\subset V$ and therefore $x=(1/2)(2y+2(x-y))\in V$. If $\underset{\sim}{F}=\underset{\sim}{C}$ we define G to be $\{t\in\beta T \mid x_r^{\beta}(t)\ , x_i^{\beta}(t) < d/4\}$, $y_r=\max(x_r,(d/2)1)+\min(x_r,(-d/4)1)$, and $y_i=\max(x_i,(d/4)1)+\text{minx}(x_i,(-d/4)1)$. After observing that $4(x_r-y_r)$ and $4(x_i-y_i)$ belong to $d\bar{V}_{p_T}\subset V$, it follows as in the real case that $2x_r$ and $2x_i$ belong to V. Thus $x=(1/2)(2x_r)+(i/2)(2x_i)\in V$.

To establish (2) it certainly suffices to show that the intersection of two support sets of V, X and Y, is a support set of V. Set $K=X\cap Y$ and suppose that x^β vanishes on W, an open neighborhood of K in βT. Now X and Y-W are closed and disjoint in the normal space βT so there are disjoint open sets U and P such that $X\subset U$ and $Y-W\subset P$. Since βT is normal, open sets W_1 and W_2 exist such that $K\subset W_1\subset cl_\beta W_1\subset U$ and $Y-W\subset W_2\subset cl_\beta W_2\subset P$. By the normality of βT there exists $z\in C_b(T,\underset{\sim}{R})$ such that $z^\beta(cl_\beta W_1)=\{1\}$ and $z^\beta(cl_\beta W_2)=\{0\}$. Clearly 2xz vanishes on $(W\cup W_2)\cap T$. Since $W\cup W_2$ is open in βT and T is dense in βT,

$$W\cup W_2 = (W\cup W_2)\cap cl_\beta T\subset cl_\beta((W\cup W_2)\cap T).$$

Thus by the continuity of $(2xz)^\beta$, $(2xz)^\beta$ vanishes on $cl_\beta(W\cup W_2)\cap T)$ and therefore also on $W\cup W_2$. Now we can use the facts that Y is a support set of V and $W\cup W_2$ is a neighborhood of Y together with (1) to conclude that $2xz\in V$. In the same way the extension $(2x(1-z))$ vanishes on W_1 and, therefore, $2x(1-z)\in V$. Finally $x=(1/2)(2xz)+(1/2)(2x(1-z))\in V$ and it follows by (1) that K is a support set of V.

Having established (1) and (2), we are now ready to prove that K(V), the support of V, is a support set of V. To this end suppose that x^β vanishes on an open neighborhood W of K(V) in βT. Since βT is a compact Hausdorff space and K(V) is the intersection of all support sets of V, there are support sets $K_1,\dots,K_n$ such that $\cap K_i\subset W$. By (2), $\cap K_i$ is a support set of V so that x must belong to V since x^β vanishes on the open neighborhood W of $\cap K_i$. Thus (1) may be invoked again to conclude that K(V) is a support set of V.

The final thing to be shown is that $K(V)\subset T$. Let $t\in \beta T-T$; we shall prove that $t\notin K(V)$. By Theorem 1.5-1(b) and the repleteness of T, there is a decreasing sequence (W_n) of closed neighborhoods of t such that $(\cap_n W_n)\cap T =\emptyset$. We claim that at least one of the sets βT - int W_n is a support set of V. As $t\in$ int W_n, it follows - after establishing the claim - that $t\notin K(V)$. Suppose that none of the sets βT-int W_n is a support set of V. Then, for each $n\in\underset{\sim}{N}$, there is an element $x_n\in C(T,\underset{\sim}{F},c)$ such that $x_n^\beta(\beta T\text{-int } W_n)=0$ and $x_n\notin V$. Let $y=\sup_n n|x_n|$. To see that y is continuous on T, fix a positive integer m and consider any $n > m$. Clearly $T-W_m\subset \beta T\text{-int } W_n$ and, therefore, x_n vanishes on $T-W_m$. Thus $y=\max(|x_1|,2|x_2|,\dots,m|x_m|)$ on $T-W_m$ so that $y|_{T-W_m}$ is continuous on $T-W_m$. As $(T-W_m)$ is an increasing sequence of open sets, whose union is T, we see that y is continuous on T. Since V absorbs all order segments, there is a positive number k such that $[-y,y]\subset kV$.

Now if $\underset{\sim}{F}=\underset{\sim}{R}$ then for each n, $nx_n \in [-y,y] \subset kV$ and $x_n \in V$ for all $n \geq k$. On the other hand, if $\underset{\sim}{F}=\underset{\sim}{C}$, then $2n|x_{2n}| \leq y$ for each n and it follows that the real and imaginary parts of $2x_{2n}$ belong to V for all $n \geq k$. Hence, by the absolute convexity of V, $x_{2n} \in V$ for all $n \geq k$. Thus in both cases, the contradiction that $x_p \in V$ for some $p \in \underset{\sim}{N}$ has been established. We conclude that one of the sets $\beta T\text{-int } W_n$ is a support set of V from which it follows that $t \notin K(V)$.

In summary we have shown that K(V) is a support set of V, contained in T so, for some $a > 0$, $a\overline{V}_{p_{K(V)}} \subset V$ and the proof is complete. ∇

Recall (Sec. 1.5) that a subset E of the completely regular Hausdorff space T is relatively pseudocompact if the restriction $x|_E$ of each $x \in C(T,\underset{\sim}{R})$ is bounded and ((1.5-4)) that E is relatively pseudocompact iff $cl_\beta E \subset \upsilon T$. If T is not an NS-space, then there is some closed noncompact subset E of T such that each $x \in C(T,\underset{\sim}{R})$ is bounded on E, i.e. E is relatively pseudocompact, or, equivalently, $cl_\beta E \subset \upsilon T$. If T were replete, then $cl_\beta E \subset T$ which implies that E is compact, contrary to assumption. Thus if T is not an NS-space, T is not replete. In view of Theorems 2.5-1 and 2.6-1, then:

<u>(2.6-1) $C(T,\underset{\sim}{F},c)$ BORNOLOGICAL → $C(T,\underset{\sim}{F},c)$ BARRELED</u> Let T be a completely regular Hausdorff space. If $C(T,\underset{\sim}{F},c)$ is bornological then it is barreled.

As was noted at the beginning of the section the characterizations of barreledness and bornologicity of $C(T,\underset{\sim}{F},c)$ given in Theorems 2.5-1 and 2.6-1 make it possible to construct an example of a barreled space which is not bornological.

<u>Example 2.6-1</u> <u>AN NS-SPACE WHICH IS NOT REPLETE</u> Let μ be the initial ordinal of the fourth class* and $[0,\mu)$ the collection of all ordinals smaller than μ equipped with the interval topology.** T is defined to be the subspace of $[0,\mu)$ obtained by deleting from $[0,\mu)$ all non-isolated points (i.e. limit ordinals) with a countable neighborhood base.

<u>(a) T is an NS-space</u>. We claim that for any non-compact subset S of T there is an $x \in C(T,\underset{\sim}{R})$ which is unbounded on S. As any such S must be infinite, a countable subset C of S must exist. The existence of the desired unbounded function will follow once we show that C is closed and

---

*The finite ordinals comprise the first class, the denumerable ordinals the second class, those with the same cardinality as Ω, the first uncountable ordinal, constitute the third class, and the fourth class consists of all those with the same cardinality as the first ordinal larger than all those of the third class.

**The <u>interval topology</u> of $[0,\Omega)$ has as a base intervals of the form $(a,b]$ where $a,b \in [0,\Omega)$ and $a < b$.

discrete and T is normal, for then an unbounded continuous real-valued function on C exists which can be extended to T. The normality of T can be obtained by virtually the same argument used to prove that any of the ordinal spaces $[0,a)$ is normal (see, for example, Dugundji 1966, Chap. VII, Sec. 3, Ex. 2). The closedness as well as the discreteness of C will follow once we show that an arbitrary countable intersection of open subsets of T is open in T. Indeed if this is the case, then any countable union of one-point sets in the T_1 space T is closed. Furthermore, writing $C=\{t_n\}$ there are open subsets U_{nm} of T with the property that $t_n \in U_{nm}$ and $t_m \notin U_{nm}$ for $n \neq m$. Thus the open set $\bigcap_{m\neq n} U_{nm}$ contains t_n and excludes each t_m for $m \neq n$. Thus C is discrete in the subspace topology provided countable intersections of open sets are open.†

Let (G_n) be a sequence of open sets and $b \in \bigcap G_n$. If b is isolated, then it is an interior point of $\bigcap G_n$. If, on the other hand, b is not isolated then there is an increasing sequence (a_n) of ordinals $a_n \in [0,b)$ such that $(a_n, b] \cap T \subset G_n$ for each $n \in \underset{\sim}{N}$. Letting $a = \sup a_n$, it follows from the fact that there is no countable neighborhood base at b that $a < b$ and so $(a,b] \cap T \subset \bigcap G_n$. Thus $\bigcap G_n$ is open and, by the above discussion, T is an NS-space.

<u>(b) T is not replete</u>. Let $S = T \cup \{\mu\}$ inherit the subspace topology of $[0,\mu]$ with the interval topology. Since $a+1$ is isolated and belongs to $(a,\mu] \cap T$ for each $a \in [0,\mu)$, μ is a limit point of T in S and therefore T is dense in S. Once it has been shown that each $x \in C(T,\underset{\sim}{R})$ can be extended to an $x' \in C(S,\underset{\sim}{R})$, it will follow, by Theorem 1.1-1, that βT and βS are equivalent compactifications of T. In fact we may then identify βT and βS and view S as a topological subspace of βT.

We contend that each $x \in C(T,\underset{\sim}{R})$ is actually constant on a tail of T, i.e. on a set of the form $[a,\mu) \cap T$, and is therefore extendible to S. In proving this, a few technical results are needed which we establish now.

A subset F of T is said to be <u>cofinal</u> <u>in</u> <u>T</u> if for each $s \in T$ there is an $a \in F$ such that $a \geq s$.

(i) If the cardinality $\#F$ of F is $\leq \aleph_1$, the cardinality of the first uncountable ordinal Ω, then F is not cofinal in T.

---

†The condition on T that each countable intersection of open sets is open is equivalent to requiring that each prime ideal of $C(T,\underset{\sim}{R})$ be maximal, i.e. that T be a P-space, (Gillman and Henriksen 1954, p. 345; cf. also Gillman and Jerison 1960, p. 63).

Proof (i): It follows from the fact that $a+1 \in T$ for each $a \in [0,\mu)$ that T is cofinal in $[0,\mu)$; thus any set F which is cofinal in T is also cofinal in $[0,\mu)$. Hence if F is cofinal, $[0,\mu) = \bigcup_{a\in F}[0,a)$. Since the cardinality of $[0,a)$ and the cardinality of a are the same, $\#\mu \leq \#F\aleph_1$, and (i) follows.

(ii) The supremum of each subset E of T of order type Ω (i.e. order-isomorphic to be $[0,\Omega)$) belongs to T.

Proof (ii): If E is such a set and a=sup E, then $a \notin E$ since $[0,\Omega)$ does not contain $\sup[0,\Omega)=\Omega$. Thus there is an order-isomorphism of $E\cup\{a\}$ onto $[0,\Omega]$ mapping a into Ω. The assumption that $a \notin T$ is equivalent to assuming the existence of an increasing sequence of elements from E which converges to a. Thus if $a \notin T$, there is an increasing sequence of ordinals from $[0,\Omega)$ convergent to Ω which is contradictory.

(iii) If A and B are closed disjoint subsets of T then one of them must be bounded, i.e. not cofinal.

Proof (iii): Under the assumption that A and B are closed, disjoint, and cofinal, we use transfinite recursion to construct an element in $A\cap B$ to reach a contradiction. Suppose that f is a map taking $[0,a)$ into $A\cup B$ where $a < \Omega$. Then, as $f([0,a))$ is not cofinal in T by (i) while A and B are cofinal in T, there must be elements of A and elements of B larger than $\sup f([0,a))$. Let a_f and b_f be the smallest elements of A and B larger than or equal to $\sup(f([0,a))\cap A)$ and $\sup(f([0,a))\cap B)$ respectively. Since $A\cap B=\emptyset$, $a_f \neq b_f$ so that one must be larger than the other. If $a_f > b_f$, then define $h_a(f)$ to be the smallest element of B larger than a_f; in the event that b_f is the larger, $h_a(f)$ is to be the smallest element of A larger than b_f. Now by the principle of transfinite recursion,* there is a unique function $F:[0,\Omega) \to A\cup B$ such that $F(a)=h_a(F_{|[0,a)})$ for each $a<\Omega$. Now we contend that F is an order-isomorphism. To see this, let P be the

*The principles of transfinite recursion and induction are (Dugundji 1966, p. 40, 5.2 and 5.1): (a) Recursion. If W is a well-ordered set and E an arbitrary class such that for all $x\in W$ there is a rule R_x that associates with each map $f:\{y\in W \mid y < x\} \to E$ a unique $R_x(f)\in E$, then there is a unique map $F:W \to E$ such that $F(x)=R_x(F_{|\{y\in W \mid y < x\}})$ for each $x\in W$. (b) Induction. If Q is a subset of the well-ordered set W such that $\{y\in W \mid y < x\}\subset Q \to x\in Q$ for each $x\in W$, then $Q=W$.

collection of all $a \in [0,\Omega]$ such that $F_{|[0,a)}$ is an order-isomorphism. We use transfinite induction to show that $P=[0,\Omega]$. Suppose that $[0,a) \subset P$, i.e. $F_{|[0,a')}$ is an order-isomorphism for each $a' < a$, and let b and c be elements of $[0,a)$ such that $b < c$. Then $F(c)=h_c(F_{|[0,c)})$. Set a_f and b_f equal to the smallest elements of A and B respectively larger than $\sup[F([0,c)) \cap A]$ and $\sup[F([0,c)) \cap B]$. As $F(b)$ is $\leq \sup(F([0,c)) \cap A)$ or $\sup(F([0,c)) \cap B)$, it follows that $F(c)$ is larger than $F(b)$. Hence $F_{|[0,a)}$ is an order-isomorphism, $a \in P$, and $P=[0,\Omega]$. By transfinite induction and the definition of h_a it follows that F is an order-isomorphism. Thus, by (ii), $\sup F([0,\Omega)) \in T$.

Next we claim that $\sup F([0,\Omega)) \in A \cap B$. Since F is a net defined on the directed set $[0,\Omega)$ and $\sup F([0,\Omega))$ is its limit, it is only necessary to show that the net is frequently in each of the sets A and B. Let $a \in [0,\Omega)$. If $F(a) \notin A$, then $F(a)$ is the smallest element of B which is larger than the first element of A larger than $\sup(F([0,a)) \cap A)$. To see that F is frequently in A, consider any $a \in [0,\Omega)$. We wish to show that there is some $a' \in [a,\Omega)$ such that $F(a') \in A$. If $F(a) \notin A$, then it will be shown that $F(a+1) \in A$. If $F(a) \notin A$, then $\sup(F([0,a)) \cap A)=\sup(F([0,a+1)) \cap A)$. Thus x_{a+1}, the smallest element of $A \geq \sup(F([0,a)) \cap A)$, coincides with x_a, the smallest element of $A \geq \sup(F([0,a)) \cap A)$. Moreover since $F(a) \in B$, then $F(a)$ is the smallest element b_{a+1} of $\{b \in B \mid b \geq \sup(F([0,a+1)) \cap B)\}$. Hence $b_{a+1}= F(a) > x_a = x_{a+1}$ and, by the definition of h_{a+1}, $F(a+1) \in A$. Thus F is frequently in A and the same is true of B.

Returning to the task of showing that each function $x \in C_b(T,\mathbb{R})$ is constant on a tail of T, we note that for each $t \in T$ the set $cl\{x(a) \mid a > t\}$ is a nonempty compact set. Consequently $\cap_{t \in T} cl\{x(a) \mid a > t\}$ is nonempty and an element r can be extracted from it. It follows that $G_n=\{a \in T \mid |x(a)-r| \leq 1/n\}$ is closed and cofinal in T for each $n \in \mathbb{N}$. Since the set $F_n=\{a \in T \mid |x(a)-r| \geq 1/n\}$ is closed and disjoint from the closed cofinal sets G_{2n} for each $n \in \mathbb{N}$, F_n must be bounded by (iii). Let $a_n = \sup F_n$ and choose $a \in T$ such that $a > \sup a_n$. Certainly then $x(t)=r$ for each $t \geq a$ (otherwise $t \in F_n$ for some n).

Appealing to the discussion at the beginning of (b), we may conclude that $\mu \in \beta T - T$. To see that $\mu \in \upsilon T$, it is enough to show that $(\cap V_n) \cap T \neq \emptyset$ for any sequence (V_n) of neighborhoods of μ in βT. Let (b_n) be an increasing sequence from T such that $(b_n,\mu] \subset V_n \cap S$. Since T is cofinal in $[0,\mu)$ and $\{b_n\}$ is not cofinal in T, $(\sup b_n,\mu] \cap T \neq \emptyset$ and $(\sup b_n,\mu] \subset \cap_n V_n$. This completes the proof. ∇

2.7 Separability of C(T,$\underset{\sim}{F}$,c) M. and S. Krein (1940) were the first to mention the characterization of separability of C(T,$\underset{\sim}{F}$,c) for compact T that appears in (2.7-1). Warner (1958) generalized this to the general completely regular Hausdorff space T and this result, as well as some of its consequences, appears below.

(2.7-1) SEPARABILITY OF C(T,$\underset{\sim}{F}$,c) WHEN T IS COMPACT If T is a compact Hausdorff space, then C(T,$\underset{\sim}{F}$,c) is separable iff T is metrizable.

Proof Suppose that C(T,$\underset{\sim}{F}$,c) is separable so that a countable dense subset $\{x_n | n \epsilon \underset{\sim}{N}\}$ exists in C(T,$\underset{\sim}{F}$,c). Let $U(y,a) = \{(s,t) \, | \, |y(s)-y(t)| < a\}$ be a typical subbasic entourage in the uniformity $\mathcal{U}$(T,$\underset{\sim}{F}$) generated by the continuous $\underset{\sim}{F}$-valued functions on T. One of the x_n must be such that the uniform $p_T(x_n-y) < a/3$, given $y \epsilon C(T,\underset{\sim}{F})$. Clearly $U(x_n,r) \subset U(y,a)$ for any positive rational $r < a/3$. Thus the uniformity has a countable subbase, and T is metrizable.

Conversely, suppose that T is metrizable. Then there exists a countable base of entourages $(U(x_1^n, \ldots, x_{k_n}^n, a_n))_{r \epsilon \underset{\sim}{N}}$, $a_n > 0$, for the unique (since T is compact) uniformity $\mathcal{U}$(T,$\underset{\sim}{F}$). Since T is Hausdorff, the uniformity is separating, i.e. given $(t,s) \epsilon T \times T$, $t \neq s$, at least one of the basic entourages fails to contain (t,s). Thus, for some $n \epsilon \underset{\sim}{N}$, and some $1 \leq j \leq k_n$, $x_j^n(t) \neq x_j^n(s)$ so that the collection $S = \cup_{n \epsilon \underset{\sim}{N}} \{x_1^n, \ldots, x_{k_n}^n\}$ separates the points of T. Hence, by the Stone-Weierstrass theorem, the collection of all linear combinations of monomials of elements from S with rational or Gaussian (i.e. of the form a+ib where a and b are rational) coefficients - depending upon whether $\underset{\sim}{F} = \underset{\sim}{R}$ or $\underset{\sim}{C}$ - is dense in C(T,$\underset{\sim}{F}$,c). ∇

As an immediate application of this result, we remark that for a completely regular Hausdorff space T, $C_b(T,\underset{\sim}{F})$, with sup norm topology, is separable iff T is compact and metrizable. Indeed, if $C_b(T,\underset{\sim}{F})$ is separable, then so is C(βT,$\underset{\sim}{F}$,c). Thus by (2.7-1), βT is metrizable and therefore T is compact. (Indeed if βT is metrizable and $p \epsilon \beta T - T$, then there is a sequence (t_n) of distinct points from T convergent to p. Set $Z_e = \{t_{2n} | n \epsilon \underset{\sim}{N}\}$ and $Z_o = \{t_{2n+1} | n \epsilon \underset{\sim}{N}\}$. As $Z_e \cup \{p\}$ and $Z_o \cup \{p\}$ are closed in βT, the disjoint sets Z_e and Z_o are closed in T. But, by (1.2-2), the class of closed sets is the same as the class of zero sets in any metric space so that Z_e and Z_o are disjoint zero sets in T whose closures are not disjoint, each containing p, in βT, in contradiction to Theorem 1.2-1. Thus βT=T and T is compact.

Before going on to the more general result on separability of C(T,$\underset{\sim}{F}$,^{r} , for non-compact T, let us examine more closely what follows from assuming $C(T,\underset{\sim}{F},c)$ to be separable. Let $\{x_n | n\in\underset{\sim}{N}\}$ be dense in $C(T,\underset{\sim}{F},c)$. The initial uniformity generated by $\{x_n\}$ is certainly weaker than the uniformity $\mathcal{C}(T,\underset{\sim}{F})$, so the former topology must also be weaker than the latter. Let T' denote T endowed with the weaker topology. Since $\{x_n\}$ is dense in $C(T,\underset{\sim}{F},c)$, these functions must separate points in T. Under the map $t\to\hat{t}$ where $\hat{t}(n)=x_n(t)$, T' is homeomorphic to a subspace of the product space $\underset{\sim}{R}^{\underset{\sim}{N}}$. This countable product of real lines is separable and metrizable; hence so is T'. Thus whenever $C(T,\underset{\sim}{F},c)$ is separable, T admits a separable metrizable topology which is weaker than the original topology. As we shall now see, this condition is also sufficient.

Theorem 2.7-1 SEPARABILITY OF $C(T,\underset{\sim}{F},c)$ Let T be a completely regular Hausdorff space. Then $C(T,\underset{\sim}{F},c)$ is separable iff T admits a separable metrizable topology which is weaker than the original topology.

Proof It only remains to verify sufficiency. Let T' denote T equipped with a weaker separable metrizable topology. We assert that $C(T',\underset{\sim}{F},c)$ is separable. By Urysohn's metrization theorem (Kelley 1955, Theorem 4.17, p. 125) T' is homeomorphic to a subspace M of the compact metrizable product space $[0,1]^{\underset{\sim}{N}}=S$. Certainly it suffices to show that $C(M,\underset{\sim}{F},c)$ is separable. Since $C(S,\underset{\sim}{F},c)$ is separable by (2.7-1), it contains a countable dense subset $\{z_n\}$. Let $w_n=z_{n|M}$. If F is compact in M and $w\in C(M,\underset{\sim}{F})$ then, by Tietze's extension theorem, there is a function $z\in C(S,\underset{\sim}{F})$ that agrees with w on F. Now for each $a > 0$ there exists z_n such that $p_F(z_n-z) < a$. But $w_n-w=z_n-z$ on F so $p_F(w_n-w) < a$ and $\{w_n\}$ is dense in $C(M,\underset{\sim}{F},c)$. It follows that $C(T',\underset{\sim}{F},c)$ is separable.

To see that $C(T,\underset{\sim}{F},c)$ is separable, let $\{x_n\}$ be dense in $C(T',\underset{\sim}{F},c)$ and let K be a compact subset of T. Clearly K is also compact in T'. Next we claim that T and T' induce the same topology on K. Indeed the injection map taking the compact space K into the Hausdorff space T' is 1-1 and continuous and therefore a homeomorphism. Thus each restriction $x_{|K}$ of $x\in C(T,\underset{\sim}{F})$ can be extended, by the Tietze extension theorem, to T'. Letting y denote the extension, it follows that $p_K(x_n-y) < a$ for some $n\in\underset{\sim}{N}$ corresponding to the given positive number a. Since $x_n\in C(T,\underset{\sim}{F})$ and $p_K(x_n-x) = p_K(x_n-y) < a$, $\{x_n\}$ is also dense in $C(T,\underset{\sim}{F},c)$. ∇

An immediate consequence of this theorem and Urysohn's metrization theorem is that the cardinality of T must be less than or equal to that of the continuum whenever $C(T,\underset{\sim}{F},c)$ is separable. To illustrate the converse is false, we may take T to be the compact ordinal space $[0,\Omega]$ where Ω is the first uncountable ordinal. Now the cardinality of T is $\leq$ that of the continuum while T is not metrizable since no countable base exists at Ω so that $C(T,\underset{\sim}{F},c)$ is not separable by (2.7-1).

For any discrete space T whose cardinality is $\leq$ that of the continuum, $C(T,\underset{\sim}{F},c)$ must be separable; for in this case there must be a 1-1 map from T into $\underset{\sim}{R}$ so, by transferring the Euclidean topology of $\underset{\sim}{R}$ on the image of T in $\underset{\sim}{R}$, a separable metric topology is obtained for T which is weaker than the discrete topology and it follows that $C(T,\underset{\sim}{F},c)$ is separable.

(2.7-1) showed that for compact T, metrizability of T was enough to guarantee separability of $C(T,\underset{\sim}{F},c)$. We show next that $C(T,\underset{\sim}{F},c)$ is separable even when T is a countable union of compact metrizable spaces.

<u>(2.7-2) σ-COMPACTNESS OF T AND SEPARABILITY OF $C(T,\underset{\sim}{F},c)$</u> If the completely regular Hausdorff space T is a countable union of compact metrizable sets, then $C(T,\underset{\sim}{F},c)$ is separable. If, in addition, T is locally compact, then $C(T,\underset{\sim}{F},c)$ is also metrizable.

<u>Proof</u> Let $T=\bigcup_{n\in\underset{\sim}{N}}K_n$ where each K_n is a compact metrizable subspace of T. Since the union of a neighborhood-finite collection of closed metrizable subspaces is metrizable (Dugundji 1966, p.194f., and Ex. 9.4, p. 207), any finite union of compact metrizable spaces is compact and metrizable. Thus we may assume that (K_n) is an increasing sequence. Since $C(K_n,\underset{\sim}{F},c)$ is separable by (2.7-1) there is a countable dense subset $\{x_{ni} \mid i\in\underset{\sim}{N}\}\subset C(K_n,\underset{\sim}{F},c)$ for each $n\in\underset{\sim}{N}$. By the Tietze extension theorem, each x_{ni} has a continuous extension y_{ni} to all of T. To see that $\{y_{ni} \mid (n,i)\in\underset{\sim}{N}\times\underset{\sim}{N}\}$ separates points in T, let t and s be distinct points of T. Choosing n large enough so that $t, s\in K_n$, it follows from the fact that K_n is Hausdorff that $y_{ni}(s)\neq y_{ni}(t)$ for some $i\in\underset{\sim}{N}$. Thus the uniformity generated by the countable family $\{y_{ni} \mid (n,i)\in\underset{\sim}{N}\times\underset{\sim}{N}\}$ induces a separable metrizable topology on T which is weaker than the original topology of T. The separability of $C(T,\underset{\sim}{F},c)$ now follows from Theorem 2.7-1.

To see that $C(T,\underset{\sim}{F},c)$ is also metrizable when T is locally compact it is enough to recall that a locally compact σ-compact space is hemicompact by the discussion following Def. 2.1-1. ∇

What about the converse of (2.7-2)? Consider $T=\underset{\sim}{R}$ with discrete topology. As we have already mentioned, prior to (2.7-2), $C(T,\underset{\sim}{R},c)$ is separable. On the other hand, only finite subsets of T are compact so that T is not even σ-compact. Thus the converse of (2.7-2) is false. A partial converse is obtainable however when $C(T,\underset{\sim}{F},c)$ is metrizable.

(2.7-3) SEPARABILITY OF $C(T,\underset{\sim}{F},c)$ WHEN T IS HEMICOMPACT Let T be a completely regular Hausdorff space. $C(T,\underset{\sim}{F},c)$ is separable and metrizable iff T is a hemicompact space in which each compact set is metrizable.

Proof It is only necessary to prove the necessity of the condition. If $C(T,\underset{\sim}{F},c)$ is separable, then T admits a separable metrizable topology which is weaker than the original topology. Thus the injection map of a typical compact subset of T into T with the weaker separable metrizable topology is a continuous injection, and therefore a homeomorphism. ∇

In the table below the arrows denote implication, T a completely regular Hausdorff space, "weakly" separable metrizable indicates the existence of a separable metrizable topology weaker than the original topology, and "σ-metrizably compact" means that the space can be written as a countable union of compact metrizable subspaces.

Summary

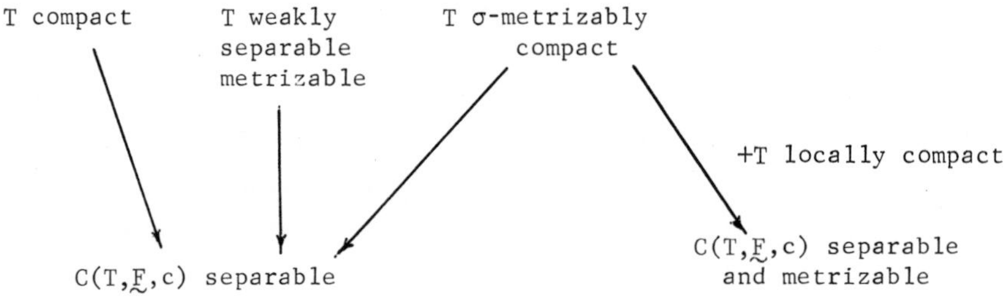

2.8 The bornology of $C(T,\underset{\sim}{F},c)$ The bornology of a topological vector space X is the collection $\mathcal{B}$ of all bounded subsets of X. A base for the bornology is a collection $\mathcal{A}$ of bounded sets such that each $B \in \mathcal{B}$ is a subset of some $A \in \mathcal{A}$. This section deals with the bornology of $C(T,\underset{\sim}{F},c)$. In particular, the following questions are investigated:

(1) Under what conditions on T are all the bounded sets of $C(T,\underset{\sim}{F},c)$ precompact, relatively compact?

(2) When is there a countable base for the bornology of $C(T,\underset{\sim}{F},c)$?

The first question is dealt with in our first two results, while the second is answered in (2.8-3). In answering (1) the critical topological properties of T are pseudofiniteness and discreteness. As will be shown, the pseudofinite spaces are the spaces for which all the bounded subsets of $C(T,\underset{\sim}{F},c)$ are precompact, while the necessary and sufficient condition for $C(T,\underset{\sim}{F},c)$ to be Montel is that T be discrete. Questions concerning the countability properties of the bornology are less sharply answered in that they are not characterized solely by topological properties of T: There is a countable base for $\mathcal{B}$ on $C(T,\underset{\sim}{F},c)$, for example, iff T is pseudocompact and $C(T,\underset{\sim}{F},c)$ is sequentially complete.

(2.8-1) WHEN DOES $C(T,\underset{\sim}{F},c)$ CARRY ITS WEAK TOPOLOGY? Let T be a completely regular Hausdorff space, and let $X=C(T,\underset{\sim}{F},c)$. Then the following statements are equivalent:

(a) Each bounded subset of $C(T,\underset{\sim}{F},c)$ is precompact;

(b) T is pseudofinite;

(c) $C(T,\underset{\sim}{F},c)$ carries its weak topology $\sigma(X,X')$.

Proof (a) → (b): Let K be a compact subset of T. By (a), the bounded set $\overline{V}_{p_T}$ is precompact so that there are a finite number of functions $x_1,\ldots,x_n \in C(T,\underset{\sim}{F})$ such that $\overline{V}_{p_T} \subset \bigcup_{i=1}^{n}(x_i+V_{p_K})$. We claim that K can contain at most n-1 distinct points. Indeed if $t_1,\ldots,t_n$ are distinct points of K, then by the complete regularity of T, there is a function $x \in \overline{V}_{p_T}$ such that $x(t_i)=-1$ if Re $x_i(t_i) \geq 0$ and $x(t_i)=1$ if Re $x_i(t_i) < 0$ for $i=1,\ldots,n$. But then $p_K(x-x_i) \geq 1$ for each i, $1 \leq i \leq n$, so that $x \notin \bigcup_{i=1}^{n}(x_i+V_{p_K})$. Thus each compact set K must be finite.

(b) → (c): If each compact subset of T is finite, the point-open and compact-open topologies coincide on $C(T,\underset{\sim}{F})$. Since all the evaluation maps t^* (where $t^*(x)=x(t)$ as x runs through $C(T,\underset{\sim}{F})$) are continuous linear functionals on $C(T,\underset{\sim}{F},c)=X$, $\sigma(X,X')$ is generally finer than the point-open topology. Thus, given that T is pseudofinite, all three topologies coincide.

(c) → (a): It is generally true that the weakly bounded subsets of any locally convex Hausdorff space are weakly precompact. ∇

In addition to answering the question of when $C(T,\underset{\sim}{F},c)$ carries its weak topology, the proof that (b) → (c) reveals that if the compact-open and weak topologies coincide on $C(T,\underset{\sim}{F})$ then the common topology must coincide with the point-open topology.

Reflexivity of $C(T,\underset{\sim}{F},c)$ was first characterized for compact T by S.B. Myers (1949). Our next result, due to Warner (1958), subsumes Myers' result.

(2.8-2) REFLEXIVITY OF $C(T,\underset{\sim}{F},c)$ Let T be a completely regular Hausdorff space. Then the following statements are equivalent*:

* Let X be a locally convex Hausdorff space. X is semireflexive if each strongly continuous linear functional on X' is generated by an element of X. X is reflexive if the natural map between X and X'' is a surjective topological isomorphism when X'' carries its strong topology. Equivalently, X is reflexive if it is semireflexive and barreled. X is semi-Montel if each bounded subset of X is relatively compact, Montel if it is semi-Montel and infrabarreled.

(a) $C(T,\underset{\sim}{F},c)$ is a Montel space;

(b) $C(T,\underset{\sim}{F},c)$ is reflexive;

(c) $C(T,\underset{\sim}{F},c)$ is semireflexive;

(d) T is discrete.

Proof The implications (a) → (b) → (c) are generally true, for any locally convex Hausdorff space.

(c) → (d): First we show that T is pseudofinite. Suppose that an infinite compact set $K \subset T$ exists and let (t_n) be a countable collection of distinct points of K. Since K is compact, the sequence (t_n) has a cluster point t which we may assume to be different from each t_n. By the complete regularity of T, there are functions $x_n \in C(T,\underset{\sim}{F})$ such that $x_n(t_m)=1$ for $m \leq n$, $x_n(t)=0$, and $p_T(x_n) \leq 1$. Thus $\{x_n\}$ is bounded. Since a locally convex Hausdorff space is semireflexive iff it is semi-Montel in its weak topology (Horvath 1966, Chap. 3, Sec. 8, Prop. 1) it follows that the weak closure of $\{x_n\}$ is compact, so that (x_n) has a weak cluster point x. Since x lies in the weak closure of $\{x_n\}, n \geq m,$ for each positive ϵ and $m \in \underset{\sim}{N}$ there is an integer $n \geq m$ such that

$$|t_m^*(x_n) - t_m^*(x)| = |x_n(t_m) - x(t_m)| < \epsilon .$$

Thus $x(t_m)=1$. Since x is continuous, it follows that $x(t)=1$. But $x_n(t)=0$ for each $n \in \underset{\sim}{N}$ and x is a weak cluster point of (x_n), so $x(t)=0$. It follows that each compact set is finite.

Having just shown that T is pseufofinite, (2.8-1) may be invoked together with the remarks immediately following it to conclude that $C(T,\underset{\sim}{F},c)$ carries its weak topology (which, in this case, coincides with the point-open topology). In light of this and the semireflexivity of $C(T,\underset{\sim}{F},c)$, we see that $C(T,\underset{\sim}{F},c)$ is semi-Montel and, therefore, quasi-complete. By Th. (2.1-2), $C(T,\underset{\sim}{F},c)$ is complete. Since $C(T,\underset{\sim}{F},c)$ carries the point-open topology - the subspace topology induced by the product topology - and $C(T,\underset{\sim}{F})$ is dense in $\underset{\sim}{F}^T$, $C(T,\underset{\sim}{F})=\underset{\sim}{F}^T$. It follows that T is discrete.

(d) → (a): Since T is discrete, $C(T,\underset{\sim}{F},c)$ coincides with $\underset{\sim}{F}^T$ carrying the product topology. Since a product of Montel spaces is a Montel space, the proof is complete. ∇

(2.8-3) WHEN IS THE BORNOLOGY OF $C(T,\underset{\sim}{F},c)$ COUNTABLY GENERATED? Let T be a completely regular Hausdorff space. Then the following statements are equivalent:

(a) There exists a countable base for the bornology of $C(T,\underset{\sim}{F},c)$;

(b) Each Cauchy sequence from $C(T,\underset{\sim}{F},c)$ converges uniformly on T;

(c) T is pseudocompact and $C(T,\underset{\sim}{F},c)$ is sequentially complete;

(d) The family of sets $(n\bar{V}_{p_T})_{n\in\underset{\sim}{N}}$ is a base for the bornology of $C(T,\underset{\sim}{F},c)$ (i.e. $\bar{V}_{p_T}$ is bornivorous).

Proof (a) → (b): Suppose that (x_n), $x_n \in C(T,\underset{\sim}{F})$, is not uniformly convergent on T. Then (x_n) also fails to be uniformly Cauchy on T. Thus for some $\epsilon > 0$ there are strictly increasing sequences of integers (s_n) and (r_n) such that $r_n < s_n < r_{n+1}$ and $p_T(x_{s_n} - x_{r_n}) > \epsilon$. As the functions x_{s_n} and x_{r_n} are continuous, open neighborhoods V_n of some t_n exist so that $|x_{s_n}(t) - x_{r_n}(t)| > \epsilon$ for each $t \in V_n$. By using (a) we shall show that (x_n) is not Cauchy in $C(T,\underset{\sim}{F},c)$. Furthermore we claim that in order to do this, it is only necessary to show that there is a compact set K meeting an infinite number of the V_n's. Indeed if such a compact set K exists then there is a strictly increasing sequence of integers (n_k) such that $K \cap V_{n_k} \neq \emptyset$. Thus $p_K(x_{s_{n_k}} - x_{r_{n_k}}) > \epsilon$ for each k and (x_n) is not Cauchy in $C(T,\underset{\sim}{F},c)$.

To prove that such a compact set exists, let us assume to the contrary, i.e. that for each compact set K there is an integer n_k such that $V_n \cap K = \emptyset$ for all $n \geq n_k$. By (a) there is a countable base (A_n) for the bornology of $C(T,\underset{\sim}{F},c)$. Certainly $C_n = \sup\{|x(t_n)| \; | x \in A_n\}$ is finite for each n. Since T is completely regular, for each $n \in \underset{\sim}{N}$, an element $y_n \in C(T,\underset{\sim}{F})$ can be found which vanishes on $T - V_n$ and evaluates to $C_n + 1$ at t_n. Now the set $B = \{y_n\}$ cannot be contained in any one A_n as $y_n \notin A_n$. On the other hand we claim that B is bounded, which would contradict the fact that (A_n) is a base for the bornology of $C(T,\underset{\sim}{F},c)$. To see that B is bounded it is enough to note that $\sup_n p_K(y_n) = \max_{1 \leq n \leq n_k} p_K(y_n)$ for each compact K. This contradiction establishes the result.

(b) → (c): Certainly $C(T,\underset{\sim}{F},c)$ is sequentially complete whenever (b) holds. To see that T is pseudocompact, we need to show that each $x \in C(T,\underset{\sim}{R})$ is bounded. Let

$$x_n(t) = \begin{cases} n & \text{if } x(t) \geq n \\ x(t) & \text{if } |x(t)| \leq n \\ -n & \text{if } x(t) \leq -n \end{cases}$$

If K is compact then, for some $m \in \underset{\sim}{N}$, $|x(t)| < m$ for each $t \in K$ and therefore, $x_n(t) = x_k(t)$ for each $t \in K$ whenever $n, k \geq m$. Thus (x_n) is Cauchy in $C(T,\underset{\sim}{F},c)$ and, by hypothesis, converges uniformly on T. Since x is the pointwise limit of (x_n) on T, it is also the uniform limit. Hence, for sufficiently large n, $p_T(x_n - x) < 1$ and the boundedness of x follows from that of x_n.

(c) → (d): Suppose that (d) does not hold, i.e., that there is a bounded set B which is not contained in any $n\overline{V}_{p_T}$. Then for each n there is an $x_n \in B$ and a $t_n \in T$ such that $|x_n(t_n)| \geq n^3$. We claim that the sequence of partial sums of the series $\Sigma_{k\in \underset{\sim}{N}} k^{-2}|x_k|$ is Cauchy in $C(T,\underset{\sim}{F},c)$ so that the series converges to an element x of $C(T,\underset{\sim}{F},c)$. This follows from the observation that for each compact K, $p_K(\Sigma_{k=m}^{p} k^{-2}|x_k|) \leq M\, \Sigma_{k=m}^{p} k^{-2}$ where $M = \sup_{x\in B} p_K(x)$. As $x(t_n) \geq n^{-2}n^3 = n$ for each n, x is an unbounded continuous function on T and we conclude that (c) fails to hold.

The proof that (d) → (a) is trivial. ∇

The situation examined in the previous theorem is considerably simpler for the class of replete spaces. Indeed if T is replete then T is pseudocompact only if T is compact ((1.5-3)) and thus it is necessary for T to be compact whenever the bornology of $C(T,\underset{\sim}{F},c)$ is countable. Conversely, whenever T is compact, both conditions of (c) are satisfied so that the bornology of $C(T,\underset{\sim}{F},c)$ is countably generated.

We conclude the chapter with a consequence of (2.8-3), a characterization of those pseudocompact spaces T which are compact in terms of topological vector spaces properties of $C(T,\underset{\sim}{F},c)$ (cf. (1.5-3)).

<u>(2.8-4)</u> <u>PSEUDOCOMPACTNESS VS. COMPACTNESS</u> A pseudocompact (completely regular Hausdorff) T is compact iff $C(T,\underset{\sim}{F},c)$ is infrabarreled and sequentially complete.

<u>Proof</u> The necessity of both conditions follows immediately from the observation that $C(T,\underset{\sim}{F},c)$ is a Banach space whenever T is compact. To see that they are also sufficient, we note first that by the previous result, each bounded set is contained in some integral multiple of $\overline{V}_{p_T}$. Thus since $C(T,\underset{\sim}{F},c)$ is infrabarreled, $\overline{V}_{p_T}$ is a neighborhood of the origin and the norm p_T generates the compact open topology. $C(T,\underset{\sim}{F},c)$ is thereby a Frechet space so that by Theorem 2.1-1, T is hemicompact. Consequently T is a Lindelöf space and it follows by Theorem 1.5-3 that T is replete. The result now follows by an application of (1.5-3) which asserts that a pseudocompact replete space is compact. ∇

2.1 C(T,$\underset{\sim}{F}$,c) BORNOLOGICAL IFF ULTRABORNOLOGICAL (de Wilde and Schmets 1971). A locally convex Hausdorff space X is bornological iff it is an inductive limit of normed spaces $(X_\mu)_{\mu\in M}$, i.e. iff there exist linear maps $A_\mu : X_\mu \to X$ such that the linear span of $\bigcup_{\mu\in M} A_\mu(X_\mu)$ is X and a set $U \subset X$ is a neighborhood of 0 iff for each $\mu \in M$ $A_\mu^{-1}(0)$ is a neighborhood of 0 in X_μ. X is ultrabornological if the spaces X_μ can be chosen such that they are Banach spaces.

The point of this exercise is to show that C(T,$\underset{\sim}{F}$,c) is bornological iff it is ultrabornological.

(a) Bornivores in C(T,$\underset{\sim}{F}$,c) For $x \in C(T,\underset{\sim}{F},c)$, $x \geq 0$, let $N(x) = \{y \in C(T,\underset{\sim}{F}) \mid |y| \leq x\}$. For replete T, show that an absolutely convex subset of C(T,$\underset{\sim}{F}$,c) is bornivorous (absorbs all bounded sets) iff it absorbs each N(x). (Hint: Cf. proof of Theorem 2.6-1.)

(b) Bounded linear functionals on C(T,$\underset{\sim}{F}$,c) For T replete, show that a linear functional f on C(T,$\underset{\sim}{F}$,c) is bounded iff f(N(x)) is bounded for each nonnegative x in C(T,$\underset{\sim}{F}$,c).

In (c) and (d) some technicalities needed for the proof of (f) are discussed.

(c) If every absolutely convergent series in a normed space X is convergent, then X is a Banach space. (Hint: Let (x_n) be a Cauchy sequence in X and let (m_K) be an increasing sequence of positive integers such that $\|x_m - x_{m_k}\| \leq 2^{-k}$ for all $m \geq m_k$. The series $x_{m_1} + \Sigma_{k\in\underset{\sim}{N}}(x_{m_{k+1}} - x_{m_k})$ is convergent, and $(x_{m_N}) = (x_{m_1} + \Sigma_{k=1}^{N}(x_{m_{k+1}} - x_{m_k}))$ is a subsequence of (x_n).)

(d) Any absolutely convergent series $\Sigma_{n\in\underset{\sim}{N}} x_n$ in a normed linear space X can be written $\Sigma_{n\in\underset{\sim}{N}} c_n y_n$ where $y_n \to 0$ and $\Sigma_{n\in\underset{\sim}{N}} |c_n|$ converges. (Hint: Let (N_i) be an increasing sequence of positive integers such that $\Sigma_{m \geq N_i} \|x_m\| \leq 2^{-2i}$ for each $i \geq 1$. For each $m \geq 0$ such that $N_i \leq m \leq N_{i+1}$ let $c_m = \max(2^i \|x_m\|, 2^{-m})$.) Let $x_n = c_n y_n$. As $\|y_n\| \leq 2^{-i}$ for some i and each n, and $i \to \infty$ as $n \to \infty$, it follows that $y_n \to 0$. Also $\Sigma_{n\in\underset{\sim}{N}} |c_n| \leq \Sigma_{m\in\underset{\sim}{N}} 2^{-m} + \Sigma_{i\in\underset{\sim}{N}} 2^i \Sigma_{m\in\underset{\sim}{N}} \|x_m\|$ and therefore $\Sigma_{n\in\underset{\sim}{N}} |c_n|$ is convergent.

Return attention now to C(T,$\underset{\sim}{F}$,c). Given N(x), let X_x denote the space spanned by N(x) in C(T,$\underset{\sim}{F}$) and $\|\ \|_x$ the gauge of N(x). If $y \in (1/n)N(x)$ for each $n \in \underset{\sim}{N}$ then clearly y=0. Hence $\|\ \|_x$ is a norm on X_x.

(e) X_x is a Banach space. Hint: As $\|y\|_x = \inf\{r > 0 \mid |y| \in rN(x)\} = \sup_{x(t)\neq 0} |y(t)|/x(t)$, $N(x) = \{y \in X_x \mid \|y\|_x \leq 1\}$. Now by parts (c) and (d) it is sufficient to show that given any sequence (x_m) from X_x, with $x_m \to 0$, $y = \Sigma_{m\in\underset{\sim}{N}} 2^{-m} x_m \in C(T,\underset{\sim}{F})$. Clearly the series converges pointwise to y and we show that $y \in C(T,\underset{\sim}{F})$ as follows. For $\epsilon > 0$, choose a neighborhood V of $t_o \in T$ such that y is bounded on V, and a positive integer M such that $\sup_{t\in V}|\Sigma_{m \geq M} 2^{-m}(x_m(t)-x_m(t_o))| \leq \epsilon/2$. Now there exists a neighborhood V' of t such that $V' \subset V$ and $\sup_{t\in V'}|\Sigma_{m=1}^{M} 2^{-m}(x_m(t)-x_m(t_o)|$. Thus $\sup_{t\in V'}|y(t)-y(t_o)| \leq \epsilon$ and $y \in C(T,\underset{\sim}{F})$. Now it remains to show that $y \in X_x$.

(f) If T is replete, then $C(T,\underset{\sim}{F},c)$ is ultrabornological. Hint: It is clear that $C(T,\underset{\sim}{F}) = \cup_{x \geq 0} X_x$ and as X_x is a Banach space we need only show that an absolutely convex absorbing set $U \subset C(T,\underset{\sim}{F},c)$ is a neighborhood of 0 if and only if $U \cap X_x$ is a neighborhood of 0 in X_x for all nonnegative functions $x \in C(T,\underset{\sim}{F})$. Using (a), this follows iff U is bornivorous. But $C(T,\underset{\sim}{F},c)$ is bornological whenever T is replete so that the bornivore U must be a neighborhood of 0.

<u>2.2 k-Spaces</u> A topological space T is a k-space if a set $U \subset T$ is open whenever $U \cap G$ is open in G for each compact subset G of T. Clearly any function x mapping a k-space T into a topological space Z is continuous whenever each of its restrictions to a compact subset of T is continuous. A space T having the property that a Z-valued function defined on T is continuous whenever each restriction to a compact subset is continuous is referred to as a <u>k_Z-space</u> (cf. the definition of a k_R-space in Sec. 2.2). Thus if T is a k-space then it is a k_Z-space for each topological space Z. The converse is also true:

<u>(a) $k \leftrightarrow k_Z$ FOR ALL Z</u> A topological space T is a k-space iff it is a k_Z-space for all topological spaces Z. <u>Hint for sufficiency:</u> Consider the function $I:(T,\mathcal{T}) \to (T,\mathcal{T}_k)$ where $\mathcal{T}$ is the original topology of T and $\mathcal{T}_k$ is the k-extension topology (defined before (2.3-1)).

<u>(b) LOCAL COMPACTNESS OR 1st COUNTABILITY $\to$ k</u> All locally compact spaces and all 1st countable spaces are k-spaces. As there are locally compact spaces which are not 1st countable and 1st countable spaces which are not locally compact the class of k-spaces is strictly smaller than either the class of locally compact spaces or the class of 1st countable spaces.

Though a k-space need not be locally compact it must have a locally compact "ancestor":

<u>(c) A k-SPACE IS A QUOTIENT OF A LOCALLY COMPACT SPACE</u> A topological space is a k-space iff it is a quotient of a locally compact space. Hint. <u>Necessity</u>: Let $\mathcal{G}$ denote the class of all compact subsets of T, $G' = \{(t,G) | t \in G\}$ for each $G \in \mathcal{G}$, and $T' = \{G' | G \in \mathcal{G}\}$. If equipped with the final topology generated by the injection maps

$$i_G : G \to T' \qquad (G \in \mathcal{G}),$$
$$t \to (t,G)$$

T' is referred to as the free union of $\mathcal{G}$, i.e. if the topology of G is transferred to G' in the natural way a subset of T' is open iff its intersection with each G' is open in G' and so T' is clearly locally compact. The relation $\sim$ defined on T' by:

$$(t,G) \sim (s,H) \text{ iff } t = s,$$

is an equivalence relation on T' and the mapping

$$h : T \to T'/R$$
$$t \to h(t) = \{(t,G) \mid t \in G\}$$

is a bijection. It remains to show that h is a homeomorphism when T'/R carries the quotient topology, i.e. the final topology generated by the map

$$p : T' \to T'/R$$
$$(t,G) \to h(t) = \{(t,H) \mid t \in H\}.$$

The result follows from the observation that for each $U \subset T$, $p^{-1}(h(U) = \cup\{U' \cap G' \mid G \in \mathcal{G}\}$, and the fact that U is open in T iff $U \cap G$ is open in G for each $G \in \mathcal{G}$.

Sufficiency: If there exists a locally compact space S, and equivalence relation R and a homeomorphism h taking S/R onto T then the mapping

$$f : S \to S/R \to T$$
$$s \to Rs \to h(Rs)$$

has the property that a set $U \subset T$ is open iff $f^{-1}(U)$ is open in S. Thus to prove that T is a k-space it suffices to show that $f^{-1}(U)$ is open whenever $U \cap G$ is open in G for compact $G \subset T$. Since S is locally compact so that there exists a covering of S by relatively compact open sets $\{V_\alpha\}_{\alpha \in \Lambda}$, the desired conclusion follows from the sequence of observations:

(1) $U \cap f(\mathrm{cl} V_\alpha)$ is open in the compact set $f(\mathrm{cl} V_\alpha)$,

(2) there exists an open set $W \subset T$ such that $U \cap f(\mathrm{cl} V_\alpha) = W \cap f(\mathrm{cl} V_\alpha)$,

(3) $f^{-1}(U) \cap V_\alpha = f^{-1}(W) \cap V_\alpha$.

(d) SUBSPACES A subspace of a k-space need not be a k-space. Indeed if T is any completely regular Hausdorff space which is not a k-space, e.g. $T = W^A$ where A is uncountable (discussed in Example 2.3-1), then T is certainly a subspace of the k-space βT. What about closed subspaces of k-spaces? They are also k-spaces.

(e) QUOTIENTS A quotient space of a k-space is a k-space.

Hint: Use (a).

(f) PRODUCTS (Bagley and Young 1966). We already know by Example 2.3-1 that an infinite product of k-spaces needn't be a k-space. What about finite products? Unfortunately a product of two k-spaces need not be k_R even if one of the spaces is metrizable.

(f1) Let T be a completely regular Hausdorff hemicompact k-space which is not locally compact. Then $C(T,\underset{\sim}{R},c)$ is metrizable (and a k-space by (b)) while $C(T,\underset{\sim}{R},c)\times T$ is not a k_R-space.

Hint: It suffices to show that the evaluation map

$$e: C(T,\underset{\sim}{R},c)\times T \to \underset{\sim}{R}$$
$$(x,t) \to x(t)$$

is continuous on each compact subset of $C(T,\underset{\sim}{R},c)\times T$ but is not continuous on all of $C(T,\underset{\sim}{R},c)\times T$. To prove the first assertion it is enough to consider e on compact subsets of the form $F\times K$ where F is compact in $C(T,\underset{\sim}{R},c)$ and K is compact in T. Fix $(x_o,t_o)\in F\times K$. As $F|_K=\{x|_K \,|x\in F\}$ is a compact subset of $C(K,\underset{\sim}{R},c)$ and thereby equicontinuous, for $\epsilon>0$ there exists a neighborhood U of t_o in T such that $|x(t)-x(t_o)|<\epsilon/2$ for all $t\in U\cap K$ and all $x\in F$. It follows that $|x(t)-x_o(t_o)|<\epsilon$ whenever $(x,t)\in((x_o+\epsilon/2\, V_{p_K})$ $\times U)\cap F\times K$.

To establish discontinuity of e we produce a net $(x_\alpha,t_\alpha)_{\alpha\in\wedge}$ in $C(T,\underset{\sim}{R},c)$ $\times$ T convergent to $(0,t_o)$ (where 0 denotes the function on T which sends each $t\in T$ into $0\in\underset{\sim}{R}$ and t_o is fixed in T) such that $x_\alpha(t_\alpha)=1\not\to 0=0(t_o)$. As T is not locally compact there exists $t_o\in T$ with no relatively compact neighborhood. Let $\mathcal{U}(t_o)$ denote the neighborhood system at t_o. Since T is hemicompact we may choose an increasing sequence (K_n) of compact sets such that $\{1/n\, V_{p_{K_n}} |n\in\underset{\sim}{N}\}$ is a neighborhood base at 0 in $C(T,\underset{\sim}{R},c)$. For each $n\in\underset{\sim}{N}$ and $U\in\mathcal{U}(t_o)$ there exists $t_{n,u}\in U$ such that $t_{n,u}(K_n)=0$ and $x_{n,u}(t_{n,u})=1$. It follows that $(x_{n,u},t_{n,u})$ are the elements of a net on $N\times\mathcal{U}(t_o)$ convergent to $(0,t_o)$.

Thus requiring that one of a pair of k-space be 1st countable is not enough to guarantee that the product be a k-space. However

(f2) a product of two 1st countable spaces is a k-space;

(f3) a product of two Hausdorff k-spaces, one of which is locally compact, is a k-space.

Hint: So that (a) may be used in establishing this result some preliminary results are needed. Let C(S,Z,c) denote the space of continuous Z-valued functions defined on S with the compact-open topology, i.e. the topology having as a subbasis all sets of the form $[K,V]=\{x\in C(S,Z)\,|\,x(K)\subset V\}$ where K is a compact subset of S and V is open in Z.

Let T, S, and Z be topological spaces and $x\in C(T\times S,Z)$. If t is fixed in T then $\hat{x}(t)(s)=x(t,s)$ $(s\in S)$ defines an element of $C(S,Z)$. Thus $\hat{x}$ maps elements of T into continuous Z-valued functions on S. Conversely suppose we are given a function $\hat{x}:T\to C(S,Z)$. Then $x(t,s)=\hat{x}(t)(s)$ $((t,s)\in T\times S)$ defines a Z-valued function on T×S which is continuous in the variable s for each fixed $t\in T$.

(f4) If x is a given continuous function then $\hat{x}$ is also continuous.
Hint: Suppose $\hat{x}(t)\in[G,V]=\{y\in C(X,Z)\,|\,y(G)\subset V\}$ where G is compact in S and V is open in Z. We must find a neighborhood U_t of t in T such that $x(U_t)\subset[G,V]$, or equivalently, $x(t',g)=\hat{x}(t')(g)\in V$ for all $t'\in U_t$ and $g\in G$. Now $x(t,g)=\hat{x}(t)(g)\in V$ for each $g\in G$ so by the continuity of x, open sets $U_{t,g}\subset T$ and $W_g\subset S$ exist such that $x(U_{t,g}\times W_g)\subset V$. As G is compact and is covered by finitely many W_g's, say $W_{g_1},\ldots,W_{g_n}$, $U_t=\bigcap_{i=1}^{n}U_{t,g_i}$ serves the desired end.

(f5) If $\hat{x}: T\to C(S,Z,c)$ is continuous and S is locally compact and Hausdorff then the associated function x (defined above) is also continuous.
Hint: Suppose $(t_\alpha,s_\alpha)_{\alpha\in\Lambda}$ is a net in T×S convergent to a point $(t,s)\in T\times S$ and V is a neighborhood of $x(t,s)$ in Z. We shall prove that $x(t_\alpha,s_\alpha)$ is eventually in V. As $\hat{x}(t)\in C(S,Z)$ and S is locally compact Hausdorff there exists a relatively compact neighborhood W_s of s such that $\hat{x}(t)(\mathrm{cl}W_s)\subset V$. Thus $\hat{x}(t)\in[\mathrm{cl}W_s,V]$. Now $\hat{x}$ is continuous so there exists a neighborhood U_t of t in T such that $\hat{x}(U_t)\subset[\mathrm{cl}W_s,V]$. Since $(t_\alpha,s_\alpha)\to(t,s)$ there exists an $\alpha\in\Lambda$ such that for $\alpha\le\beta$, $(t_\beta,s_\beta)\in U_s\times W_s$. Hence $x(t_\beta,s_\beta)=\hat{x}(t_\beta)(s_\beta)\in V$ for all $\beta\ge\alpha$.

Returning to the proof of (f3) we claim that $x:T\times S\to Z$ is continuous whenever $x|_{K\times S}$ is continuous for each compact set $K\subset T$. To see this suppose that $x|_{K\times S}$ is continuous. Then $\hat{x}|_K$ is also continuous by (i). As T is a k-space it follows that $\hat{x}$ is continuous and so by (ii) x is continuous. Thus it remains to show that $x|_{K\times S}$ is continuous whenever $x|_{K\times G}$ is continuous for all compact $G\subset S$. But this fact follows by exactly the same argument since K is locally compact Hausdorff and S is a k-space (by part (b)).

<u>(g) (Noble 1967) T A k-SPACE $\not\Rightarrow$ υT A k-SPACE</u> In Example 2.3-1 it was established that $\underset{\sim}{W}^A$, the collection of all non-negative integer-valued functions defined on the <u>uncountable</u> set A, is not a k-space. Moreover we saw that the set S consisting of all elements of $\underset{\sim}{W}^A$ which vanish at all but at most

a countable number of elements from A is a k_R-space with $\upsilon S = \underset{\sim}{W}^A$. Here we claim: S is a k-space whose repletion is not a k-space.

Hint: To see that S is a k-space let s be fixed in S and U be a set containing s which is not a neighborhood of s. We shall exhibit a compact set K such that $CU \cap K$ is not closed in K. Since U is not a neighborhood of s there exists an element $s_1 \in S$ such that $s_1 \notin U$. As s_1, $s \in S$ they differ on a countable subset $A_1 = \{a_{1j} \mid j \in \underset{\sim}{N}\}$ (if A_1 is finite $a_{1j} = a_{1J}$ for some J and all $j \geq J$). Let $F_1 = \{a_{11}\}$; then the neighborhood $\{t \in S \mid t(a_{11}) = s(a_{11})\} \not\subset U$ so we can find $s_2 \notin U$ such that $s_2(a_{11}) = s(a_{11})$. Let $A_2 = \{a_{2j} \mid j \in \underset{\sim}{N}\}$ denote the elements of A on which s_2 and s differ and set $F_2 = \{a_{ij} \mid 1 \leq i,j \leq 2\}$. Then the neighborhood $\{t \in S \mid t(a) = s(a)$ for each $a \in F_2\} \not\subset U$ so there exists $s_3 \notin U$ such that $s_3 = t$ on F_2. Continuing by induction we obtain a sequence (s_n) from S all of whose elements lie in CU and, denoting the collection of elements of A on which s_n and s differ by $A_n = \{a_{nj} \mid j \in \underset{\sim}{N}\}$, $s_{n+1} = s$ on $F_n = \{a_{ij} \mid 1 \leq i,j \leq n\}$. Set $C = \cup F_n$. If $a \notin C$ then $s_n(a) = s(a)$ for each n so that $s_n(a) \to s(a)$. If $a \in C$ there exists $m > 0$ such that $a \in F_n$ for each $n \geq m$. Thus $s_{n+1}(a) = s(a)$ whenever $n \geq m$ and $s_n(a) \to s(a)$. Therefore $s_n \to s$ in S, $K = \{s_n \mid n \in \underset{\sim}{N}\} \cup \{s\}$ is compact and $CU \cap K = \{s_n \mid n \in \underset{\sim}{N}\}$ is not closed in K as $s \notin CU \cap K$.

(h) ASCOLI THEOREMS (Kelley 1955, pp. 223-249; Bagley & Young 1966) The point of this exercise is to obtain generalizations [(iii) and (iv) below] of Ascoli's Theorem as presented in Kelley 1955 (Theorem 21, p.236), as well as the version in Kelley's Theorem 7.17 (p. 233).

If F is a family of maps from a topological space S into a topological space T, then any topology $\mathcal{T}$ for F which makes the evaluation map e sending (x,s) into x(s) from FxS into T is called jointly continuous. Two results on joint continuity are needed for the Ascoli Theorems.

(i) If S and T are Hausdorff spaces, $F \subset C(S,T)$, $\mathcal{T}$ is a topology for F which is finer than the compact-open topology $\mathcal{T}_c$ and which makes $(F,\mathcal{T})xS$ a k-space, then $\mathcal{T}$ is jointly continuous.

(Hint: Let C be a closed subset of T, K a compact subset of $(F,\mathcal{T})xS$, and (x,s) be a point outside of $M = K \cap e^{-1}(C)$. If $(x,s) \notin K$ then obviously $(x,s) \notin$ cl M. Suppose $(x,s) \in K$ and $(x,s) \notin e^{-1}(C)$. Let U=T-C and let K_S be the projection of K into S. There is a compact neighborhood N of S relative to K_S such that $x(N) \subset U$ and $e([N,U]xN) \subset U$ where $[N,U] = \{y \in F \mid y(N) \subset U\} \in \mathcal{T}_c \subset \mathcal{T}$. Thus $([N,U]xN) \cap e^{-1}(C) = \emptyset$. It follows that (x,s) is not in the closure of M relative to $(F,\mathcal{T})xK_S$. But since $M \subset K \subset FxK_S$, it follows that $(x,s) \notin$ cl M. Since FxS is a k-space, $e^{-1}(C)$ is closed and the proof is complete.)

(ii) If $(F,\mathcal{T}_c)$ is locally compact, S a Hausdorff k-space, and T a Hausdorff space then $\mathcal{T}_c$ is jointly continuous.

(Hint: Use (f3) and (i) above.)

Using (ii), the generalizations of Kelley's Ascoli Theorems mentioned above (Kelley's Theorem 7.17 and 7.21 respectively) may now be obtained using the same proofs that are in Kelley.

(iii) ASCOLI THEOREM Let S be a Hausdorff k-space, T a Hausdorff uniform space and $F \subset C(S,T)$. Then $(F,\mathcal{T}_c)$ is compact iff (1) $(F,\mathcal{T}_c)$ is closed in C(S,T,c) where C(S,T,c) denotes C(S,T) with compact-open topology, (2) F(s) has compact closure for each $s \in S$ and (3) F is equicontinuous.

(iv) ASCOLI THEOREM Let S be a Hausdorff k-space, T a regular Hausdorff space and $F \subset C(S,T)$. $(F,\mathcal{T}_c)$ is compact iff (1) $(F,\mathcal{T}_c)$ is closed in C(S,T,c), (2) F(s) has compact closure for each $s \in S$ and (3) F is evenly continuous. (To say that F is evenly continuous means that for each $s \in S$, each $t \in T$, and each neighborhood U of t there is a neighborhood V of s and a neighborhood W of t such that $x(V) \subset U$ whenever $x(s) \in W$.)

2.3 k-SPACES AND FULL COMPLETENESS OF $C(T,\underset{\sim}{F},c)$ (Ptak 1953; see also comment following (4.12-8).) Let T be a completely regular Hausdorff space. $C(T,\underset{\sim}{F},c)$ is complete iff T is a k-space, whenever T is a hemicompact space ((2.3-4)). This equivalence is not general for, as we have seen in Example 2.3-1, there are completely regular Hausdorff spaces which are not k-spaces but for which $C(T,\underset{\sim}{F},c)$ is complete. What condition on the LCHS $C(T,\underset{\sim}{F},c)$ will force T to be a k-space? Ptak (1953) has shown that T is a k-space whenever $C(T,\underset{\sim}{F},c)$ is fully complete. Unfortunately the converse is not true. Recall that a LCHS X is fully complete if every continuous linear transformation A taking X onto a LCHS Y which is almost open (i.e. cl A(V) is a neighborhood of 0 in Y for each neighborhood V of 0 in X) is an open map.

(a) If $C(T,\underset{\sim}{F},c)$ is fully complete then T is a k-space

(i) The image of a fully complete space under a topological homomorphism (i.e. a continuous and open linear map) is fully complete. (Suppose the topological homomorphism A takes the fully complete LCHS X onto the LCHS Y and let B be an almost open continuous linear map from Y onto the LCHS Z. If U is a neighborhood of 0 in Y then $A^{-1}(U)$ is a neighborhood of 0 in X. Therefore, because X is fully complete and $B \cdot A$ is almost open, $B(U)=(B \cdot A)(A^{-1}(U))$ is open in Z.)

(ii) If S is a closed subset of T and $C(T,\underset{\sim}{F},c)$ is fully complete, then so is $C(S,\underset{\sim}{F},c)$. Furthermore if $C(T,\underset{\sim}{F},c)$ is fully complete, then T is normal.

Consider the restriction map

$$R: C(T,\underset{\sim}{F},c) \longrightarrow C(S,\underset{\sim}{F},c)$$
$$x \longrightarrow x|_S$$

R is clearly linear. Next we claim that the range of R is dense in $C(S,\underset{\sim}{F},c)$. Indeed, if $y\in C(S,\underset{\sim}{F})$ and $K\subset S$ is compact then $y|_K$ has a continuous extension $\hat{y}$ defined on all of T. Thus $\hat{y}|_K=y|_K$ and so in any basic neighborhood of y, say $y+\epsilon V_{p_K}=\{y+z \mid z\in C(S,\underset{\sim}{F}),\ p_K(z) < \epsilon\}$, there exists an element of the range of R.

It only remains to show that R is a topological homomorphism for then, by (i), the range of R is fully complete. As R is certainly continuous our concern is to establish openness of R. But $C(T,\underset{\sim}{F},c)$ is fully complete so it suffices to show that R is almost open. Let K be compact in T and consider the set cl $R(V_{p_K})$. In the event that $K\cap S\neq\emptyset$ we contend that cl $R(V_{p_K})\supset\{y\in C(S,\underset{\sim}{F}) \mid p_{K\cap S}(y) \leq 1/2\}$ so that cl $R(V_{p_K})$ is a neighborhood of 0 in $C(S,\underset{\sim}{F},c)$. Suppose that $y\in C(S,\underset{\sim}{F})$ such that $p_{K\cap S}(y) \leq 1/2$ and $H\subset S$ is compact. Now by what has already been shown, there exists $u\in C(T,\underset{\sim}{F})$ such that u agrees with y on $(K\cap S)\cup H$. Thus the set $W=\{t\in K \mid |u(t)| \geq 1\}$ is disjoint from H. Choose $z\in C(T,\underset{\sim}{F})$ such that $0 \leq z \leq 1$, $z(H)=\{1\}$, and $z(W)=\{0\}$. Set $x'=zu$ and $x'|_S=x$. If $t\in K$ then there are two possibilities. First suppose that $t\in W$. Then $z(t)=0$ and $|x'(t)|=0 \leq 1$. If $t\notin W$ then $|x'(t)| = |u(t)||z(t)| < 1$. As x agrees with y on H (since $z(H)=\{1\}$ and $u=y$ on H), $y\in$ cl $R(V_{p_K})$.

If $K\cap S=\emptyset$ then we may choose a function $z\in C(T,\underset{\sim}{F})$ such that $0 \leq z \leq 1$, $z(K)=\{0\}$, and $z(S)=\{1\}$. Now if y is any element of $C(S,\underset{\sim}{F})$ and $H\subset S$ is compact there exists $u\in C(T,\underset{\sim}{F})$ with the property that $u|_H=y$ then $x'=zu\in V_{p_K}$ and $x=x'|_S$ agrees with y on H. Thus cl $R(V_{p_K})=C(S,\underset{\sim}{F})$ is a neighborhood of 0 in $C(S,\underset{\sim}{F})$.

Hence R is a topological homomorphism onto a fully complete dense subspace of $C(S,\underset{\sim}{F})$ so that R is, in fact, surjective, and $C(S,\underset{\sim}{F})$ is fully complete. Moreover, by the ontoness of R, each continuous real-valued function defined on S has a continuous extension to all of T. Consequently T is normal by the Tietze extension theorem.

<u>(iii) If $C(T,\underset{\sim}{F},c)$ is fully complete then T is a k-space</u> Suppose that $E\subset T$ and $E\cap K$ is closed for each compact $K\subset T$. Show that $S=$ cl $E=E$. To this end consider the topology on $C(S,\underset{\sim}{F})$ generated by the collection of seminorms $p_{K\cap E}(\cdot)=\sup|(\cdot)(K\cap E)|$ as K runs through the compact subsets of

S for which $K\cap E\neq\emptyset$ and let $C(S,\underset{\sim}{F},c_o)$ denote $C(S,\underset{\sim}{F})$ with this topology. To see that this space is Hausdorff, let $x\in C(S,\underset{\sim}{F})$, $x\neq 0$. Then there exists $s\in S$ such that $x(s)\neq 0$ and so there exists a neighborhood V_s of s in S in which x never vanishes. Now E is dense in S so $E\cap V_s\neq\emptyset$ and $p_{\{t\}}(x)\neq 0$ for any $t\in E\cap V_s$.

As the compact-open topology is finer than the c_o topology of $C(S,\underset{\sim}{F})$ the bijective mapping

$$I: C(S,\underset{\sim}{F},c) \longrightarrow C(S,\underset{\sim}{F},c_o)$$
$$x \longrightarrow x$$

is continuous. Since $C(S,\underset{\sim}{F},c)$ is fully complete (by (ii)) it is only necessary to show that I is almost open in order to prove that the compact-open and c_o topologies coincide. To see this show that $cl_o V_{p_K}$ (where cl_o denotes closure in the c_o topology) is a c_o-neighborhood of the origin for each compact $K\subset S$. Indeed, if $K\cap E\neq\emptyset$ we claim that $1/2\, V_{p_{K\cap E}} \subset cl_o V_{p_K}$. Let $x\in 1/2 V_{p_{K\cap E}}$ and suppose that $H\subset S$ is compact and $H\cap E\neq\emptyset$. Using the procedure outlined below, show that there exists a function $x'\in C(S,\underset{\sim}{F})$ such that x' agrees with x on $H\cap E$ and $p_K(x') < 1$ to establish the desired inclusion. Let $W=\{t\in K \mid |x(t)| \geq 1\}$; note that W is compact and, since $\sup|x(K\cap E)| < 1/2$, $H\cap E$ is also compact so there exists $z\in C(S,\underset{\sim}{F})$ such that $0 \leq z \leq 1$, $z(H\cap E)=\{1\}$ and $z(W)=\{0\}$. Set $x'=x\cdot z$. If $t\in H\cap E$ then $z(t)=1$ so $x'(t)=x(t)$ and x' and x agree on $H\cap E$. Next suppose that $t\in K$. If $t\in W$ then $z(t)=0$ and therefore $|x'(t)|=0 < 1$. If $t\notin W$ then $|x(t)| \leq 1$ and, since $|z(t)| \leq 1$, $|x'(t)| < 1$. Thus $cl_o V_{p_K}$ is a c_o-neighborhood of 0 provided $K\cap E \neq\emptyset$. What if $K\cap E=\emptyset$? Then by using an argument similar the one above show that $cl_o V_{p_K}=C(S,\underset{\sim}{F})$ so that in any event $cl_o V_{p_K}$ is c_o-neighborhood of 0. Thus, since $C(S,\underset{\sim}{F},c)$ is fully complete, I is a topological isomorphism and the compact-open and c_o-topologies agree on $C(S,\underset{\sim}{F})$.

Now show that E=S. If $t\in S-E$ then the compact-open neighborhood $V_{p_{\{t\}}}$, being also a c_o-neighborhood of the origin contains a neighborhood $V_{p_{K\cap E}}$ where K is compact in T and $K\cap E\neq\emptyset$. Since $K\cap E$ is closed and $t\notin K\cap E$ there exists $y\in C(S,\underset{\sim}{F})$ such that $0 \leq y \leq 1$, $y(t)=1$, and $y(K\cap E)=\{0\}$. Hence $y\in V_{p_{K\cap E}}$ while $y\notin V_{p_{\{t\}}}$ - a contradiction. It follows that S=E and T is a k-space.

(b) <u>There are (completely regular Hausdorff) k-spaces T for which $C(T,\underset{\sim}{F},c)$ is not fully complete.</u>

Hint: Consider the ordinal spaces $[0,\omega]$ and $[0,\Omega]$ where ω is the first infinite ordinal and Ω the first uncountable ordinal. Then $[0,\omega]X[0,\Omega]$ is a compact Hausdorff space and the open subset $T=[0,\omega]X[0,\Omega]-\{(\omega,\Omega)\}$ is a locally compact completely regular Hausdorff space. Hence T is a k-space by Exercise 2.2(b). On the other hand T is not normal (see Dugundji 1966, p. 145, Ex. 4) so, by (ii) of part (a), $C(T,\underset{\sim}{F},c)$ is not fully complete.

2.4 $C(T,\underset{\sim}{R})'$ AND BAIRE MEASURES Throughout this exercise we take T to be a completely regular Hausdorff space. We know from (2.4-4) that each positive linear function h on $C_b(T,\underset{\sim}{R})$ has the form $h(\cdot)=\int(\cdot)\,d\mu$ where μ is a regular additive non-negative set function defined on $\mathcal{A}_z$, the algebra generated by $\mathcal{Z}$, the zero sets of T. Recall how μ was defined: for any $Z\in\mathcal{Z}$, $\mu(Z)=\sup\{h(x)\,|x\in C_b(T,\underset{\sim}{R}),\ 0\le x\le k_{CZ}\}$ and for arbitrary $A\in\mathcal{P}(T)$, $\mu(A)=\inf\{\mu(CZ)\,|Z\in\mathcal{Z},A\subset CZ\}$. In (a) below we show that μ is, in fact, countably additive on the σ-algebra $\mathcal{B}_a$ generated by $\mathcal{Z}$ (the Baire sets) and that, if h is obtained as the restriction of a positive linear functional on $C(T,\underset{\sim}{R})$ the representation as an integral given above holds for all $x\in C(T,\underset{\sim}{R})$.

(a) POSITIVE LINEAR FUNCTIONS ON $C(T,\underset{\sim}{R})$ AND BAIRE MEASURES Let h be a positive linear function on $C(T,\underset{\sim}{R})$ and μ be the set function defined above. Then μ is a regular measure on $\mathcal{B}_a$ and

$$h(x)=\int x\,d\mu \qquad (x\in C(T,\underset{\sim}{R})).$$

For each $x\in C(T,\underset{\sim}{R})$ there exists $E\in\mathcal{B}_a$ such that x is bounded on E and $\mu(CE)=0$.

Sketch of Proof Let $\mathcal{P}$ denote the collection of complements of zero sets. First show that

(i) for each $G\in\mathcal{P}$ and $\epsilon>0$ there exists $H\in\mathcal{P}$ such that cl $H\subset G$ and $\mu(G)-\epsilon<\mu(H)$: Choose $w\in C_b(T,\underset{\sim}{R})$ such that $0\le w\le k_G$ and $h(w)>\mu(G)-\epsilon/2$. Now $H=w^{-1}(\epsilon/2,1]\in\mathcal{P}$ by (1.2-3) and cl $H\subset G$. If $\mu(H)\le\mu(G)-\epsilon$ then

$$\begin{aligned} h(w)-\epsilon/2 &= h(w-(\epsilon/2)1)\le h((w-(\epsilon/2)1)\vee 0)\le\mu(H)\\ &\le\mu(G)-\epsilon<h(w)-\epsilon/2 \end{aligned}$$

-a contradiction.

To establish countable subadditivity on $\mathcal{P}$ prove:

(ii) if (G_n) is a sequence from $\mathcal{P}$ such that $G_n\supset G_{n+1}$ for all n and $\bigcap G_n=\emptyset$ then $\lim_n\mu(G_n)=0$: Consider first the situation where $\bigcap \mathrm{cl}(G_n)=\emptyset$. As μ is monotone on $\mathcal{A}_z$, $(\mu(G_n))$ is a decreasing sequence of real numbers; as such it possesses a limit $a\ge 0$. If we assume a to be positive then for each $n>0$, there exists $x_n\in C_b(T,\underset{\sim}{R})$ such that $0\le x_n\le k_{G_n}$ and $h(x_n)>a/2$. Let $x=\sum_n x_n$. Since $\bigcap G_n=\emptyset$, each $t\in T$ can belong to only a finite number of the sets G_n, so that x is a real-valued non-negative function on T. Moreover, since $\bigcap \mathrm{cl}\, G_n = \emptyset$, each $t \in T$ belongs to $U_n = C(\mathrm{cl}\, G_n)$ for some n. But U_n is open and only a finite number of the x_k's are non-zero on U_n so x is continuous at t for each $t \in T$. Applying h to x we obtain the contradiction:

$$h(x) = \sum_{k=1}^{n} h(x_k) + h(\sum_{k>n} x_k) > \frac{na}{2} \quad (n > 0)$$

Next consider a sequence (G_n) for which $\bigcap G_n = \emptyset$. For any given $\epsilon > 0$, by (i), there exists $V_n \in \mathcal{P}$ such that $\mathrm{cl}(V_n) \subset G_n$ and $\mu(G_n) - \epsilon/2^n < \mu(V_n)$ for $n > 0$. Thus if we set $H_1 = V_1$ and inductively define $H_n = V_n \cap H_{n-1}$ it follows that $\mathrm{cl}\, H_n \subset G_n$ and $H_n \supset H_{n+1}$ for each $n > 0$. Furthermore, with the aid of the observation:

$$G_n - H_n \subset (G_n - V_n) \cup (G_{n-1} - H_{n-1}) \quad (n > 0)$$

and the fact that μ is subadditive on $\mathcal{a}_z$ (since it is additive and monotone on $\mathcal{a}_z$), it is readily established that $\mu(G_n) < \mu(H_n) + \epsilon \sum_{k=1}^{n} 1/2^k < \mu(H_n) + \epsilon$ for each positive n. As $\bigcap \mathrm{cl}\, H_n \subset \bigcap G_n = \emptyset$, so that $\lim_n \mu(H_n) = 0$, it follows that $\lim_n \mu(G_n) \leq \epsilon$. Hence $\lim_n \mu(G_n) = 0$.

(iii) μ is countably subadditive on $\mathcal{P}$, i.e. if $G_n \in \mathcal{P}$ for each $n > 0$ then $\mu(\bigcup G_n) \leq \Sigma_n \mu(G_n)$: First we claim that for each G_n there exists a sequence (Z_{nm}) of zero sets such that $G_n = \bigcup_m Z_{nm}$. Indeed, by (1.2-1(d) there exists a $0 \leq y_n \leq 1$ from $C(T, \underset{\sim}{R})$ such that $G_n = y_n^{-1}(0,1]$. Then choosing $Z_{n,m} = y_n^{-1}[1/m, 1]$ - a zero set by (1.2-3) - the claim follows. Now $G = \bigcup_n G_n = \bigcup_m W_m$ where each W_m equals some Z_{nm}. Set $M_k = \bigcup_{m \leq k} W_m$. Since $G \in \mathcal{P}$ by (1.2-1(e)) and each $M_k \in \mathcal{Z}$ by (1.2-1(c)) the sequence $(G \cap CM_k)$ satisfies the hypothesis of (ii). Thus for any positive ϵ there exists k_o such that $\mu(G \cap CM_{k_o}) < \epsilon$. But M_{k_o} is contained in some finite union $\bigcup_{n \leq N} G_n$ so that

$$\mu(G) \leq \mu(\bigcup_{n \leq N} G_n) + \mu(G \cap CM_{k_o}) < \Sigma_{n \leq N}\, \mu(G_n) + \epsilon.$$

(iv) μ is countably subadditive on $\mathcal{P}(T)$: Let $A = \bigcup_n A_n$ and choose $G_n \supset A_n$ such $\mu(G_n) < \mu(A_n) + \epsilon/2^n$ for each $n > 0$. Set $G = \bigcup_n G_n$ and note that $G \supset A$. Thus

$$\mu(A) \leq \mu(G) \leq \Sigma_n\, \mu(G_n) < \Sigma_n\, \mu(A_n) + \epsilon.$$

In summary then μ is non-negative, monotone, countably subadditive and $\mu(\emptyset) = 0$; i.e. μ is an outer measure on $\mathcal{P}(T)$. Hence the collection of μ-measure sets, i.e. the sets $E \in \mathcal{P}(T)$ for which $\mu(A) \geq \mu(A \cap E) + \mu(A \cap CE)$ for each $A \in \mathcal{P}(T)$, form a σ-algebra of subsets of T on which μ is countably additive (see Dunford and Schwartz 1958, III 5.4, p. 134).

(v) μ is a regular measure on $\mathcal{B}_a$ - the Baire sets: To see that μ is a measure on $\mathcal{B}$ it is enough to show that each $G \in \mathcal{P}$ is μ-measurable. Let $A \subset T$, $\epsilon > 0$, and choose $H \in \mathcal{P}$ such that $H \supset A$ and $\mu(A) > \mu(H) - \epsilon$. Then, since μ is additive on $\mathcal{a}_z$ and monotone on $\mathcal{P}(T)$,

$$\mu(A) + \epsilon > \mu(H) = \mu(H \cap G) + \mu(H \cap CG) \geq \mu(A \cap G) + \mu(A \cap CG)$$

and the result follows. As for regularity, given $A \in \mathcal{B}_a$ there exists $G \supset A$ and $H \supset CA$ such that $\mu(G) - \mu(A) < \epsilon/2$ and $\mu(H) - \mu(CA) < \epsilon/2$. But $H - CA = A - CH$ so that

$$\mu(A) - \mu(CH) = \mu(A - CH) = \mu(H - CA) = \mu(H) - \mu(CA) < \epsilon/2.$$

Hence $CH \subset A \subset G$ and $\mu(G - CH) < \epsilon$.

To finish the proof of (a) it remains to show that

(vi) for each $x \in C(T, \underset{\sim}{R})$ there exist real numbers a, b such that $\mu(x^{-1}[a\ b]) = \mu(T)$: It suffices to prove the result for $x \geq 0$. Suppose that $\mu(x^{-1}[0,r]) < \mu(T)$ for all $r > 0$. It follows then that a sequence of real numbers

$$0 < a_1 < b_1 < a_2 < b_2 < \dots < a_n < b_n < \dots$$

exists such that $\lim_n a_n = \lim_n b_n = \infty$ and $\mu(x^{-1}[a_n, b_n]) > 0$ for each $n > 0$. Letting $c_n = 1/\mu(x^{-1}[a_n, b_n])$ for each $n > 0$ we define $g \in C(\underset{\sim}{R}, \underset{\sim}{R})$ as follows:

$$g(t) = \begin{cases} 0 & t < 0 \\ (c_1/a_1)t & t \in [0, a_1] \\ \sum_{k=1}^{n} c_k & t \in [a_n, b_n],\ n \in \underset{\sim}{N} \\ [c_{n+1}/(a_{n+1} - b_n)]\,(t - b_n) + \sum_{k=1}^{n} c_k, & t \in [b_n, a_{n+1}],\ n \in \underset{\sim}{N} \end{cases}$$

Then the function $y = g \cdot x \in C(T, \underset{\sim}{R})$. Let

$$g_n(t) = \begin{cases} g(t) & t \leq a_n \\ \sum_{k=1}^{n} c_k & t \geq a_n \end{cases}$$

so that $y_n = g_n \cdot x \in C_b(T, \underset{\sim}{R})$, $y_n \leq y$, and $y_n(t) = \sum_{k=1}^{m} c_k$ for $t \in x^{-1}[a_m, b_m]$, and $m \leq n$. As h is positive

$$h(y) \geq h(y_n) = \int y_n d\mu \geq \int_{\bigcup_{m=1}^{n} x^{-1}[a_m,b_m]} y_n \, d\mu$$

$$= \sum_{m=1}^{n} \left(\sum_{k=1}^{m} c_k\right) \mu(x^{-1}[a_m,b_m]) > n$$

for each $n > 0$. This contradiction establishes the result.

As a consequence of (vi) we see that the functional

$$x \to \hat{h}(x) = \int x \, d\mu$$

is real-valued on $C(T,\underset{\sim}{R})$. Furthermore suppose that $x \geq 0$. Then, again by (vi), there exists $r > 0$ such that $\mu(x^{-1}[0,r])=\mu(T)$ so that $\hat{h}(x)=\int_{x^{-1}[0,r]} x d\mu$. Let $x_r=\min(x,r)$. Clearly $\hat{h}(x)=\hat{h}(x_r)$. On the other hand $x_r \in C_b(T,\underset{\sim}{R})$ so that $h(x) \geq h(x_r)=\int x_r d\mu=\hat{h}(x_r)=\hat{h}(x)$. We may therefore conclude that $h-\hat{h}$ is a positive linear functional on $C(T,\underset{\sim}{R})$ which vanishes on $C_b(T,\underset{\sim}{R})$. Thus to finish the proof of (a) we need only prove that

(vii) a positive linear functional x^* on $C(T,\underset{\sim}{R})$ which vanishes on $C_b(T,\underset{\sim}{R})$ vanishes identically on $C(T,\underset{\sim}{R})$: Since each $x \in C(T,\underset{\sim}{R})$ may be expressed in the form $x=(x \vee 0)-((-x)\vee 0)$ it is enough to prove that $x^*(x)=0$ whenever $x \geq 0$. If $x \in C_b(T,\underset{\sim}{R})$ there is nothing to show, so let us assume that x is unbounded on T. Choose an increasing sequence $(a_n)_{n \geq 1}$ of real numbers from $x(T)$ and let $t_n \in T$ be such that $x(t_n)=a_n$ for each $n > 0$. Next define the bounded functions

$$x_n(t) = \begin{cases} 0 \text{ for } t \in x^{-1}[0,a_{n-1}] \\ x(t) - a_{n-1} \text{ for } t \in x^{-1}(a_{n-1},a_n] \\ a_n - a_{n-1} \text{ for } t \in x^{-1}(a_n, \infty) \end{cases}$$

where $a_0=0$ and $n \in \underset{\sim}{N}$. Each such function is continuous on T. If $x(t) \in (a_{n-1}, a_n]$ then $x_k(t)=a_k-a_{k-1}$ for $k < n$, $x_n(t)=x(t)-a_{n-1}$, and $x_k(t)=0$ for $k > n$. Thus, for any such t, $x(t)=\Sigma_k x_k(t)$. If $x(t)=0$ then $x_k(t)=0$ for all k so that once again $x(t)=\Sigma_k x_k(t)$. Therefore $\Sigma_k x_k$ converses pointwise to x on T. Next set $y_n=a_n x_n$ for each positive n. Clearly $\Sigma_n y_n(s)=0$ for all s in the open set $x^{-1}[0,a_1)$ so that $y(s)=\Sigma_n y_n(s)$ exists and is continuous on $x^{-1}[0,a_1)$. For $s \in x^{-1}(a_{n-1},a_{n+1})$ for some $n \geq 1$ we see that

$$\Sigma_k y_k(s) = \begin{cases} \Sigma_{k \le n-1}\, a_k(a_k - a_{k-1}) + a_n(x(t) - a_{n-1}) & \text{for } s \in x^{-1}(a_{n-1}, a_n] \\ \Sigma_{k \le n}\, a_k(a_k - a_{k-1}) + a_{n+1}(x(t) - a_n) & \text{for } s \in x^{-1}[a_n, a_{n+1}) \end{cases}$$

so that $y(s) = \Sigma_k\, y_k(s)$ exists and is continuous on $x^{-1}(a_{n-1}, a_{n+1})$. Now

$$y - \Sigma_{k \le n}\, y_k = \Sigma_{k > n}\, y_k = \Sigma_{k > n}\, a_k\, x_k \ge \Sigma_{k > n}\, a_n\, x_k = a_n(x - \Sigma_{k \le n} x_k)$$

and, as x^* vanishes on $C_b(T,\underset{\sim}{R})$,

$$x^*(y) = x^*(y - \Sigma_{k \le n}\, y_k) > a_n\, x^*(x - \Sigma_{k \le n}\, x_k) = a_n\, x^*(x)$$

for each positive n. Thus $x^*(x)=0$.

In (2.4-2) and (2.4-3) we saw that a linear functional on $C_b(T,\underset{\sim}{R})$ is continuous in the uniform norm iff it is the difference of two positive linear functionals. Close scrutiny of that result reveals that we actually proved that a linear functional is order-bounded (i.e. it sends sets of the form $[x_1,x_2]=\{x \,|\, x_1 \le x \le x_2\}$ into bounded sets of numbers) iff it is a difference of two positive linear functionals. The same proof may be used to obtain the corresponding result for linear functionals on $C(T,\underset{\sim}{R})$.

(b) ORDER-BOUNDED LINEAR FUNCTIONALS A linear functional x^* on $C(T,\underset{\sim}{R})$ is order-bounded iff $x^* = x_p^* - x_n^*$ where x_p^* and x_n^* are positive on $C(T,\underset{\sim}{R})$.

Now it is a simple matter to see that (a) may be extended to

(c) ORDER-BOUNDED LINEAR FUNCTIONALS ON $C(T,\underset{\sim}{R})$ AND REGULAR SET FUNCTIONS

If x^* is order-bounded on $C(T,\underset{\sim}{R})$ then there exists a unique regular countably additive set function μ defined on $\mathcal{B}_a$ such that

$$x^*(x) = \int x\, d\mu \qquad (x \in C(T,\underset{\sim}{R})).$$

Furthermore for each $x \in C(T,\underset{\sim}{R})$ there exists $E \in \mathcal{B}_a$ such that x is bounded on E and $|\mu|(CE)=0$ (where $|\mu|$ represents the total variation of μ).

Proof of uniqueness If μ_1 and μ_2 are two regular countably additive set functions defined on $\mathcal{B}_a$ yielding the same finite integral for each $x \in C(T,\underset{\sim}{R})$ then $\mu_1 - \mu_2 = \mu$ is regular and countably additive on $\mathcal{B}_a$ and $\int x\, d\mu = 0$ for each $x \in C(T,\underset{\sim}{R})$. Let $A \in \mathcal{B}_a$ and choose G, $H \in \mathcal{P}$ such that $CH \subset A \subset G$ and $|\mu|(G-CH) < \epsilon$. By the normality of T there exists a function $x \in C_b(T,\underset{\sim}{R})$ such that $0 \le x \le 1$, $x(CH)=\{1\}$ and $x(CG)=\{0\}$. Then

$$|\mu(A)| = |\mu(A) - \int x\, d\mu| = |\int k_A\, d\mu - \int x\, d\mu|$$

$$= |\int_H (k_A - x) d\mu| = |\int_{H \cap G} x\, d\mu| \le \mu\,(G \cap H)$$

Thus $\mu(A)=0$ for all $A\in \mathcal{B}_a$.

Since our aim is to obtain a representation theorem for the elements of $C(T,\underset{\sim}{R},c)'$ it is natural at this point to inquire as to the circumstances under which all order-bounded linear functionals are continuous.

(d) T REPLETE $\rightarrow$(ORDER-BOUNDED $\leftrightarrow$ CONTINUOUS): If T is replete then the linear functional x^* on $C(T,\underset{\sim}{R})$ is order-bounded iff x^* is continuous in the compact-open topology. Thus, in this case, the elements of $C(T,\underset{\sim}{R},c)'$ are characterized in (c).

Proof As any order segment $[x,y]=\{z\in C(T,\underset{\sim}{R})\,|\,x\le z\le y\}$ is bounded in the compact-open topology (see proof of Theorem 2.6-1), continuity implies order-boundedness. To obtain the converse we recall that in the proof of Theorem 2.6-1 it was shown that any absolutely convex set V which absorbs all order-segments is a neighborhood of the origin in $C(T,\underset{\sim}{R},c)$ provided T is replete. Since $x^*([x,y])\subset Mx^{*-1}((-\epsilon,\epsilon))$, $\epsilon>0$, x^* is continuous.

(e) BOUNDED $\not\rightarrow$ CONTINUITY FOR T NOT REPLETE Of the homomorphisms of $C(T,\underset{\sim}{R})$ only the evaluation maps $T^*=\{t^*\,|\,t\in T\}$ are continuous in the compact-open topology. Any homomorphism fixed to a point of $\upsilon T-T$, i.e. a homomorphism of the type $p^*(x)=x^{\upsilon}(p)$, $x\in C(T,\underset{\sim}{R})$, $p\in\upsilon T-T$, is discontinuous. On the other hand any such homomorphism is bounded as the following argument shows: If $x\le y$ then by the denseness of T in υT, $x^{\upsilon}\le y^{\upsilon}$. Thus for any $x\le z\le y$ it follows that

$$p^*(x)=x^{\upsilon}(p)\le p^*(z)=z^{\upsilon}(p)\le p^*(y)=y^{\upsilon}(p)$$

and p^* of any order segment is a bounded set of numbers.

What then do the elements of $C(T,\underset{\sim}{R},c)'$ look like when T is not replete?

(f) $C(T,\underset{\sim}{R},c)'$ AND RESTRICTIONS OF REGULAR SET FUNCTIONS Let T be a completely regular Hausdorff space. If $x'\in C(T,\underset{\sim}{R},c)'$ then there exists a set function $\mu=\mu_1-\mu_2$ defined on $\mathcal{P}(T)$, where μ_1 and μ_2 are outer measures which when restricted to $\mathcal{B}_a$ are regular measures, and a compact set $K\subset T$ such that

$$x'(x)=\int_K x\,d\mu \quad \text{for} \quad x\in C(T,\underset{\sim}{R})$$

and μ is countably additive on $\mathcal{B}_a\cap K=\{A\cap K\,|\,A\in\mathcal{B}_a\}$.

Proof Just as in (2.4-3) we may decompose x' into a difference of positive linear functionals $x'=x_p'-x_n'$ where for $x\ge 0$, $x_p'(x)=\sup\{x'(y)\,|\,y\in C_b(T,\underset{\sim}{R}),\ 0\le y\le x\}$ and for arbitrary $x=(x\vee 0)-(x\wedge 0)$, $x_p'(x)=x_p'(x\vee 0)-x'(x\wedge 0)$. As x' is continuous there exists a compact set $K\subset T$ and a constant $M>0$

such that $|x'(x)| \leq M\, p_K(x)$. It follows that $|x_p'(x)| \leq |x_p'(x \vee 0)| + |x_p'(x \wedge 0)| \leq 2M\, p_K(x)$ so that x_p' and x_n' are both continuous. Thus by (a) and its proof there are outer measures μ_p and μ_n defined on $\mathcal{P}(T)$ which when restricted to $\mathcal{B}_a$ are regular measures such that $x_p'(x)=\int x\, d\mu_p$ and $x_n'(x)=\int x\, d\mu_n$ for each $x \in C(T,\underset{\sim}{R})$. Next we claim that compact sets K_p and K_n exist such that $\mu_p(K_p)=\mu_p(T)$ and $\mu_n(K_n)=\mu_n(T)$: If this isn't the case, say for μ_p, then for each compact $K \subset T$, $\mu_p(K) < \mu_p(T)$. Thus there exists $G_K \in \mathcal{P}$ such that $G_K \supset K$ and $\mu_p(G_K) < \mu_p(T)$. As K and $\mathrm{cl}_\beta CG_K$ are disjoint in βT there exists $x_K \in C(T,\underset{\sim}{R})$ such that $0 \leq x_K \leq 1$, $x_K(K)=\{0\}$ and $x_K(CG_K)=\{1/\mu_p(CG_K)\}$. Hence $x_p'(x_K)=\int x_K\, d\mu_p \geq \int_{CG_K} x_K\, d\mu_p = 1$. On the other hand if the compact subsets of T are ordered by set inclusion, $(x_K)_K$ is a net in $C(T,\underset{\sim}{R},c)$ clearly converges to the zero function, thereby contradicting the continuity of x'_p. Letting $K=K_p \cup K_n$ it is seen that $\mu_p(K)=\mu_p(T)$ and $\mu_n(K)=\mu_n(T)$.

At this point it is clear that it suffices to consider <u>positive</u> x', i.e. $x'=x_p'$ and $\mu=\mu_p$. The next thing to do is to transfer x' over to $C(K,\underset{\sim}{R})$. Each $x \in C(K,\underset{\sim}{R})$ has a continuous extension to T (extend x to βT and then restrict it to T) so that it is natural to attempt to define $x_K(x)=x'(y)$ where y is some continuous extension of x up to T. To see that this definition is meaningful we need only show that $x'(z)=0$ for any $z \in C(T,\underset{\sim}{R})$ which vanishes on K. If z is such a function then $G=z^{-1}(-\epsilon,\epsilon) \in \mathcal{P}$ and contains K. Thus $\mu(T-G_\epsilon)=0$, $\mu(G_\epsilon)=\mu(K)$ and therefore

$$|x'(z)| = \left|\int_{G_\epsilon} z\, d\mu + \int_{T-G_\epsilon} z\, d\mu\right| = \left|\int_{G_\epsilon} z\, d\mu\right| \leq \epsilon\, \mu(K)$$

for each $\epsilon > 0$.

Now x_K' is clearly positive so there exists an outer measure μ_K on $\mathcal{P}(K)$ which is a regular measure on the Baire subsets of K such that $x_K'(x)=\int x\, d\mu_K$ for each $x \in C(K,\underset{\sim}{R})$. We contend that (iii) the Baire subsets of K and $\mathcal{B}_a \cap K$ coincide: Denoting the class of all complements of zero sets in K by $\mathcal{P}_K$ observe that $\mathcal{P} \cap K=\{G \cap K \mid G \in \mathcal{P}\} \subset \mathcal{P}_K$. Conversely, if $G \in \mathcal{P}_K$ then there exists a non-negative $x \in C(K,\underset{\sim}{R})$ such that $x^{-1}(0,\infty)=G$. Thus, if y is a continuous extension of x to T, $y^{-1}(0,\infty) \cap K=G \in \mathcal{P} \cap K$. Now $\mathcal{B}_a$ is the σ-ring generated by $\mathcal{P}$ so it follows that $\mathcal{B}_a \cap K$ is the σ-ring generated by $\mathcal{P}_K=\mathcal{P} \cap K$ (Halmos, 1950, p. 25, Theorem E).

The next thing to be shown is that μ and μ_K agree on the power set $\mathcal{P}(K)$. As the first step in that direction we claim that (iv) μ and μ_K agree on $\mathcal{P}_K=\mathcal{P} \cap K$: Let $G \in \mathcal{P}_K$ so that for each $\epsilon > 0$ there exists $y \in C(K,\underset{\sim}{R})$

such that $0 \leq y \leq k_G$ (on K) and $\mu_K(G)-\epsilon < x_K'(y)$. As $G \subset T$ there also exists an $H \in \mathcal{P}$ such that $\mu(H)-\epsilon < \mu(G)$. Let F denote the compact set $\{t \in K \mid |y(t)| \geq \epsilon\}$. Since $cl_\beta CH$ and F are disjoint in βT there exists $u \in C(T,\underset{\sim}{R})$ such that $0 \leq u \leq 1$, $u(F)=\{1\}$ and $u(CH)=\{0\}$. Let $\hat{y}$ be any continuous extension of y to T, $\hat{x}=u\,\hat{y}$ and $\hat{x}|_K=x$. It follows that $y \leq x + \epsilon$ and $0 \leq x \leq k_H$. Thus

$$\begin{aligned}\mu_K(G) - \epsilon &\leq x_K'(y) \leq x_K'(x + \epsilon) = x_K'(x) + \epsilon\,\mu_K(K)\\ &= x'(\hat{x}) + \epsilon\,\mu_K(x)\\ &\leq \mu(H) + \epsilon\,\mu_K(x) < \mu(G) + \epsilon(1 + \mu_K(K)).\end{aligned}$$

As ϵ is arbitrary $\mu_K(G) \leq \mu(G)$. To obtain the reverse inequality choose $J \in \mathcal{P}$ such that $J \cap K=G$. For $\epsilon > 0$ choose $\hat{z} \in C(T,\underset{\sim}{R})$ such that $0 \leq \hat{z} \leq k_j$ and $\mu(J)-\epsilon < x'(\hat{z})$. Setting $z=\hat{z}|_K$ it follows that $0 \leq z \leq k_G$ on F and

$$\mu_K(G) \geq x_K'(z) = x'(z) > \mu(J) - \epsilon \geq \mu(G) - \epsilon.$$

Hence $\mu_K(G) \geq \mu(G)$ and finally $\mu_K=\mu$ on $\mathcal{P}_K$.

(v) μ and μ_K agree on $\mathcal{P}(K)$: Let $A \subset K$, and choose $G \in \mathcal{P}_K$ such that $\mu_K(G)-\epsilon < \mu_K(A)$. If $H \in \mathcal{P}$ and $H \cap K=G$ then $\mu(H)=\mu(H \cap K)=\mu(G)$ so that

$$\mu(A) - \epsilon \leq \mu(H) - \epsilon = \mu(G) - \epsilon = \mu_K(G) - \epsilon < \mu_K(A).$$

Thus $\mu(A) \leq \mu_K(A)$. On the other hand for $\epsilon > 0$ there exists $J \in \mathcal{P}$ such that $J \supset A$ and $\mu(J)-\epsilon < \mu(A)$. Then, noting that $\mu(J \cap K)=\mu(J)$, $J \cap K \in \mathcal{P}_K$, and $J \cap K \supset A$, we see that

$$\mu_K(A) - \epsilon \leq \mu_K(J \cap K) - \epsilon = \mu(J) - \epsilon < \mu(A)$$

for each $\epsilon > 0$. Hence $\mu_K(A) \leq \mu(A)$.

Since μ_K is countably additive on the Baire subsets of K (= $\mathcal{B}_a \cap K$ by (iii)), μ is countably additive on $\mathcal{B}_a \cap K$. Furthermore for each $x \in C(T,\underset{\sim}{R})$

$$x'(x) = x_K'(x|_K) = \int x|_K\, d\mu_K = \int_K x\, d\mu. \ \nabla$$

<u>2.5 FIELD-VALUED CONTINUOUS FUNCTIONS</u> (Cf. Exercises 1.9 and 1.13). Let T be an ultraregular space and $\underset{\sim}{K}$ a complete nonarchimedean nontrivially valued field. The main point of this exercise is to characterize when $C(T,\underset{\sim}{K},c)$ is "barreled" and "bornological" (parts (e) and (f)) after introducing suitable meanings for those terms.

A <u>bounded measure</u> μ on T is a finitely additive $\underset{\sim}{K}$-valued set function on the algebra $\mathfrak{S}$ of clopen subsets of T such that for some positive number M, $|\mu(S)| \leq M$ for all $S \in \mathfrak{S}$. The linear space (usual operations) of all such measures μ is denoted by $B(T,\underset{\sim}{K})$. If μ assumes only the values 0 and 1 and is monotone (i.e. if $\mu(S)=0$, then $\mu(S')=0$ for all $S'\subset S$), then μ is called a monotone 0-1 measure. The set of all such measures is denoted by $M_o(T)$. We note that any 0-1 measure is monotone if the characteristic of $\underset{\sim}{K}$, ch $\underset{\sim}{K}$, is not 2; if ch $\underset{\sim}{K}=2$, then there are non-monotone 0-1 measures.

(a) $C_c(T,\underset{\sim}{K})'$ and $B(T,\underset{\sim}{K})$ Let $C_c(T,\underset{\sim}{K})$ be the subalgebra of functions in $C(T,\underset{\sim}{K})$ with relatively compact range, endowed with the topology of uniform convergence on T, and let $C_c(T,\underset{\sim}{K})'$ denote the continuous dual of $C_c(T,\underset{\sim}{K})$. Show that there is a 1-1 correspondence between $B(T,\underset{\sim}{K})$ and $C_c(T,\underset{\sim}{K})'$. (Hint: If $f\in C_c(T,\underset{\sim}{K})'$, let $\mu_f(S)=f(k_S)$ where k_S is the characteristic function of $S\in\mathfrak{S}$. Conversely, let $D(T,\underset{\sim}{K})$ be the linear span of the functions $(k_S)_{S\in\mathfrak{S}}$. For $\mu\in B(T,\underset{\sim}{K})$ and $\Sigma\, a_i k_{S_i}\in D(T,\underset{\sim}{K})$ let $f_\mu(\Sigma\, a_i k_{S_i})=\Sigma\, a_i\mu(S_i)$. Show that f_μ is continuous on $D(T,\underset{\sim}{K})$ and extend it to $C_c(T,\underset{\sim}{K})$ by continuity.)

For $\mu\in B(T,\underset{\sim}{K})$ let D_μ be the collection of sets $S\in\mathfrak{S}$ such that for all clopen subsets S' of S, $\mu(S')=0$. The support, supp μ, of μ (cf. (2.4-7)) is defined to be $C(\bigcup D_\mu)$. If supp μ is compact, μ is said to have compact support, and the collection of all such measures is denoted by $B_c(T,\underset{\sim}{K})$. The next result establishes a 1-1 correspondence between $B_c(T,\underset{\sim}{K})$ and $C(T,\underset{\sim}{K},c)'$.

(b) $B_c(T,\underset{\sim}{K})$ and $C(T,\underset{\sim}{K},c)'$ Let T be compact, $\mu\in B(T,\underset{\sim}{K})$, $P=(S_i)$ be a finite partition of T into disjoint clopen sets, and $d_P=\max_i|\mu(S_i)|$. For $t_i\in S_i$ and $x\in C(T,\underset{\sim}{K})$, $\lim_{d_P\to 0}\Sigma_i x(t_i)\mu(S_i)$ exists and is denoted by $\int x\,d\mu$. Show that $\int\cdot\, d\mu$ is a continuous linear functional on $C(T,\underset{\sim}{K},c)$. Conversely, if $f\in C(T,\underset{\sim}{K},c)'$ then there exists $\mu_f\in B(T,\underset{\sim}{K})$ such that $f(x)=\int x\,d\mu_f$ for each $x\in C(T,\underset{\sim}{K})$. Moreover, letting $\|\mu_f\|=\sup_{S\in\mathfrak{S}}|\mu_f(S)|$, $\|f\|=\|\mu_f\|$.

For $f\in C(T,\underset{\sim}{K},c)'$, let D_f denote the collection of clopen sets S such that $f(xk_S)=0$ for all $x\in C(T,\underset{\sim}{K})$. The support of f, supp f, is defined to be $C(\bigcup D_f)$. Next, it is shown that there is a bounded measure μ such that supp μ_f=supp f and that this notion of support of a continuous linear functional has essentially the same properties as support does in the classical case (cf. (2.4-7) and (2.4-8)).

(c) Bounding sets and vanishing sets A bounding set for $f\in C(T,\underset{\sim}{K},c)'$ is a compact subset K of T for which there is some $N_K>0$ such that $|f(x)|\leq N_K p_K(x)=N_K\sup|x(K)|$ for each $x\in C(T,\underset{\sim}{K})$. A vanishing set for f is a compact set K with the property that if x vanishes on K, f must vanish on x.

(i) Show that a compact set is a bounding set iff it is a vanishing set. (Hint: Let K be a vanishing set for f, $x\in C(T,\underset{\sim}{K})$, and x' the restriction of x to K. Note that each $x\in C(K,\underset{\sim}{K})$ can be extended to an $\hat{x}\in C(T,\underset{\sim}{K})$ as follows. Let $\sup |x(K)|=M$, and construct a finite disjoint cover (W_i) of K of clopen subsets of T such that $\sup\{|x(t)-x(t')| \mid t,t'\in W_i\cap K\} \leq M/2$. For $t_i\in W_i\cap K$, let $x_1=\Sigma_{i=1}^n x(t_i)k_{W_i}$. Then $\sup|(x-x_1)(K)| \leq M/2$ while $\sup|x_1(T)|\leq M$. In similar fashion construct a function x_2 such that $\sup|(x-x_1-x_2)(K)| \leq M/4$ while $\sup|x_2(T)| \leq M/2$, and so on. Consider the map

$$f':C(K,\underset{\sim}{K},c) \rightarrow \underset{\sim}{K}$$

$$x' \rightarrow f(x)$$

Show that f' is a well-defined continuous linear functional on $C(K,\underset{\sim}{K},c)$ and, thus, that K is a bounding set for f.)

(ii) For each $f\in C(T,\underset{\sim}{K},c)'$, show that D_f is a ring of sets.

(iii) Show that supp $f=C(\bigcup D_f)$ is compact and nonempty iff f is nontrivial. (Hint: Let K be a vanishing set for F. If $S\in\mathfrak{S}$ and $S\subset CK$, then $S\in D_f$. Hence $CK\subset\bigcup D_f$ and supp $f\subset K$.)

(iv) Show that if $S\in\mathfrak{S}$ and $S\cap$ supp $f\neq\emptyset$, then there is some $x\in C(T,\underset{\sim}{K})$ such that $x(CS)=\{0\}$ and $f(x)=1$.

(v) If T is compact, then supp f is a vanishing set for f. (Hint: Suppose that supp f is a subset of the clopen set S. Since CS is compact, CS is covered by finitely many $S_i\in D_f$. Thus $CS\in D_f$ and S is a vanishing set. Thus any clopen superset of supp f is a vanishing set; this insures that supp f is a vanishing set. Indeed, if x vanishes on supp f and $S_n=\{t\in T \mid |x(t)|< 1/n\}$, then supp $f\subset S_n$ for each $n\in\underset{\sim}{N}$ and each S_n is a vanishing set. Thus $f(x)=f(xk_{S_n})+f(xk_{CS_n})=f(xk_{S_n})$. As $|f(x)| \leq N_K p_K(xk_{S_n})$ for some compact subset K of T, by the definition of S_n it follows that $|f(x)|\leq N_K(1/n)$ for all $n\in\underset{\sim}{N}$ and $f(x)=0$.)

(vi) Let $f\in C(T,\underset{\sim}{K},c)'$ and μ_f be the bounded measure defined by taking $\mu_f(S)=f(k_S)$ for each $S\in\mathfrak{S}$. Show that supp μ_f=supp f and that supp f is the minimal (with respect to set inclusion) vanishing set for f. (Hint: Let K be a vanishing set for f and let μ_f' be the bounded measure defined on the clopen subsets $S\cap K$ of K by taking $\mu_f'(S\cap K)=\mu_f(S)=f(k_S)$. That every clopen subset of K is of the form $S\cap K$ for some clopen subset S of T is guaranteed by the compactness of K. To see that μ_f' is well-defined, note that if X, $Y\in\mathfrak{S}$ and $X\cap K=Y\cap K$, then $(X-X\cap Y)\cap K=\emptyset$ so that $f(k_{X-X\cap Y})=0$; thus $f(k_X)=f(k_Y)=f(k_{X\cap Y})$. Let x' be the restriction of $x\in C(T,\underset{\sim}{K})$ to K, and let $f'(x')=f(x)=\int x'd\mu_f'$. (Also note that $(\mu_f)'=\mu_{(f')}$.) Now show that supp f=supp f', supp μ_f=supp μ_f', and supp μ_f'=supp f'. Then use (b) and (v) above to show that $f'(x')=f(x)=\int x'd\mu_f'=\int_{\text{supp }\mu_f'} x'd\mu_f'=\int_{\text{supp } f'} x'd\mu_{f'} = \int_{\text{supp } f} x'd\mu_{f'}$.)

A topological vector space X over $\underset{\sim}{K}$ is called locally K-convex if its topology is generated by a family P of nonarchimedean seminorms (i.e. for each $p \in P$, $p(x+y) < \max(p(x), p(y))$. Letting C denote the closed unit disc about 0 in $\underset{\sim}{K}$, a subset U of X is called absolutely K-convex if $CU+CU \subset U$. If p is a nonarchimedean seminorm, all sets of the form $\{x \in X \mid p(x) < \epsilon\}, \epsilon > 0$, are absolutely $\underset{\sim}{K}$-convex. X is K-barreled if every absolutely $\underset{\sim}{K}$-convex closed absorbent subset of X is a neighborhood of 0. Similarly X is K-bornological if every absolutely $\underset{\sim}{K}$-convex bornivore is a neighborhood of 0. The space $C(T,\underset{\sim}{K},c)$ is a nonarchimedean locally K-convex space.

$\underset{\sim}{K}$ is called spherically complete if every totally ordered collection of closed discs in $\underset{\sim}{K}$ has nonempty intersection and an analog of the Hahn-Banach theorem holds for X when $\underset{\sim}{K}$ has this property (Ingleton 1952, Narici, Beckenstein and Bachman 1971, p. 78).

(d) Relative $\underset{\sim}{K}$-pseudocompactness (Cf. (1.5-4)) A closed subset L of T is relatively K-pseudocompact if every $x \in C(T,\underset{\sim}{K})$ is bounded on L. Show that a closed set L is relatively $\underset{\sim}{K}$-pseudocompact iff $\text{cl}_{\beta_o T} L \subset \upsilon_o T$ where $\beta_o T$ and $\upsilon_o T$ are as in Exercises 1.9(f) and 1.13(e). Thus the "$\underset{\sim}{K}$" in "relatively $\underset{\sim}{K}$-pseudocompact" is superfluous and may be omitted.

(e) $\underset{\sim}{K}$-barreledness of $C(T,\underset{\sim}{K},c)$ If $\underset{\sim}{K}$ is spherically complete, show that $C(T,\underset{\sim}{K},c)$ is $\underset{\sim}{K}$-barreled iff every relatively $\underset{\sim}{K}$-pseudocompact subset of T is compact.

(f) $\underset{\sim}{K}$-bornologicity of $C(T,\underset{\sim}{K},c)$ If $\underset{\sim}{K}$ has non-Ulam cardinal, then $C(T,\underset{\sim}{K},c)$ is $\underset{\sim}{K}$-bornological iff $\upsilon_o T=T$.

(g) If $\underset{\sim}{K}$ is a spherically complete field of non-Ulam cardinal, then $\underset{\sim}{K}$-bornologicity of $C(T,\underset{\sim}{K},c)$ implies $\underset{\sim}{K}$-barreledness of $C(T,\underset{\sim}{K},c)$.

THREE

Lattices and Wallman Compactifications

Lattices, in general, possess two descriptions: As an ordered space and a more algebraic description as a set equipped with two operations. The equivalence of these descriptions is discussed in the first section. Examples of lattices abound; a few are of special concern to us. (The prototype lattices are collections of subsets of a given set.) Later we focus particular attention on the lattice of closed subsets of a topological space. Another of special interest is the lattice of zerosets of a topological space.

Using various lattices $\mathcal{L}$ of subsets of a topological space T one may always construct a nearby compact space $w(T,\mathcal{L})$ which is a compactification of T under certain conditions. More generally, there is a compact space wL associated with any lattice L and this is discussed in Sec. 3.2; the compact space wL is the collection of all L-ultrafilters endowed with a certain topology. Given a topological space T we have already seen an instance where a compactification has been obtained from an associated space of ultrafilters: In Chap. 1 it was seen that the Stone-Cech compactification βT of T was the collection of all ultrafilters from the lattice $\mathcal{Z}$ of zerosets of T.

We specialize to the setting of main interest to us in Sec. 3.3 and restrict consideration to lattices $\mathcal{L}$ of subsets of a topological space T. Conditions on $\mathcal{L}$ are developed which ensure that the map

$$t \to \mathcal{F}_t = \{A \in \mathcal{L} \mid t \in A\}$$

is a 1-1 map of T into $w(T,\mathcal{L})$. It must be guaranteed that $\mathcal{F}_t$ is an $\mathcal{L}$-ultrafilter and that distinct t's produce different ultrafilters. When these conditions are satisfied, we can ask whether the compact space $w(T,\mathcal{L})$ is a compactification of T. It turns out ((3.3-2)) that the map $t \to \mathcal{F}_t$ is a homeomorphism iff $\mathcal{L}$ is a base for the closed sets in T, while $\{\mathcal{F}_t \mid t \in T\}$ is always dense in $w(T,\mathcal{L})$.

In Sec. 3.4 sufficient conditions are obtained for $w(T,\mathcal{L})$ to be βT. If T is any completely regular Hausdorff space e.g. and one takes the lattice $\mathcal{L}$ to be the lattice $\mathcal{Z}$ of zerosets of T, it happens that $w(T,\mathcal{Z}) = \beta T$.

Wallman compactifications constitute a more general breed of compactification than the Stone-Cech. Every Stone-Cech compactification may be realized as a Wallman compactification, as mentioned in the proceeding paragraph, and so can any one-point compactification of a locally compact Hausdorff space be realized as a Wallman compactification. One-point compactifications, however, are certainly only rarely Stone-Cech compactifications, (bounded continuous functions on $\underset{\sim}{R}$ for example may generally not be continuously extended to the one-point compactification of $\underset{\sim}{R}$). This raises the question: Perhaps any compactification is of the Wallman type? This question, which has to do with felicitously choosing a lattice of subsets, hasn't been answered yet. Even so, some progress has been made and in Sec. 3.5: Given a Hausdorff compactification S of a topological space T and a lattice $\mathcal{L}$ of subsets of T, necessary and sufficient conditions are obtained for $w(T,\mathcal{L})$ to equal S in Theorem 3.5-1.

The subject of Sec. 3.6 is 'when' two lattices $\mathcal{H}$ and $\mathcal{L}$ of subsets of a topological space T determine the same Wallman compactification. These results are used in Chap. 5 (principally in Sec. 5.4) where we investigate how spaces $\mathcal{M}$ of maximal ideals of certain kinds of topological algebras X may be realized as Wallman compactifications $w(\mathcal{M}_c,\mathcal{L})$ of the space $\mathcal{M}_c$ of closed maximal ideals of X.

<u>3.1 Lattices</u> A partially ordered set containing the supremum and infimum of each finite set is a <u>lattice</u>. Though we introduce lattices here - actually distributive lattice with 0 and 1 - we scarcely investigate them, providing only what is needed for use in the sequel. Some of our nomenclature is unconventional - "filter" for "additive ideal without 0" for example - because of the desire to use language which is topologically suggestive. For more complete information about lattices per se, the reader is referred to the standard work of Birkhoff (1948) and to Grätzer (1971).

The power set $\mathcal{P}(T)$ of a set T, i.e. the collection of all subsets of T, equipped with the relation

$$A \leq B \text{ iff } A \subset B$$

is a partially ordered set in which each pair of elements A,B has a greatest lower bound $A \cap B$ and a least upper bound $A \cup B$. Motivated by this example we define a lattice to be any partially ordered set $(L, \leq)$ in which each pair of elements $a,b \in L$ has a greatest lower bound $a \wedge b$, the "meet" of a and b, and a least upper bound $a \vee b$, the "join" of a and b. L is a <u>distributive lattice</u> provided

$$a\wedge(b\vee c) = (a\wedge b)\vee(a\wedge c)$$

and

$$a\vee(b\wedge c) = (a\vee b)\wedge(a\vee c)$$

for all $a,b,c\in L$. Actually these rules are equivalent to each other. The power set lattice $\mathcal{P}(T)$ is certainly distributive. Moreover $\mathcal{P}(T)$ has a least element $\emptyset$ and a greatest element T. When a lattice L possesses a least or greatest element it is unique and is denoted by 0 and 1 respectively. Corresponding to each element $A\in\mathcal{P}(T)$ there is an element CA, the complement of A, having the properties:

$$A\wedge CA = A\cap CA = \emptyset, \quad A\vee CA = A\cup CA = T.$$

If each element a of a general lattice L with 0 and 1 possesses a complement **b**, i.e. $a\wedge b=0$, $a\vee b=1$, L is said to be complemented and a complemented distributive lattice is referred to as a Boolean lattice.

A more algebraic description of these ideas is also useful. First we replace the symbols $\vee$ and $\wedge$ by + and . respectively, i.e. for a, b in the lattice L we define $a+b=a\vee b$ and $ab=a\wedge b$. Then it clearly follows that L is a commutative semigroup with respect to each of these operations. Furthermore if L is a lattice with 0 and 1 then these semigroups possess identities. Some things concerning these new operations are apparent; in particular the idempotent properties: For any $a\in L$

$$a + a = aa = a.$$

Moreover

$$a \leq b \text{ iff } a + b = b \text{ iff } ab = a.$$

In a distributive lattice the distributive laws take the form:

$$a(b + c) = ab + ac$$
$$a + bc = (a + b)(a + c),$$

while in a complemented lattice to each element a of the lattice there corresponds and element b such that

$$a + b = 1 \quad \text{and} \quad ab = 0.$$

Having just seen how to convert a $\leq$ formulation of a lattice into a (+,.) phrasing, we now consider the reverse question: Given a set L provided with operations + and ., when is it possible to recover the ordering? What conditions must + and . satisfy?

(3.1-1) ALGEBRAIC DESCRIPTIONS OF LATTICES Let L be a non-empty set provided with two operators $+$ and $\cdot$. If

(1) L is a commutative semigroup with respect to each operation,

(2) $a + a = aa = a$ for each $a \in L$, and

(3) $a + b = b$ iff $ab = a$

then the relation

$$a \leq b \text{ iff } a + b = b$$

defines a partial order on L with respect to which L is a lattice. If $a, b \in L$ then $a \vee b = a+b$ and $a \wedge b = ab$. Moreover if each of the commutative semigroups possesses an identity then the identities serve as 0 and 1 for the lattice. L is a distributive lattice if

(4) $a(b + c) = ab + ac$ and $ab + c = (a + c)(b + c)$ for all $a,b,c \in L$.

Finally if L contains 0 and 1 and

(5) for each $a \in L$ there exists $b \in L$ such that $a + b = 1$ while $ab = 0$

L is a Boolean lattice.

Proof Clearly $\leq$ defines a partial order on L. To see that $a \vee b = a+b$ it is necessary to show that $a, b \leq a+b$. Indeed

$$a + (a + b) = (a + a) + b = a + b$$

so that $a \leq a+b$. In the same way $b \leq a+b$. Suppose that $a, b \leq c$ and consider

$$(a + b) + c = a + (b + c) = a + c = c.$$

Thus $a+b \leq c$ and $a+b$ is therefore the join of a and b. An analogous argument yields $a \wedge b = ab$. The remaining claims are obvious. ▽

Example 3.1-1 THE POWER SET BOOLEAN LATTICE Let T be a set and L the collection of all subsets of T. For $A,B \in L$ let

(1) $A \cdot B = A \cap B$

(2) $A + B = A \cup B$.

L with these operations is, as was previously noted, a Boolean lattice. Rather than consider all subsets of T, it is of course possible to restrict attention to subclasses of $\mathcal{P}(T)$ which contain a least upper bound and a greatest lower bound of each pair of its elements with respect to the inclusion ordering. It is not necessary for $A \wedge B$ and $A \vee B$ to be $A \cap B$ as happens for example if L consists just of sets $\emptyset$, T, A, and B where $A \cup B \neq T$, $A \cap B \neq \emptyset$, and $A \neq B$.

Example 3.1-2 THE LATTICE OF CLOSED SETS Let T be a T_1 topological space and L be the collection of closed subsets of T. Then L is a distributive lattice with 0 and 1 with respect to $\cup$ and $\cap$. It is clear that the complement of any closed set is closed iff the topology on T is discrete. Hence, L cannot be a Boolean lattice unless T is discrete.

A different-looking example of a lattice is given next.

<u>Example 3.1-3 THE LATTICE OF CONTINUOUS FUNCTIONS</u> Let T be a topological space. Then $C(T,\underset{\sim}{R})$ is a lattice with respect to the partial order

$$x \leq y \text{ iff } x(t) \leq y(t) \text{ for each } t \in T \ (x, y \in C(T,\underset{\sim}{R})).$$

Since $\min(x,y) = \frac{1}{2}((x+y)-|x-y|))$ and $\max(x,y) = \frac{1}{2}(x+y+|x-y|)$ are continuous and

$$(x \wedge y)\ (t) = \min(x(t),\ y(t)),$$
$$(x \vee y)\ (t) = \max(x(t),\ y(t))$$

$x \wedge y$ and $x \vee y$ belong to $C(T,\underset{\sim}{R})$. It is evident that the distributive laws hold. On the other hand $C(T,\underset{\sim}{R})$ contains neither a least nor a greatest element. If instead of taking all of $C(T,\underset{\sim}{R})$ we choose $L=[x,y]= \{z \in C(T,\underset{\sim}{R}) \,|\, x \leq z \leq y\}$, where x and y are continuous real-valued functions on T such that $x \leq y$, then L is a distributive lattice with $0=x$ and $1=y$. In most cases however, L is not complemented, e.g. if x and y are the continuous functions assuming the real number values 0 and 1 respectively, the only elements in L with complements are x and y.

In Chap. 1 it was shown that in the class of replete spaces T, the algebra $C(T,\underset{\sim}{R})$ completely characterizes T, i.e. if T and S are replete spaces and $C(T,\underset{\sim}{R})$ and $C(S,\underset{\sim}{R})$ are isomorphic algebras, then T and S are homeomorphic ((1.6-3)).

As noted in Sec. 1.6 the lattice properties of $C(T,\underset{\sim}{R})$ are often enough to characterize T, e.g. if T and S are compact Hausdorff spaces and $C(T,\underset{\sim}{R})$ and $C(S,\underset{\sim}{R})$ are order-isomorphisms (there exists a 1-1 mapping ψ taking $C(T,\underset{\sim}{R})$ onto $C(S,\underset{\sim}{R})$ such that $x \leq y$ implies $\psi(x) \leq \psi(y)$) then T and S are homeomorphic (Kaplansky 1947).

Since the distributive function lattices of the previous example differ so drastically in appearance from the set lattices given in Examples 3.1-1 and 3.1-2 it is perhaps surprising to find that any lattice can be "represented" by a set lattice. More precisely, if L is a lattice then there exists a set T and a collection of its subsets which, when supplied with the ordering defined by set inclusion, is order-isomorphic to L. Indeed by taking T=L and $\mathcal{L}$ to be the collection of all subsets of T of the form

$$L_a = \{b \in L \,|\, b \leq a\} \quad (a \in L)$$

the mapping $a \to L_a$ is seen to be an order-isomorphism between L and $\mathcal{L}$. Even more can be said - if L is a distributive lattice then L is order-isomorphic to a set lattice in which the meet and join of a pair of sets are given by $\cap$ and $\cup$ respectively (Nobeling 1954, p. 28, and Hermes 1955, p. 106).

The lattice of the following example is quite similar in form to the lattice of closed sets mentioned in Example 3.1-2. After linking lattices and compactifications, it will be seen (Example 3.3-3) that the lattice of all closed sets leads to the Wallman compactification. Sublattices of the lattice of all closed sets yield other compactifications. The lattice of zero sets of a completely regular Hausdorff space T, for example, produces βT ((3.4-3)).

Example 3.1-4 THE ZERO SET LATTICE Let T be a completely regular Hausdorff space and $\mathcal{Z}$ the collection of all zero sets of functions in $C(T,\mathbb{R})$. By the results of Sec. 1.2, $\mathcal{Z}$ is a distributive lattice with 0 and 1 with respect to $\cup$ and $\cap$:

$$z(x) \cup z(y) = z(x^2 + y^2) \in \mathcal{Z}$$
$$z(x) \cap z(y) = z(xy) \in \mathcal{Z}$$

The notions to be considered next are abstractions of concepts particular to the power set lattice. First we have a generalization of "filter".

Definition 3.1-1 LATTICE FILTERS AND ULTRAFILTERS A subset F of a distributive lattice L with 0 and 1 is a lattice filter (or simply "filter") if

(1) $0 \notin F$,

(2) F is stable under finite products (i.e. if a, $b \in F$ then $ab \in F$) and

(3) (the "upper bound" condition) if $a \in F$ and $b \geq a$ then $b \in F$.

A lattice filter F is a lattice ultrafilter (or just "ultrafilter") if no other filter contains F as a proper subset.

Analogous to the power set situation, ultrafilters can be characterized in terms of a "primality" condition.

(3.1-2) ULTRAFILTERS Let F be a lattice filter, in the distributive lattice L with 0 and 1. Then if F is an ultrafilter and $a+b \in F$, either $a \in F$ or $b \in F$. Conversely, if L is a Boolean lattice and

$$a + b \in F \to a \in F \text{ or } b \in F \quad (a,\ b \in L)$$

then F is an ultrafilter.

Proof Suppose that F is an ultrafilter, $a+b \in F$, and $b \notin F$. Then, letting $H=\{c \in L \mid c+b \in F\}$, it follows that H is a filter: Indeed $0 \notin H$ as $b \notin F$. If $c,d \in H$

then, since $cd+b=(c+b)(d+b)\in F$, $cd\in H$. Moreover, if $c\in H$ and $d \geq c$ then, as $d+b \geq c+b$, it follows that $d+b\in F$ and $d\in H$. Clearly $a\in H$ and $F\subset H$ so the maximality of F implies that $a\in H=F$.

Conversely, if F is a filter in the Boolean lattice L and F contains a or b whenever $a+b\in F$, suppose that $H\supset F$ and $a\in H$ while $a\notin F$. As $1\in F$, $a\notin F$, and $1=a+a'$ where a' is the complement of a, it follows that $a'\in F$. But then both a and a' belong to H so that $0=aa'\in H$ - a contradiction. Thus H=F and F is an ultrafilter. ∇

Example 3.1-5 LATTICE FILTERS VS. SET FILTERS In a power set lattice $\mathcal{P}(T)$ a collection of subsets is a lattice filter iff it is a filter in the ordinary set - theoretic sense. In a lattice $\mathcal{L}$ of subsets of a set T (with respect to $\cup$ and $\cap$) a lattice filter is not, in general, a set filter, as it need not contain all supersets of elements of the filter - just those belonging to $\mathcal{L}$.

The primality condition given in (3.1-2) completely characterizes ultrafilters in Boolean lattices. Below we obtain a characterization of them in any distributive lattice with 0 and 1 (Theorem 3.1-1) by essentially strengthening condition (3) of the definition of a lattice filter. First, however, we define a notion which is somewhat weaker than that of a lattice filter.

Definition 3.1-2 LATTICE FILTER SUBBASE A subset S of a distributive lattice L with 0 and 1 is a lattice filter subbase if it has the finite intersection property, i.e. all finite products of elements of S are non-zero.

Every lattice filter subbase S is contained in a lattice filter for clearly the collection F(S) of all elements b of L which are greater than or equal to some finite product of elements from S is a lattice filter and, in fact, is the smallest lattice filter of L containing S. As promised, we now show that by strengthening the third condition of Definition 3.1-1, a characterization of ultrafilters is obtained.

Theorem 3.1-1 ULTRAFILTERS AND THE "MEET" CONDITION Let L be a distributive lattice L with 0 and 1. Then $F\subset L$ is a lattice ultrafilter iff

(1) F is a lattice filter subbase,

(2) F is stable under finite products, and

(3) (the "meet" condition) for any $c\in L$, if $ca\neq 0$ for each $a\in F$ then $c\in F$.

Remark Clearly conditions (1) and (2) are equivalent to conditions (1) and (2) of the definition of a lattice filter (Definition 3.1-1). Moreover the meet condition is stronger than the upper bound condition of Definition 3.1-1: Whenever $b \geq a$ for some $a\in F$ then, for each $d\in F$, $bd\neq 0$ - indeed if bd=0 then ad=bad=a(bd)=0 - a contradiction. Hence, by the meet condition the upper bound condition of Definition 3.1-1 is fulfilled: $b\in F$.

Proof Clearly (1) and (2) are necessary for F to be a lattice ultrafilter. As for the meet condition suppose that $ca\neq 0$ for each $a\in F$. As the smallest lattice filter H generated by the lattice filter subbase $F\cup\{c\}$ contains the ultrafilter F, H=F and, therefore, $c\in F$.

Conversely, as was noted in the remark above, conditions (1), (2) and (3) force F to be a lattice filter. If H is a lattice filter $\supset$ F and $c\in H$ then $ca\neq 0$ for each $a\in F$. Thus $c\in F$ by (3), H=F, and F is therefore a lattice ultrafilter. ∇

The collection of all lattice filter subbases of a given distributive lattice L with 0 and 1 is partially ordered by set inclusion and, via the Zorn's lemma argument, lattice filter subbases which are maximal with respect to set inclusion can be shown to exist. Such objects are referred to as maximal lattice filter subbases. It is perhaps surprising that maximal lattice filter subbases are in fact lattice ultrafilters viz. :

(3.1-3) MAXIMAL FILTER SUBBASE "=" ULTRAFILTER In a distributive lattice L with 0 and 1 a lattice filter subbase is maximal iff it is a lattice ultrafilter.

Proof If H is a lattice ultrafilter and S is a lattice filter subbase containing H then the smallest lattice filter F(S) containing S contains H. Thus by the maximality of H,

$$H\subset S\subset F(S) = H$$

and S=H thereby proving that H is a maximal lattice filter subbase.

Conversely, suppose that S is a maximal lattice filter subbase. Then, as S' - the collection of all finite products of S - is a lattice filter subbase containing S, S'=S. By Theorem 3.1-1, it remains to show that $c\in S$ whenever $ca\neq 0$ for each $a\in S$. But this is clearly the case as $S\cup\{c\}=S'\cup\{c\}$ is a lattice filter subbase containing S. ∇

Section 3.2 LATTICES AND ASSOCIATED COMPACTIFICATIONS In this section we show that given a distributive lattice L with 0 and 1 there is an associated compact space wL: wL consists of all lattice ultrafilters. As observed in Sec. 3.1 given a topological space T there are certain lattices which one naturally tends to consider, e.g. the lattice $\mathcal{C}$ of all closed subsets of T and, for T completely regular and Hausdorff, the lattice of all zero sets of T. Determining the properties of wL in these special cases is what this section is directed toward.

To begin with we single out a collection of subsets of the family of all ultrafilters of a distributive lattice L with 0 and 1 to serve as a base of closed subsets for a topology. For each $a \in L$ we define the basic set $\mathcal{B}_a$ generated by a to be the collection of all ultrafilters to which a belongs. As no ultrafilter can contain 0, $\mathcal{B}_0 = \emptyset$. Thus for the basic sets to be a base rather than a subbase of closed sets for a topology it suffices that they be stable with respect to the formation of finite unions. Indeed, as we now show, $\mathcal{B}_a \cup \mathcal{B}_b = \mathcal{B}_{a+b}$ for each pair $a, b \in L$. If $F \in \mathcal{B}_a$ then, since for any $b \in L$ $a \leq a+b$, $a+b \in F$ and $F \in \mathcal{B}_{a+b}$. Thus $\mathcal{B}_a, \mathcal{B}_b \subset \mathcal{B}_{a+b}$ and so $\mathcal{B}_a \cup \mathcal{B}_b \subset \mathcal{B}_{a+b}$. On the other hand, as ultrafilters satisfy the primality condition of (3.1-2)

$$F \in \mathcal{B}_{a+b} \rightarrow a+b \in F \rightarrow a \in F \text{ or } b \in F \rightarrow F \in \mathcal{B}_a \cup \mathcal{B}_b.$$

Hence $\mathcal{B}_a \cup \mathcal{B}_b = \mathcal{B}_{a+b}$ and it follows that $(\mathcal{B}_a)_{a \in L}$ is a base of closed subsets for a topology on the family of ultrafilters. This topological space is referred to as the <u>Wallman space generated by L</u> and is denoted by <u>wL</u>. If $\mathcal{L}$ is a lattice of subsets of a topological space T with respect to union and intersection the associated Wallman space is denoted by $w(T, \mathcal{L})$. Certainly the family $(C\mathcal{B}_A)_{A \in \mathcal{L}}$ is a base of open sets of $w\mathcal{L}$. In the Wallman space $w(T, \mathcal{C})$, $\mathcal{C}$ being the collection of all closed subsets of the space T, these open sets are in a sense generated by the open subsets of T.

<u>Example 3.2-1 BASIC OPEN SETS IN w(T, $\mathcal{L}$)</u> Let $\mathcal{L}$ be a lattice of subsets of the set T with respect to union and intersection. Then (a) a base for the open sets of $w(T, \mathcal{L})$ is the collection of sets of the form

$$\mathcal{V}_U = \{\mathcal{F} \in w(T, \mathcal{L}) \mid A \in \mathcal{F} \rightarrow A \cap U \neq \emptyset\}$$

where U runs over all complements taken in T of sets in $\mathcal{L}$; in other words $\mathcal{V}_U$ consists of all ultrafilters $\mathcal{F}$ whose trace $\mathcal{F} \cap U$ on U is a filterbase. Moreover the sets $\mathcal{V}_U$ have the property that $U \subset W \rightarrow \mathcal{V}_U \subset \mathcal{V}_W$.

<u>Proof</u> We show that $\mathcal{V}_{CF} = C\mathcal{B}_F$ for each $F \in \mathcal{L}$. If $\mathcal{F} \in C\mathcal{B}_F$ then $F \notin \mathcal{F}$ so that an $H \in \mathcal{F}$ exists such that $H \cap F = \emptyset$ and so $H \subset CF$. As $A \cap H \neq \emptyset$ for each $A \in \mathcal{F}$, CF meets each element of $\mathcal{F}$ and $\mathcal{F} \in \mathcal{V}_{CF}$. To obtain the reverse inclusion suppose that $\mathcal{F} \in \mathcal{V}_{CF}$. Then as $F \cap CF = \emptyset$, $F \notin \mathcal{F}$ and $\mathcal{F} \notin \mathcal{B}_F$. The truth of the monotonicity assertion about the $\mathcal{V}_U$ is obvious. ▽

Thus the open subsets of a topological space T generate an open base for $w(T, \mathcal{C})$. Will a <u>base</u> $\mathcal{B}$ for the topology of T generate an open base for $w(T, \mathcal{C})$? Although not generally true (see Exercise 3.3(c)) this will be the case if $\mathcal{B}$ satisfies the normality condition of (b) below.

(b) If $\mathcal{B}$ is a base of open sets for T which separate the closed subsets of T, i.e. for each disjoint pair $F_1, F_2 \in \mathcal{C}$ there exist disjoint $U_1, U_2 \in \mathcal{B}$ such that $F_1 \subset U_1$ and $F_2 \subset U_2$, then the family $(V_U)_{U\in\mathcal{B}}$ is an open base for $w(T,\mathcal{C})$.

Proof Let $W \subset T$ be open and $\mathcal{F} \in V_W = C\mathcal{B}_{CW}$. Then there exists an $F \in \mathcal{F}$ such that $F \cap CW = \emptyset$. Choosing $U \in \mathcal{B}$ such that $F \subset U \subset W$, it follows that $\mathcal{F} \in V_U \subset V_W$. ∇

This last result hints at the existence of an interrelationship between the topological natures of T and $w(T,\mathcal{C})$. Indeed, the interrelationship is a strong one.

Returning to the general (i.e. lattice) situation we claim that for any distributive lattice L with 0 and 1, wL is a compact T_1 space. To establish T_1-ness we show that $cl_w\{F\}=\{F\}$ for each $F \in wL$ where cl_w denotes closure in wL. As $cl_w\{F\}=\cap\{\mathcal{B}_a \mid F \in \mathcal{B}_a\} = \bigcap_{a\in F} \mathcal{B}_a$, we need only show that $\bigcap_{a\in F} \mathcal{B}_a = \{F\}$. Certainly if $G \in \bigcap_{a\in F} \mathcal{B}_a$ then $F \subset G$ and therefore, by the maximality of F, G=F. In order to establish compactness another fundamental property of the basic sets is needed: Namely stability under finite intersections; in fact $\mathcal{B}_a \cap \mathcal{B}_b = \mathcal{B}_{ab}$ for each pair $a, b \in L$. First, if $F \in \mathcal{B}_a \cap \mathcal{B}_b$ then $a, b \in F$ so that $ab \in F$ and $F \in \mathcal{B}_{ab}$. On the other hand, if $ab \in F$ then, since $a, b \geq ab$, both a and b belong to F and $F \in \mathcal{B}_a \cap \mathcal{B}_b$. It clearly follows from this stable property that for any $a_1,\ldots,a_n \in L$, $\bigcap_{i=1}^{n} \mathcal{B}_{a_i} = \emptyset$ iff $a_1 \cdot\ldots\cdot a_n = 0$. Next wL is shown to be compact. A closed subset of wL is uniquely determined by some subset $\Lambda \subset L$ in that it be written in the form $\mathcal{F}_\Lambda = \bigcap_{a\in\Lambda} \mathcal{B}_a$. Let $(\mathcal{F}_\Lambda)$ be a family of closed subsets of wL with the finite intersection property. It is clear that the family $\{\mathcal{B}_a \mid a \in \cup\Lambda\}$ also enjoys the finite intersection property. Thus, as $\bigcap_{i=1}^{n} \mathcal{B}_{a_i} \neq \emptyset$, $a_1 \cdot a_2 \cdots a_n \neq 0$ for any choice of $a_1,\ldots,a_n \in \cup\Lambda$ and $\cup\Lambda$ is a lattice filter subbase. Letting F be any lattice ultrafilter containing $\cup\Lambda$, it follows that $F \in \mathcal{B}_a$ for each $a \in \cup\Lambda$ - whence $F \in \cap\mathcal{F}_\Lambda$ and $\cap\mathcal{F}_\Lambda \neq \emptyset$. Hence wL is compact.

The above discussion is summarized in the two following results. In the first, (3.2-1), we list those fundamental properties of the basic sets used in defining the Wallman topological space and in obtaining compactness and T_1-ness for ease in future reference.

(3.2-1) ELEMENTARY PROPERTIES OF THE BASIC SETS Let L be a distributive lattice with 0 and 1 and $(\mathcal{B}_a)_{a\in L}$ the associated family of basic sets. Then

(1) for each $F \in wL$, $\bigcap_{a\in F} \mathcal{B}_a = \{F\}$,

(2) for each pair $a, b \in L$

$$\mathcal{B}_a \cup \mathcal{B}_b = \mathcal{B}_{a+b} \text{ and } \mathcal{B}_a \cap \mathcal{B}_b = \mathcal{B}_{ab}, \text{ and}$$

(3) for any choice of $a_1,\ldots,a_n \in L$

$$\bigcap_{i=1}^{n} \mathcal{B}_{a_i} = \emptyset \text{ iff } a_1 \ldots a_n = 0.$$

Theorem 3.2-1 wL IS A COMPACT T_1 SPACE If L is a distributive lattice with 0 and 1 then the Wallman space wL is a compact T_1 space.

At this point the reader may be wondering when wL is Hausdorff. wL is certainly Hausdorff if for each pair of distinct elements F, G∈wL there exists a pair of disjoint complements of basic sets $C\mathcal{B}_{b'}$ and $C\mathcal{B}_{a'}$ to which F and G belong. Corresponding to each such pair F and G there exists at least one pair of elements a∈F and b∈G such that ab=0. For $C\mathcal{B}_{a'}$ and $C\mathcal{B}_{b'}$ to be disjoint, or equivalently for $\mathcal{B}_{a'+b'} = \mathcal{B}_{a'} \cup \mathcal{B}_{b'} = wL$, it is necessary and sufficient that a'+b' belong to each ultrafilter. To be certain that $F \in C\mathcal{B}_{b'}$ and $G \in C\mathcal{B}_{a'}$ it suffices that a'b=ab'=0. Thus if to each pair of nonzero elements a, b∈L such that ab=0 there corresponds a pair a', b'∈L such that a'b=ab'=0 and a'+b' belongs to each ultrafilter then wL is Hausdorff. Lattices of this type are called normal because of the similarity of the lattice condition to the topological notion of normality.

Definition 3.2-1 NORMAL LATTICES A distributive lattice L with 0 and 1 is normal if corresponding to each pair of nonzero elements a, b∈L such that ab=0 there exists a pair a', b'∈L such that

(1) a'b = ab' = 0 and

(2) a' + b'∈F for each F∈wL

If in a normal lattice we choose any pair of ultrafilters F and G containing a and b respectively then $G \notin \mathcal{B}_{a'}$ and $F \notin \mathcal{B}_{b'}$. Since $\mathcal{B}_{a'} \cup \mathcal{B}_{b'} = \mathcal{B}_{a'+b'} = wL$, $F \in \mathcal{B}_{a'}$ and $G \in \mathcal{B}_{b'}$. Hence aa'≠0 and bb'≠0. Moreover the elements a'+a and b'+b are ≥ a and b respectively and the elements a, a'+a, b, b'+b satisfy conditions (1) and (2) of the definition. Thus we may state an equivalent but stronger looking definition of normality: To each pair of nonzero elements a, b of L such that ab=0 there exists a' ≥ a and b' ≥ b such that (1) and (2) hold.

Now the discussion preceding the definition served to prove that wL is Hausdorff whenever L is normal. As it happens the converse is also true. We have:

Theorem 3.2-2 L IS NORMAL IFF wL IS T_2 Let L be a distributive lattice with 0 and 1. Then wL is Hausdorff iff L is a normal lattice.

Proof It remains to prove that L is normal when wL is Hausdorff. Since wL is compact it is also normal. If a, b∈L are nonzero, ab=0, then $\mathcal{B}_a \cap \mathcal{B}_b = \emptyset$. By the normality of wL there exists a pair of disjoint open sets $\mathcal{U}$ and $\mathcal{V}$

containing $\mathcal{B}_a$ and $\mathcal{B}_b$ respectively. Since $C\mathcal{U}$ and $C\mathcal{V}$ are both closed, subsets R and S of L exist such that $C\mathcal{U} = \bigcap_{c\in R} \mathcal{B}_c$ and $C\mathcal{V} = \bigcap_{d\in S} \mathcal{B}_d$. Moreover

$$\bigcap_{c\in R} \mathcal{B}_c \cap \mathcal{B}_a = \bigcap_{d\in S} \mathcal{B}_d \cap \mathcal{B}_b = \emptyset.$$

Now $\mathcal{B}_a$ and $\mathcal{B}_b$, being closed in the compact space wL, are compact and so there exists $c_1,\ldots,c_n \in R$ and $d_1,\ldots,d_m \in S$ such that

$$\bigcap_{i=1}^{n} \mathcal{B}_{c_i} \cap \mathcal{B}_a = \bigcap_{j=1}^{m} \mathcal{B}_{d_j} \cap \mathcal{B}_b = \emptyset.$$

If we set $b'=c_1,\ldots,c_n$ and $a'=d_1,\ldots,d_m$ then by (3.1-1)(3) $\mathcal{B}_{b'} = \bigcap_{i=1}^{n} \mathcal{B}_{c_i}$ and $\mathcal{B}_{a'} = \bigcap_{j=1}^{n} \mathcal{B}_{d_j}$ and, therefore,

$$\mathcal{B}_{b'} \cap \mathcal{B}_a = \mathcal{B}_{a'} \cap \mathcal{B}_b = \emptyset.$$

Next we claim that $b'a=a'b=0$. Indeed if $b'a\neq 0$ then $S=\{a,b'\}$ is a filter subbase and, as such is contained in some ultrafilter F. It follows that $F\in \mathcal{B}_{b'} \cap \mathcal{B}_a$ - a contradiction. Finally $C\mathcal{U} \subset \mathcal{B}_{b'}$ and $C\mathcal{V} \subset \mathcal{B}_{a'}$ from which it follows that

$$\mathcal{B}_{a'+b'} = \mathcal{B}_{a'} \cup \mathcal{B}_{b'} \supset C\mathcal{U} \cup C\mathcal{V} = wL.$$

Hence $a'+b'$ belongs to each ultrafilter and L is a normal lattice. The converse has been proved in the discussion prior to Definition 3.2-1.∇

In the remainder of this section we present some examples of normal lattices.

<u>Example 3.2-2 BOOLEAN LATTICES ARE NORMAL</u> Let L be a Boolean lattice, i.e. a complemented distributive lattice with 0 and 1, and suppose that $a, b\in L$ are nonzero and $ab=0$. Choosing $a'=a$ and b' to be the complement of a we see that $a'b=ab'=0$ and $a'+b'=1$ belongs to each ultrafilter of L. Hence L is normal.

<u>Example 3.2-3 $\mathcal{Z}$ IS NORMAL</u> Let T be a completely regular Hausdorff space and $\mathcal{Z}$ be the lattice of all zero sets in T (see Example 3.1-4). To see that $\mathcal{Z}$ is a normal lattice let Z_1 and Z_2 be disjoint zero sets. Then by Theorem 1.2-2 $cl_\beta Z_1$ and $cl_\beta Z_2$ are disjoint so that there exists a continuous function $0 \leq x^\beta \leq 1$ defined on the normal space βT such that $x^\beta(cl_\beta Z_1) = \{0\}$ while $x^\beta(cl_\beta Z_2)=\{1\}$. Thus, denoting the restriction of x^β to T by x, it follows by (1.2-3) that $CZ_2' = x^{-1}(-\infty,1/4)$ and $CZ_1' = x^{-1}(3/4,\infty)$ are disjoint complements of zero sets containing Z_1 and Z_2 respectively. Thus $Z_1' \supset Z_1$

and $Z_2' \supset Z_2$, and $Z_1' \cup Z_2'$, being equal to T, belongs to each z-ultrafilter.

In the next example another reason for using the terminology "normal lattice" comes to light.

Example 3.2-4 T IS $T_1 \to$ (T IS NORMAL $\leftrightarrow \mathcal{C}$ IS NORMAL) Let T be a T_1 topological space. Then if T is normal, in essentially the same way as the lattice of zero sets was shown to be normal, it can be shown that $\mathcal{C}$ is a normal lattice. Conversely suppose $\mathcal{C}$ is normal. Then by the stronger form of normality mentioned after Definition 3.2-1 if F and E are disjoint closed sets there exists closed sets $F' \supset F$ and $E' \supset E$ such that

$$F' \cap E = E' \cap F = \emptyset$$

and $F' \cup E'$ belongs to each $\mathcal{C}$-ultrafilter. Next we claim that $T = F' \cup E'$. If not then there exists $t \notin F' \cup E'$. Since T is T_1, the set $\{t\}$ is closed. Letting $\mathcal{F}$ be any $\mathcal{C}$-ultrafilter containing the $\mathcal{C}$-filter $\{A \in \mathcal{C} | \{t\} \subset A\}$, it follows, since $\{t\} \cap (E' \cup F') = \emptyset$, that $E' \cup F' \notin \mathcal{F}$. As $E' \cup F'$ must belong to each $\mathcal{C}$-ultrafilter no such t can exist and $T = E' \cup F'$. Thus CE' and CF' are disjoint open sets containing F and E respectively so that T is a normal topological space.

3.3 WALLMAN COMPACTIFICATIONS OF TOPOLOGICAL SPACES In the representation of the Stone-Cech compactification of a completely regular Hausdorff space T as the collection of all z-ultrafilters T is imbedded in βT via the mapping

$$\varphi : T \longrightarrow \beta T$$

$$t \longrightarrow \mathcal{F}_t = \{Z \in \mathcal{Z} | t \in Z\}.$$

At this time we wish to imbed a set T in the collection $w(T, \mathcal{L})$ of all $\mathcal{L}$-ultrafilters where $\mathcal{L}$ is a distributive lattice with 0 and 1 of subsets with respect to $\cup$ and $\cap$. Motivated by the above we consider the mapping φ taking $t \in T$ into $\mathcal{F}_t = \{A \in \mathcal{L} | t \in A\}$. Our purpose now is to determine conditions on $\mathcal{L}$ such that φ is a 1-1 mapping of T into $w(T, \mathcal{L})$. If this is to be the case then certainly each $\mathcal{F}_t$ must be an ultrafilter. But $\mathcal{F}_t$ is always an $\mathcal{L}$-filter, so by the meet condition in order for $\mathcal{F}_t$ to be an $\mathcal{L}$-ultrafilter (for each $t \in T$) it is necessary and sufficient that for each $t \in T$ and lattice element $A \notin \mathcal{F}_t$ there exists an element $B \in \mathcal{F}_t$ such that $A \cap B = \emptyset$ (Theorem 3.1-1). In other words for each $t \in T$ and $A \in \mathcal{L}$ such that $t \notin A$ there exists $B \in \mathcal{L}$ such that $t \in B$ and $A \cap B = \emptyset$. A lattice $\mathcal{L}$ satisfying this condition is called an <u>α-lattice</u>. Thus φ maps T into $w(T, \mathcal{L})$ iff $\mathcal{L}$ is an α-lattice. What about 1-1ness of φ? Suppose that φ is 1-1; i.e. if $t \neq t'$

then $\mathcal{F}_t \neq \mathcal{F}_t'$. As $\mathcal{F}_t$ and $\mathcal{F}_t'$ are unequal there exists $A \in \mathcal{F}_t$ such that $A \notin \mathcal{F}_t'$ or $A \in \mathcal{F}_t'$ and $A \notin \mathcal{F}_t$. Thus for each pair of distinct elements $t, t' \in T$ there exists an element $A \in \mathcal{L}$ containing exactly one of the points. Such a lattice is called a <u>β-lattice</u>. Hence if φ is 1-1 then $\mathcal{L}$ is a β-lattice. Conversely if $\mathcal{L}$ is a β-lattice and $t \neq t'$ then the element A belongs to exactly one of the filters $\mathcal{F}_t$ or $\mathcal{F}_t'$ so that $\mathcal{F}_t \neq \mathcal{F}_t'$.

We summarize the foregoing discussion in the next definition and result.

<u>Definition 3.3-1 α AND β-LATTICES</u> Let $\mathcal{L}$, a collection of subsets of a set T, be a distributive lattice with 0 and 1 with respect to $\cup$ and $\cap$. Then

(a) $\mathcal{L}$ is an <u>α-lattice</u> if for each $t \in T$ and $A \in \mathcal{L}$ such that $t \notin A$ there exists $B \in \mathcal{L}$ such that $t \in B$ and $A \cap B = \emptyset$.

(b) $\mathcal{L}$ is a β-lattice if for each pair of distinct points there exists $A \in \mathcal{L}$ containing one of the points and not the other. If $\mathcal{L}$ satisfies both conditions we refer to it as an <u>$\alpha\beta$-lattice</u>.

<u>(3.3-1) IMBEDDING T IN $w(T, \mathcal{L})$</u> Let T be a set, $\mathcal{L} \subset P(T)$ be a distributive lattice with 0 and 1 with respect to $\cup$ and $\cap$, and φ be the mapping of T into the collection of $\mathcal{L}$-filters taking t into $\mathcal{F}_t = \{A \in \mathcal{L} \mid t \in A\}$. Then

(a) $\varphi(T) \subset w(T, \mathcal{L})$ iff $\mathcal{L}$ is an α-lattice, and

(b) φ is 1-1 iff $\mathcal{L}$ is a β-lattice.

<u>Example 3.3-1 $\mathcal{Z}$ IS AN $\alpha\beta$-LATTICE</u> Let T be a completely regular Hausdorff space and $\mathcal{Z}$ the lattice of all zero-sets of T. Suppose that $t \in T$ and $Z \in \mathcal{Z}$ such that $t \notin Z$. Then, as Z is closed and T is completely regular there exists an $x \in C_b(T, \underset{\sim}{R})$ such that $0 \leq x \leq 1$, $t \in x^{-1}(0)$ and $Z \subset x^{-1}(1)$. Thus $x^{-1}(0) \in \mathcal{Z}$, $x^{-1}(0) \cap Z = \emptyset$ and $t \in x^{-1}(0)$ so that $\mathcal{Z}$ is an α-lattice. Given distinct points $t, t' \in T$ we may apply the same argument to the pair t and $\{t'\}$ to obtain an element $x^{-1}(0) \in \mathcal{Z}$ such that $t \in x^{-1}(0)$ and $t' \notin x^{-1}(0)$ (equivalently $\{t'\} \cap x^{-1}(0) = \emptyset$). Hence $\mathcal{Z}$ is also a β-lattice.∇

<u>Example 3.3-2 $\mathcal{C}$ IS AN $\alpha\beta$-LATTICE $\Leftrightarrow$ T IS T_1</u> Consider the lattice $\mathcal{C}$ of all closed subsets of the topological space T.

(a) $\mathcal{C}$ is a β-lattice if and only if T is T_o.

<u>Proof</u> If t and t' are distinct points in the T_o space T then $cl\{t\} \neq cl\{t'\}$. Thus either $cl\{t\}$ or $cl\{t'\}$ is an element of $\mathcal{C}$ containing exactly one of the points. Conversely suppose there exists $A \in \mathcal{C}$ containing one of the points, say t, and not the other. Then t' belongs to the open set CA while t does not and, as t and t' are arbitrary, T is T_o. ∇

(b) $\mathcal{C}$ is an $\alpha\beta$-lattice iff T is T_1.

<u>Proof</u> If T is T_1 then T is T_o and, by (a), $\mathcal{C}$ is a β-lattice. Let $A \in \mathcal{C}$ and

$t \notin A$. As T is T_1, $\{t\} \in \mathcal{L}$. Thus $\{t\}$ is an element of $\mathcal{L}$ which contains t and is disjoint from A, and $\mathcal{L}$ is therefore an α-lattice. Conversely, suppose that $\mathcal{L}$ is an $\alpha\beta$-lattice. Then as $\mathcal{L}$ is a β-lattice, if t and t' are distinct points of T there exists $A \in \mathcal{L}$ containing one of the points and not the other. Suppose that $t \notin A$ and $t' \in A$. Since $\mathcal{L}$ is also an α-lattice there exists $B \in \mathcal{L}$ such that $t \in B$ and $A \cap B = \emptyset$. Hence $t \in CA$ and $t' \notin CA$ while $t' \in CB$ and $t \notin CB$ and, therefore, T is T_1. ∇

Since our ultimate goal is to force $w(T,\mathcal{L})$ to be a compactification of T when T carries a topology, our next result is concerned with conditions on $\mathcal{L}$ under which $\varphi(T)$ is a dense homeomorphic image of T in $w(T,\mathcal{L})$.

<u>(3.3-2) IMBEDDING THE TOPOLOGICAL SPACE T IN $w(T,\mathcal{L})$</u> Let $\mathcal{L}$ be $\alpha\beta$-lattice of subsets of the topological space T so that the mapping

$$\varphi: T \longrightarrow w(T,\mathcal{L})$$

$$t \longrightarrow \mathcal{F}_t$$

is 1-1. Then

(a) φ is continuous iff all the sets in $\mathcal{L}$ are closed in T;

(b) φ is a homeomorphism iff $\mathcal{L}$ is a base of closed sets of the space T;

(c) $\varphi(T)$ is dense in $w(T,\mathcal{L})$. Thus $w(T,\mathcal{L})$ is a T_1 compactification of T whenever $\mathcal{L}$ is a base for the closed subsets of T. Moreover $w(T,\mathcal{L})$ is the "smallest" T_1 compactification of T in the sense that if T is already compact then $\varphi(T)=w(T,\mathcal{L})$.*

<u>Proof</u> (a) Recall (Example 3.2-1) that the sets $V_{CF}=\{\mathcal{F} \in w(T,\mathcal{L}) \mid A \in \mathcal{F} \rightarrow A \cap CF \neq \emptyset\}$ for $F \in \mathcal{L}$ constitutes an open base for $w(T,\mathcal{L})$. It is easy to see that φ is continuous iff each $F \in \mathcal{L}$ is closed in T.

(b) Suppose that $\{F_\alpha\}_{\alpha\in\Lambda} \subset \mathcal{L}$. Then it follows that $\varphi(\bigcap_\alpha F_\alpha)= \bigcap_\alpha \mathcal{B}_{F_\alpha} \cap \varphi(T)$. Thus if $\mathcal{L}$ is a base for the closed subsets of T, φ is a relatively closed mapping. Since φ is also continuous by (a), φ is a homeomorphism. Conversely, if φ is a homeomorphism and F is a closed subset of T then $\varphi(F)= {}_{\alpha\in\Lambda} F_\alpha \ \varphi(T)$ for some family $\{F_\alpha\}_{\alpha\in\Lambda}$ of elements of $\mathcal{L}$. As φ is 1-1 and $\varphi(F)=\varphi({}_\alpha F_\alpha)$, $F= {}_\alpha F_\alpha$ and $\mathcal{L}$ is a base of closed subsets of T.

---

* If S is a Hausdorff compactification of T and T is compact then, as T is thereby closed in S, S=T. This needn't be the case if we merely assume that T and S are T_1 for if S is an infinite set equipped with the cofinite topology (a set is open iff it is the complement of a finite set) and T is a <u>proper</u> infinite subset of S with the cofinite topology both T and S are compact while T is dense in S.

(c) That $\varphi(T)$ is dense in $w(T,\mathcal{L})$ follows from the observation that $\varphi(t)=\mathcal{F}_t \in V_{CF}$ for each $t\in CF$ and each $F\in\mathcal{L}$. Thus each basic open subset of $w(T,\mathcal{L})$ contains points of $\varphi(T)$ and $cl_w\varphi(T)=w(T,\mathcal{L})$. In the event that T is already compact and $\mathcal{F}$ is an $\mathcal{L}$-ultrafilter then, as $\mathcal{F}$, a collection of closed subsets of T has the finite intersection property, $\bigcap\mathcal{F}\neq\emptyset$. If we choose $t\in\bigcap\mathcal{F}$ it immediately follows from the maximality of $\mathcal{F}$ that $\mathcal{F}=\mathcal{F}_t$. Hence $\varphi(T)=w(T,\mathcal{L})$.∇

Compactifications of type generated by the theorem are called "Wallman-type" compactifications. We reserve the name "Wallman compactification" for the special case discussed in our next example.

Example 3.3-3 THE WALLMAN COMPACTIFICATION OF A T_1 SPACE If T is a T_1 space then the collection $\mathcal{C}$ of all closed subsets of T is an $\alpha\beta$-lattice (Example 3.3-2) and the mapping

$$\varphi: T \longrightarrow w(T,\mathcal{C})$$
$$t \longrightarrow \mathcal{F}_t$$

is homeomorphic as $\mathcal{C}$ is certainly a base of closed sets ((3.3-2)(b)). Thus $w(T,\mathcal{C})$ is a T_1 compactification of T by (3.3-2)(c) and is referred to as the Wallman compactification of the T_1 space T. Since $\mathcal{C}$ is normal iff T is normal (Example 3.2-4) and $w(T,\mathcal{C})$ is Hausdorff iff $\mathcal{C}$ is normal (Theorem 3.2-2), the Wallman compactification is Hausdorff iff T is normal.

Example 3.3-4 $w(T,\mathcal{Z})$ IS A HAUSDORFF COMPACTIFICATION OF T If T is a completely regular Hausdorff space then we recall that $\mathcal{Z}$ is a normal $\alpha\beta$-lattice (see Example 3.2-3 and Example 3.3-1). It remains to show that $\mathcal{Z}$ is a base for the closed subsets of T in order to conclude that $w(T,\mathcal{Z})$ is a Hausdorff compactification of T. To see this, recall that the topology of T is the weakest topology with respect to which each $x\in C_b(T,\mathbb{R})$ is continuous.

Thus a subbase for the closed subsets of T consists of all sets of the form $x^{-1}(C)$ where $x\in C_b(T,\mathbb{R})$ and C is closed in $\mathbb{R}$. As this collection coincides with $\mathcal{Z}$ ((1.2-3)) and $\mathcal{Z}$ is closed with respect to the formation of finite unions $\mathcal{Z}$ is a base for the closed sets.∇

In general if X is dense in Y and $Z\subset Y$ then it need not follow that $X\cap Z$ be dense in Z. In fact this need not be the case even if Z is closed in Y. If this is so for closed sets i.e. if $cl(X\cap Z)=Z$ for all closed subsets Z of Y, then X is said to be <u>very dense</u> in Y. Observe that Y being T_1 precludes the existence of a very dense proper subspace X for if $y\in Y-X$

then $\{y\}\neq cl(X\cap\{y\})=cl\emptyset=\emptyset$. Thus T (or rather $\varphi(T)$) cannot be very dense in $w(T,\mathcal{L})$.

Our next result has a twofold purpose. The first part shows that when T is a topological space, T is fairly close to being very dense in $w(T,\mathcal{L})$, while the second part indicates in a sense just how far off T can be from being very dense in $w(T,\mathcal{L})$. The symbol cl_w denotes closure in $w(T,\mathcal{L})$.

(3.3-3) T IS "NEARLY" VERY DENSE IN $w(T,\mathcal{L})$ Let $\mathcal{L}$ be an α-lattice of subsets of the set T.

(a) Then for each $F\in\mathcal{L}$, $cl_w\varphi(F)=cl_w(\mathcal{B}_F\cap\varphi(T))=\mathcal{B}_F$

(b) If T is T_1 space, $\mathcal{L}=\mathcal{C}$, and $\{F_\alpha|\alpha\in\Lambda\}\subset\mathcal{C}$ then

$$cl_w(\bigcap_\alpha \mathcal{B}_{F_\alpha}\cap\varphi(T)) = \mathcal{B}_{\cap_\alpha F_\alpha}.$$

Proof (a) Certainly $\mathcal{B}_F\cap\varphi(T)=\varphi(F)$. Observe that $cl_w(\mathcal{B}_F\cap\varphi(T))\subset\mathcal{B}_F$. To obtain the reverse inclusion let $\mathcal{F}\in\mathcal{B}_F$ and V_{CH} be a typical basic open neighborhood of $\mathcal{F}$ in $w(T,\mathcal{L})$ where $H\in\mathcal{L}$. Then, as $\mathcal{F}\in\mathcal{B}_{CH}$ and $F\in\mathcal{F}$, $F\cap CH\neq\emptyset$. Choosing any $t\in F\cap CH$ it follows that $F\in\mathcal{F}_t$ and therefore $\mathcal{F}_t\in\mathcal{B}_F$. Since t also belongs to CH, CH meets every set in $\mathcal{F}_t$ and $\mathcal{F}_t\in V_{CH}$. Thus V_{CH} meets $\mathcal{B}_F\cap\varphi(T)$ for such $H\in\mathcal{L}$ so that $\mathcal{F}\in cl_w(\mathcal{B}_F\cap\varphi(T))$.

(b) Since $\bigcap_\alpha F_\alpha\in\mathcal{C}$ we have

$$\cap_\alpha\mathcal{B}_{F_\alpha}\cap\varphi(T) = \varphi(\bigcap_\alpha F_\alpha) = \mathcal{B}_{\cap_\alpha F_\alpha}\cap\varphi(T)$$

Thus by (a), $cl_w(\bigcap_\alpha \mathcal{B}_{F_\alpha}\cap\varphi(T))=\mathcal{B}_{\cap_\alpha F_\alpha}$. ∇

Remark As a result of (a) of the preceding result the collection of closures of the basic elements $A\in\mathcal{L}$ forms a base of closed sets for $w(T,\mathcal{L})$.

Recall that in the Stone-Cech compactification βT of a completely regular Hausdorff space T the closures of disjoint zero sets of T are disjoint (Theorem 1.2-1). An analogous result for Wallman-type compactifications is true and is obtained from part (a) of the previous result.

(3.3-4) DISJOINT LATTICE SETS HAVE DISJOINT CLOSURES IN $w(T,\mathcal{L})$ Let T be a topological space and $\mathcal{L}$ be a base of closed subsets of T which forms an $\alpha\beta$-lattice. Then for any pair F, $H\in\mathcal{L}$,

$$F\cap H = \emptyset \rightarrow cl_wF\cap cl_wH = \emptyset.$$

Proof By (3.3-3)(a), $cl_w\varphi(F)=\mathcal{B}_F$ and $cl_w\varphi(H)=\mathcal{B}_H$. Since $F\cap H=\emptyset$,

$$(cl_w\varphi(F))\cap(cl_w\varphi(H)) = \mathcal{B}_F\cap\mathcal{B}_H = \emptyset. \nabla$$

A similar property of $w(T,\mathcal{L})$ is one of the properties that distinguishes the Hausdorff compactifications of completely regular Hausdorff spaces which are of the Wallman-type, (see Theorem 3.5-1).

3.4 <u>βT AND WALLMAN COMPACTIFICATIONS</u> For $w(T,\mathcal{L})$ to be a Hausdorff compactification of the topological space T it is necessary and sufficient that $\mathcal{L}$ be a normal $\alpha\beta$-lattice which serves as a base of closed subsets of T (see Theorem 3.2-2, (3.3-1), and (3.3-2)). Viewed from a different point of view this amounts to saying that a topological space has a Hausdorff Wallman-type compactification iff there exists a base of closed subsets of T which forms a normal $\alpha\beta$-lattice of subsets of T. Such topological spaces are singled out in Definition 3.4-1. After that we develop a sufficient condition for coincidence to occur between βT and $w(T,\mathcal{L})$.

<u>Definition 3.4-1 SEMI-NORMAL SPACES</u> The topological space T is called <u>semi-normal</u> iff there exists a base of closed subsets of T which forms a normal $\alpha\beta$-lattice with respect to $\cup$ and $\cap$. Such a base is referred to as a <u>normal base</u> for T.

We already know that the completely regular Hausdorff spaces are precisely the spaces with Hausdorff compactifications so it is reasonable to suspect that the semi-normal and completely regular Hausdorff spaces are related. Indeed, as we now show, in the class of T_1 spaces, the classes of semi-normal and completely regular spaces coincide. The notion of a completely regular space is defined in an external fashion utilizing the elements of $C_b(T,\underset{\sim}{R})$. Our next result provides an internal characterization of complete regularity via the notion of a normal base.

<u>(3.4-1) INTERNAL CHARACTERIZATION OF COMPLETE REGULARITY</u> A topological space T is a completely regular Hausdorff space iff it is semi-normal.

<u>Proof</u> If T is a completely regular Hausdorff space then as we have shown in Example 3.3-4 the lattice of zero sets $\mathcal{Z}$ is a normal $\alpha\beta$-lattice which serves as a base for the closed subsets of T.

Conversely, as we have already noted, a semi-normal topological space T has a Hausdorff Wallman-type compactification $w(T,\mathcal{L})$ where $\mathcal{L}$ is a normal base for T. Since $w(T,\mathcal{L})$ is a compact Hausdorff space then the subspace T must be completely regular and Hausdorff. ∇

How does a Wallman compactification $w(T,\mathcal{L})$ of a completely regular Hausdorff space T compare with βT? Certainly they could only be equivalent when $w(T,\mathcal{L})$ is Hausdorff - i.e. iff $\mathcal{L}$ is a normal base for T. In this case then, by Theorem 1.1-1(c), $w(T,\mathcal{L})=\beta T$ if T is C_b - imbedded in $w(T,\mathcal{L})$. In the following discussion we determine those bounded continuous real-valued functions which are continuously extendible to a given Hausdorff Wallman compactification $w(T,\mathcal{L})$ of T.

Definition 3.4-2 $\mathcal{L}$-UNIFORM CONTINUITY Let $\mathcal{L}$ be a normal base for the completely regular Hausdorff space T. Then the function $x\in C(T,\mathbb{R})$ is $\mathcal{L}$-uniformly continuous if for each $\epsilon > 0$ there exist sets $A_1,\ldots,A_n\in\mathcal{L}$ such that $T=\bigcup_{i=1}^{n} CA_i$ and, using "o" for "oscillation",

$$o(x, CA_i) = \sup\{|x(s) - x(t)| \;|s, t\in CA_i\} < \epsilon$$

for each $i=1,\ldots,n$.

Certainly any $\mathcal{L}$-uniform continuous function is bounded. Furthermore, the collection of all such functions forms a closed subalgebra of $C_b(T,\mathbb{R})$ in the uniform norm. In our next result we show that any $x\in C_b(T,\mathbb{R})$ that pulls closed subsets of $\mathbb{R}$ back into members of $\mathcal{L}$ must be $\mathcal{L}$-uniformly continuous.

(3.4-2) A CLASS OF $\mathcal{L}$-UNIFORMLY CONTINUOUS FUNCTIONS Let T be a completely regular Hausdorff space and $\mathcal{L}$ a normal base for T. If $x\in C_b(T,\mathbb{R})$ and $x^{-1}(C)\in\mathcal{L}$ for each closed $C\subset\mathbb{R}$ then x is $\mathcal{L}$-uniformly continuous.

Proof Let $\epsilon > 0$ be given. As x(T) is bounded in $\mathbb{R}$ there exists a finite number of real numbers $a_1,\ldots,a_n$ such that $\bigcup_{i=1}^{n} (a_i-\epsilon/4,\ a_i+\epsilon/4)\supset x(T)$. Thus each set $A_i=x^{-1}(C(a_i-\epsilon/4,\ a_i+\epsilon/4))\in\mathcal{L}$ and $T=\bigcup_{i=1}^{n} CA_i$. Furthermore, as $CA_i=x^{-1}(a_i-\epsilon/4,\ a_i+\epsilon/4)$, $o(x,CA_i)\leq\epsilon/2<\epsilon$ for each i and, since ϵ is arbitrary, x is $\mathcal{L}$-uniformly continuous. ∇

The $\mathcal{L}$-uniformly continuous functions are precisely those which are continuously extendible to $w(T,\mathcal{L})$.

Theorem 3.4-1 CONTINUOUS EXTENDIBILITY TO $w(T,\mathcal{L})$ Let T be a completely regular Hausdorff space and $\mathcal{L}$ a normal base for T. Then $x\in C(T,\mathbb{R})$ has a continuous extension to $w(T,\mathcal{L})$ iff x is $\mathcal{L}$-uniformly continuous.

Proof Let x^w be the continuous extension of $x\in C(T,\mathbb{R})$ up to $w(T,\mathcal{L})$. Then, as $\bigcup_{\mathcal{F}\in w(T,\mathcal{L})} (x^w)^{-1}((x^w(\mathcal{F})-\epsilon/2,\ x^w(\mathcal{F})+\epsilon/2))=w(T,\mathcal{L})$ and $w(T,\mathcal{L})$ is compact there exists a finite subfamily $U_i=(x^w)^{-1}(x^w(\mathcal{F}_i)-\epsilon/2,\ x^w(\mathcal{F}_i)+\epsilon/2))$, $i=1,\ldots,n$, covering $w(T,\mathcal{L})$. Each of these sets, being open in $w(T,\mathcal{L})$, is a union of basic sets of the form V_{CA} where $A\in\mathcal{L}$. Now the oscillation of x^w on each such set is less that or equal to ϵ. Moreover $w(T,\mathcal{L})$ is the union of all the basic sets making up the U_i, $i=1,\ldots,n$, so that there exists $V_{CA_1},\ldots,V_{CA_m}$ covering $w(T,\mathcal{L})$ such that $o(x^w,\ V_{CA_j})\leq\epsilon$ for $j=1,\ldots,m$. Since with φ as in (3.3-1), $\varphi(T)\subset\bigcup_{j=1}^{m} V_{CA_j}$, $T=\bigcup_{j=1}^{m} CA_j$. If s, $t\in CA_j$ for some $1\leq j\leq m$ then

$$|x(s) - x(t)| = |x^w(\mathcal{F}_s) - x^w(\mathcal{F}_t)| \leq \epsilon$$

since $\mathcal{F}_s$, $\mathcal{F}_t \in V_{CA_j}$, and x is $\mathcal{L}$-uniformly continuous.

Conversely, suppose that x is $\mathcal{L}$-uniformly continuous and let $\mathcal{F} \in w(T,\mathcal{L})$. As $\mathcal{F}$ is a filterbase on T, $x(\mathcal{F}) = \{x(A) | A \in \mathcal{F}\}$ is a filterbase on $\underset{\sim}{R}$. Furthermore, as x(T) is bounded, $x(\mathcal{F})$ has a adherence point in $\underset{\sim}{R}$ which we denote by $x^w(\mathcal{F})$. We shall show that $x(\mathcal{F}) \to x^w(\mathcal{F})$ by proving $x(\mathcal{F})$ to be a Cauchy filterbase with the aid of the $\mathcal{L}$-uniform continuity of x. Indeed for a given $\epsilon > 0$ there exists $A_1,\ldots,A_n \in \mathcal{L}$ such that $T = \bigcup_{i=1}^{n} CA_i$ and $o(x, CA_i) < \epsilon$ for each $1 \leq i \leq n$. Since $\bigcap_{i=1}^{n} A_i = \emptyset$ there exists $1 \leq j \leq n$ such that $A_j \notin \mathcal{F}$. By the maximality of $\mathcal{F}$ there exists $A \in \mathcal{F}$ such that $A_j \cap A = \emptyset$ or, equivalently, $A \subset CA_j$. Thus for any pair t, $s \in A$, $|x(t) - x(s)| < \epsilon$ and it follows that $x(\mathcal{F})$ is Cauchy.

To complete the proof we must do two things. We must show that x^w, as defined above, extends x and is continuous on $w(T,\mathcal{L})$. To see that x^w extends x, let $t \in T$ and consider $\mathcal{F}_t$. We claim that $x^w(\mathcal{F}_t) = x(t)$. As $x(\mathcal{F}_t) \to x^w(\mathcal{F}_t)$ there exists $A \in \mathcal{F}_t$ such that for each $s \in A$, $|x(s) - x^w(\mathcal{F}_t)| < \epsilon$. Thus, since $t \in A$,

$$|x(s) - x(t)| \leq |x(s) - x^w(\mathcal{F}_t)| + |x(t) - x^w(\mathcal{F}_t)| < 2\epsilon$$

and so $x(\mathcal{F}_t) \to x(t)$. Hence $x^w(\mathcal{F}_t) = x(t)$. It remains to show that x^w is continuous on $w(T,\mathcal{L})$. Let $A_1,\ldots,A_n \in \mathcal{L}$ be such that $T = \bigcup_{i=1}^{n} CA_i$ and $o(x, CA_i) < \epsilon$ for $i=1,\ldots,n$. We contend that given any $\mathcal{F} \in w(T,\mathcal{L})$ there exists V_{CA_i} for some $1 \leq i \leq n$ such that $\mathcal{F} \in V_{CA_i}$ and $|x^w(\mathcal{G}) - x^w(\mathcal{F})| < 3\epsilon$ whenever $\mathcal{G} \in V_{CA_i}$. Since $\bigcap_{i=1}^{n} A_i = \emptyset$ if $\mathcal{F} \in w(T,\mathcal{L})$ there exists $A \in \mathcal{F}$ and an A_i, $1 \leq i \leq n$ such that $A \cap A_i = \emptyset$. Thus $A \subset CA_i$ and $\mathcal{F} \in V_{CA_i}$. Now suppose that $\mathcal{G}$ also belongs to V_{CA_i}. As $x(\mathcal{F}) \to x^w(\mathcal{F})$ and $x(\mathcal{G}) \to x^w(\mathcal{G})$ there exists $A \in \mathcal{F}$ and $B \in \mathcal{G}$ such that

$$|x(s) - x^w(\mathcal{F})| < \epsilon \quad \text{for all } s \in A$$

and

$$|x(s) - x^w(\mathcal{G})| < \epsilon \quad \text{for all } s \in B.$$

Let $t \in A \cap CA_i$ and $s \in B \cap CA_i$. Then

$$|x(t) - x^w(\mathcal{F})| < \epsilon, \quad |x(s) - x^w(\mathcal{G})| < \epsilon$$

and, since t, $s \in CA_i$,

$$|x(t) - x(s)| < \epsilon.$$

Thus it follows that $|x^w(\mathcal{G})-x^w(\mathcal{F})| < 3\epsilon$ and the proof is complete. ∇

As a result of this theorem together with (3.4-2) examples of Hausdorff Wallman-type compactifications which are equivalent to βT may be given.

(3.4-3) βT IS A WALLMAN-TYPE COMPACTIFICATION Let T be a completely regular Hausdorff space, $\mathcal{Z}$ be the collection of zero sets of T, and $\mathcal{C}$ be the collection of all closed subsets of T. Then

(a) βT and $w(T,\mathcal{Z})$ are equivalent compactifications of T;

(b) βT and $w(T,\mathcal{C})$ are equivalent compactifications of T iff T is normal.

Proof (a) Since $\mathcal{Z}$ is a normal base for T, $w(T,\mathcal{Z})$ is a Hausdorff compactification of T. By (1.2-3), $x^{-1}(C)\in\mathcal{Z}$ for each closed subset C of $\underset{\sim}{R}$ and $x\in C_b(T,\underset{\sim}{R})$. Hence, by (3.4-2) and Theorem 3.4-1, all the elements of $C_b(T,\underset{\sim}{R})$ are continuously extendible to $w(T,\mathcal{Z})$ so, by Theorem 1.1-1(c), $w(T,\mathcal{Z})$ and βT are equivalent compactifications.

(b) As $\mathcal{C}$ is a normal lattice if the T_1-space is normal (Example 3.2-4) and $\mathcal{C}$ is an $\alpha\beta$-lattice whenever T is T_1 (Example 3.3-2) it only remains to note that $x^{-1}(C)\in\mathcal{C}$ for each $x\in C_b(T,\underset{\sim}{R})$ and closed subset C of $\underset{\sim}{R}$ for then (3.4-2) and Theorem 3.4-1 allow us to conclude that $w(T,\mathcal{C})$ and βT are equivalent compactifications. Conversely, if $w(T,\mathcal{C})$ and βT are equivalent compactifications, $w(T,\mathcal{C})$ is Hausdorff and T is normal (Example 3.3-3). ∇

3.5 A CLASS OF WALLMAN-TYPE COMPACTIFICATIONS We have seen that any completely regular Hausdorff space T possesses a Wallman-type Hausdorff compactification. In particular the Stone-Cech compactification is equivalent to the Wallman-type compactification $w(T,\mathcal{Z})$ where $\mathcal{Z}$ is the normal $\alpha\beta$-lattice of zero sets. Frink (1964) raised the question of whether all Hausdorff compactifications of completely regular Hausdorff spaces are Wallman-type compactifications. In the main result of this section, due to Brooks (1967) and Alo and Shapiro, 1968a and 1968b, necessary and sufficient conditions are given for a Hausdorff compactification of a completely regular Hausdorff space to be of the Wallman type. This result is then used to show that the one-point compactification of a locally compact Hausdorff space is a Wallman-type compactification. Further results along these lines can be found in Exercise 3.7.

Theorem 3.5-1 WALLMAN-TYPE? Let T be a completely regular Hausdorff space, S a Hausdorff compactification of T, and $\mathcal{L}$ a distributive lattice with 0 and 1 of closed subsets of T with respect to $\cup$ and $\cap$. Then S and $w(T,\mathcal{L})$ are equivalent compactifications of T via a homeomorphism σ:

$w(T,\mathcal{L}) \to S$ where $\sigma(\mathcal{F}_t)=\sigma(\{A\in\mathcal{L} \mid t\in A\})=\sigma(\varphi(t))=t$ for each $t\in T$ iff

(1) for each pair A, $B\in\mathcal{L}$, $cl_S(A\cap B)=cl_S A\cap cl_S B$,

(2) for each $A\in\mathcal{L}$ and $s\in S-cl_S A$ there exists $B\in\mathcal{L}$ such that $s\in cl_S B$ and $A\cap B=\emptyset$, and

(3) for each pair of distinct points s, $t\in S$ there exists A, $B\in\mathcal{L}$ such that $s\notin cl_S A$, $t\notin cl_S B$, and $A\cup B=T$.

Remark We shall write $S=w(T,\mathcal{L})$ when they are equivalent in this way.

Proof Suppose that S is a Hausdorff compactification of the completely regular Hausdorff space T and there exists an onto homeomorphism σ: $w(T,\mathcal{L}) \to S$ such that $\sigma(\mathcal{F}_t)=\sigma(\varphi(t))=t$ for each $t\in T$. Then from the fact [(3.3-3)(a)] that $cl_w(\varphi(A))=\mathcal{B}_A=\{\mathcal{F}\in w(T,\mathcal{L}) \mid A\in\mathcal{F}\}$ for each $A\in\mathcal{L}$ it follows that

$$(*) \quad \sigma(\mathcal{B}_A) = \sigma(cl_w(\varphi(A)) = cl_S(A)$$

for each $A\in\mathcal{L}$. Thus for A, $B\in\mathcal{L}$,

$$cl_S A\cap cl_S B = \sigma(\mathcal{B}_A)\cap\sigma(\mathcal{B}_B) = \sigma(\mathcal{B}_A\cap\mathcal{B}_B) = \sigma(\mathcal{B}_{A\cap B}) = cl_S(A\cap B),$$

thereby establishing (1). As the sets $\mathcal{B}_A$, $A\in\mathcal{L}$, constitute a base of closed sets in $w(T,\mathcal{L})$ we see by (*) that the sets $cl_S A$ are a base of closed sets for S. Therefore, if $A\in\mathcal{L}$ and $s\in S-cl_S A$ we may write the closed set $\{s\}=\bigcap_\alpha cl_S B_\alpha$ where $B_\alpha\in\mathcal{L}$ for each α. Since $cl_S A$ is compact and fails to meet $\bigcap_\alpha cl_S B_\alpha$ there are a finite number of sets $B_{\alpha_1},\ldots,B_{\alpha_n}$ such that $cl_S A\cap(\bigcap_{i=1}^n cl_S B_{\alpha_i})=\emptyset$. Since (1) holds so that $cl_S(\bigcap_{i=1}^n B_{\alpha_i})=\bigcap_{i=1}^n cl_S B_{\alpha_i}$, if we choose $B=\bigcap_{i=1}^n B_{\alpha_i}\in\mathcal{L}$ property (2) follows. As for (3) we may use the argument similar to the one used to prove (2) to obtain the existence of a pair of disjoint sets A, $B\in\mathcal{L}$ such that $t\in cl_S A$ and $s\in cl_S B$. Now it may happen that $A\cup B\neq T$. However, since T is a homeomorphism it follows that $w(T,\mathcal{L})$ is a Hausdorff compactification of T with respect to the imbedding φ: $T\to w(T,\mathcal{L})$, $t\to\mathcal{F}_t$. Thus $\mathcal{L}$ is a normal $\alpha\beta$-lattice and corresponding to A and B there exist supersets A', $B'\in\mathcal{L}$ such that $A'\cup B'=T$ and $A'\cap B=B'\cap A=\emptyset$. Thus by (1) $cl_S A'\cap cl_S B=cl_S B'\cap cl_S A=\emptyset$ so that $t\notin cl_S B'$ and $s\notin cl_S A'$. This establishes (3).

Conversely, suppose that conditions (1), (2) and (3) hold. First we claim that $\mathcal{L}$ is a normal $\alpha\beta$-lattice. Indeed properties (2) and (3) imply that $\mathcal{L}$ is an $\alpha\beta$-lattice. We now show that $\mathcal{L}$ is normal. If C, $D\in\mathcal{L}$ with $C\cap D=\emptyset$ then, by (1), $cl_S C\cap cl_S D=\emptyset$. With r, $s\in C$ and $t\in D$ we may use (3) to construct sets A_{st}, $B_{st}\in\mathcal{L}$ such that $cl_S C\subset cl_S A_{st}$, $cl_S D\subset cl_S B_{st}$; r, $s\notin cl_S B_{st}$, $t\notin cl_S A_{st}$, and $A_{st}\cup B_{st}=T$. Since $\bigcap_{(s,t)} cl_S A_{st}\subset C(cl_S D)$, there exist (s_i,t_i), $i=1,\ldots,n$, such that $\bigcap_i cl_S A_{s_i t_i}\subset C(cl_S D)$. Now $(\bigcap_i A_{s_i t_i})\cup(\bigcap_i B_{s_i t_i})=T$,

$A_r = \bigcap_i A_{s_i t_i} \subset CD$, and $r \notin \bigcup_i B_{s_i t_i} = B_r$ for each $r \in C$. We may repeat the same argument on the B_r to form sets $A, B \in \mathcal{L}$ with the properties $A \cap C = \emptyset$, $B \cap D = \emptyset$, and $A \cup B = T$.

Next we define a mapping $\sigma: w(T, \mathcal{L}) \to S$ which turns out to be an onto homeomorphism. Let $\mathcal{F} \in w(T, \mathcal{L})$. Then as $\mathcal{F}$ has the finite intersection property so does the collection of closed sets $\{cl_S F \mid F \in \mathcal{F}\}$. Since S is compact $\bigcap_{F \in \mathcal{F}} cl_S F \neq \emptyset$. Moreover this intersection consists of just one point for if we make the assumption to the contrary, then there exist distinct s, $t \in \bigcap_{F \in \mathcal{F}} cl_S F$. Thus by (3) there exists $A, B \in \mathcal{L}$ such that $s \notin cl_S A$, $t \notin cl_S B$, and $A \cup B = T$. Since $S = cl_S T = cl_S A \cup cl_S B$, $s \in cl_S B$ so that $cl_S B$ meets $cl_S F$ for each $F \in \mathcal{F}$. Hence $\emptyset \neq cl_S B \cap cl_S F = cl_S(B \cap F)$ for each $F \in \mathcal{F}$ and it follows that B meets all elements of the ultrafilter $\mathcal{F}$; hence $B \in \mathcal{F}$. However since $B \in \mathcal{F}$, t must belong to $cl_S B$ and we have arrived at a contradiction. Therefore $\bigcap_{F \in \mathcal{F}} cl_S F$ is a singleton and we define

$$\sigma(\mathcal{F}) = s \quad \text{where} \quad \{s\} = \bigcap_{F \in \mathcal{F}} cl_S F.$$

To see that σ is 1-1 suppose that $\mathcal{F}$ and $\mathcal{G}$ are distinct ultrafilters so that $\mathcal{G}$ contains an element G not in $\mathcal{F}$. As $\mathcal{F}$ is an ultrafilter there exists $F \in \mathcal{F}$ such that $F \cap G = \emptyset$. Thus $\sigma(\mathcal{F}) \notin cl_S G$ while $\sigma(\mathcal{G}) \in cl_S G$ so that $\sigma(\mathcal{F}) \neq \sigma(\mathcal{G})$.

Next we contend that σ is onto. To this end, suppose that $s \in S$ and let $\mathcal{F} = \{F \in \mathcal{L} \mid s \in cl_S F\}$. Now $\mathcal{F}$ is a filter subbase; in fact if $F, G \in \mathcal{F}$ then $s \in cl_S F \cap cl_S G = cl_S(F \cap G)$ by (1), whence $F \cap G \in \mathcal{F}$. To see that $\mathcal{F}$ is an ultrafilter suppose that $B \notin \mathcal{F}$, i.e. $s \notin cl_S B$. By (2) there exists $A \in \mathcal{L}$, $s \notin cl_S A$ such that $A \cap B = \emptyset$. Hence we have produced an element $A \in \mathcal{F}$ which fails to meet B and it follows by Theorem 3.1-1 that $\mathcal{F}$ is an ultrafilter. Clearly $\sigma(\mathcal{F}) =$ s and, as s is arbitrary, σ is onto.

The space S is known to be compact. If we show that $w(T, \mathcal{L})$ is Hausdorff and σ^{-1} is continuous then it follows that σ is a homeomorphism. To establish the continuity of σ^{-1} it is enough to show that $\sigma(\mathcal{B}_A) = cl_S A$. Indeed if $\mathcal{F} \in \mathcal{B}_A$ then $A \in \mathcal{F}$ so that $\sigma(\mathcal{F}) \in \bigcap_{F \in \mathcal{F}} cl_S F \subset cl_S A$. On the other hand if $\mathcal{F} \notin \mathcal{B}_A$ then $A \notin \mathcal{F}$ and there exists $F \in \mathcal{F}$ that fails to meet A. Thus $cl_S A \cap cl_S F = cl_S(A \cap F) = \emptyset$ and $\sigma(\mathcal{F}) \notin cl_S A$. Hence σ^{-1} is continuous.

It is interesting to note that this theorem provides us with an alternate argument for the fact that βT and $w(T, \mathcal{Z})$ are equivalent whenever T is a completely regular Hausdorff space. Recall first that a Hausdorff compactification S of T is equivalent to βT iff $cl_S z(y) \cap cl_S z(y) = cl_S(z(x) \cap z(y))$

for any pair $z(x)$, $z(y) \in \mathcal{Z}$ (Theorem 1.3-2). As $w(T,\mathcal{Z})$ is a Hausdorff compactification of T (Theorem 3.2-2 and Example 3.2-3), it follows by (1) of the previous result that $cl_W z(x) \cap cl_W z(y) = cl_W (z(x) \cap z(y))$ whenever $z(x)$, $z(y) \in \mathcal{Z}$ whence $w(T,\mathcal{Z})$ and βT are equivalent.

The one-point compactification of a locally compact Hausdorff space is of the Wallman type as Theorem 3.5-2 (and the discussion after it) below shows. In order to more easily describe the lattice used in the creation of the appropriate Wallman compactification of Theorem 3.5-2, it is useful to single out the class of "locally constant" functions named in Definition 3.5-1.

Definition 3.5-1 LOCALLY CONSTANT FUNCTIONS Let S be a Hausdorff compactification of the completely regular Hausdorff space T and let E denote the collection of continuously extendible (to S) $x \in C_b(T,\underset{\sim}{R})$ for which the extension x^S is constant on some neighborhood of each point $p \in S-T$. Elements of E are called locally constant.

Theorem 3.5-2 A CLASS OF WALLMAN-TYPE COMPACTIFICATIONS Hausdorff compactifications S of locally compact Hausdorff spaces T are of Wallman-type if S-T is 0-dimensional, (i.e. there is a base of clopen (=closed+open) sets for the topology of S-T). Specifically $S=w(T,z(E))$ where $z(E)$ denotes the lattice (distributive with 0 and 1 with respect to $\cup$ and $\cap$) of zero sets $z(x)$ as x runs through E.

Proof We show that the lattice conditions (1) - (3) of Theorem 3.5-1 are satisfied by $z(E)$: namely,

(1) for all $z(x)$, $z(y) \in z(E)$, $cl_S(z(x) \cap z(y)) = cl_S z(x) \cap cl_S z(y)$;

(2) for each $z(x)$ and $s \notin cl_S(z(x))$, there exists $z(y) \in z(E)$ such that $s \in cl_S z(y)$ and $z(x) \cap z(y) = \emptyset$;

(3) pairs of distinct points s, $t \in S$ there exists x, $y \in E$ such that $t \notin cl_S z(x)$, $s \notin cl_S z(y)$, and $z(x) \cup z(y) = T$.

Condition (1) That $cl_S(z(x) \cap z(y)) \subset cl_S z(x) \cap cl_S z(y)$ is clear. Conversely, if $s \in cl_S z(x) \cap cl_S z(y)$, there are two cases to consider: $s \in T$ and $s \notin T$. If $s \in T$, then $s \in z(x) \cap z(y)$ since $z(x)$ and $z(y)$ are closed subsets of T. Thus $s \in cl_S(z(x) \cap z(y))$. If $s \notin T$, then, by continuity, $x^S(s)=y^S(s)=0$. Since x and y are each constant in some neighborhood U of s in S, $U \cap T \subset z(x) \cap z(y)$. Since T is dense in S, then it follows that each neighborhood of s in S meets $z(x) \cap z(y)$; hence $s \in cl_S(z(x) \cap z(y))$.

Condition (2) We show that if $s \notin cl_S z(x)$, $x^S(s) \neq 0$. If $s \in T$, then $s \notin z(x)$. If $s \notin T$, and $x^S(s)=0$ then there is a neighborhood U of s in S on which x^S is 0. Then, as in the proof above that $z(E)$ satisfies condition (1), the contradictory result that $s \in cl_S\, z(x)$ follows. With $y^S = x^S - x^S(s)$, we have

$$y^S(s) = 0,\ y \in E, \text{ and } z(y) \cap z(x) = \emptyset.$$

The first two imply, as in the argument for condition (1) above, that $s \in cl_S\, z(y)$.

Condition 3 Here we use local compactness for the first time. We use it to show that S-T is a closed subset of S.

For S-T to fail to be closed, there must be a point $t \in T$ which belongs to $cl_S(S-T)$. For such a t, let U be an open neighborhood of t in S such that $U \cap T$ is the interior (in T) of a compact neighborhood U_t of t in T. The compact neighborhood U_t is closed in S, so that its complement CU_t is open in S and contains S-T. Now we have the contradictory implications: since $t \in cl_S(S-T)$

$$\emptyset \neq (S-T) \cap U \subset CU_t \cap U;$$

while, since $U \cap CU_t$ is a nonempty open subset of S, it must meet T, i.e.

$$\emptyset \neq CU_t \cap (U \cap T) \subset CU_t \cap U_t = \emptyset.$$

Thus S-T is closed in S.

As for condition (3) proper, we now consider three cases:

(a) $s, t \in T$;

(b) $t \in T$, $s \notin T$;

(c) $s, t \notin T$.

Case (a): $s, t \in T$. Since S-T is closed, there are pairwise disjoint open neighborhoods U_s, U_t and V of s, t and S-T respectively in S. Consequently there exists functions $x, y \in C(T, \mathbb{R})$ such that:

$$x(t) = 1, \qquad x^S(S-U_t) = \{0\}$$
$$y(s) = 1, \qquad y^S(S-U_s) = \{0\}$$

Since $S-U_t$ and $S-U_s$ each contain V and $V \supset S-T$, both x and y belong to E. If $t \in cl_S\, z(x)$, then, since $t \in T$, x(t) would have to be 0. Thus $t \notin cl_S\, z(x)$ and, similarly, $s \notin cl_S\, z(y)$. Finally, since

$$T = (T-U_s) \cup (T-U_t) \subset z(x) \cup z(y),$$

condition (3) is seen to be satisfied in this case.

Case (b): $t \in T$, $s \notin T$. If $S-T=\{s\}$, as would happen for the one-point compactification for example, then an argument similar to that used in the proof in case (a) demonstrates that condition (3) is satisfied.

If S-T contains more than one point then, as S-T is 0-dimensional, there are clopen subsets F, G of S-T which are closed in S, such that $s \in G$ and

$S-T=F\cup G$. The required functions $x, y\in E$ shall be constructed such that $x^S(S-T)=\{0\}$ while $y^S(F)=\{0\}$ and $y^S(G)=\{1\}$. As $\{t\}$, F and G are pairwise disjoint there exist a pairwise disjoint open neighborhoods U, V, and W containing t, F, and G respectively. By the normality of S, an additional open set N exists such that $G\subset N\subset cl_S N\subset W$. Now we choose $x, y\in C_b(T,\underset{\sim}{R})$ with continuous extensions $x^S, y^S\in C(S,\underset{\sim}{R})$ such that

$$x(t) = 1,\ x^S(S-U) = \{0\}$$

and

$$y^S(cl_S N) = \{1\},\ y^S(S-W) = \{0\}.$$

Certainly $t\notin cl_S\, z(x)$ and $s\notin cl_S\, z(y)$. To see that $x\in E$ we need only note that $S-U\supset S-T$ and x is thereby locally constant on S-T.

As for y^S being locally constant, it clearly suffices to show that $y^S(V)=\{0\}$ while $y^S(N)=\{1\}$. Indeed, since $S-W\supset V\supset F$ and $cl_S N\supset G$, these facts follow. Finally we claim that $z(x)\cup z(y)=T$. But this is an immediate consequence of the relations $(S-U)\cap T\subset z(x)$, $(S-W)\cap T\subset z(y)$, and $U\cap W=\emptyset$ so that case (b) has been established.

Case (c): $t, s\notin T$. Again using 0-dimensionality there exist disjoint clopen subsets, F and G, of S-T (which are therefore closed in S) such that $t\in F$ and $s\in G$. Thus we may choose disjoint neighborhoods $U\supset F$ and $V\supset G$, N and W such that $U\supset cl_S N\supset N\supset F$ and $V\supset cl_S W\supset W\supset G$, and $x, y\in C_b(T,\underset{\sim}{R})$ such that

$$x^S(cl_S N) = \{1\},\quad x(S-U) = \{0\}$$

and

$$y^S(cl_S W) = \{1\},\quad y^S(S-V) = \{0\}$$

Now $S-U\supset G$ so that $x^S(G)=\{0\}$ and x^S is locally constant on S-T. Similarly y^S is locally constant on S-T. That condition (3) is satisfied follows as in case (b). ∇

Remark As is well known [Narici, Beckenstein, and Bachman 1971, pp. 151-153] a locally compact Hausdorff space is 0-dimensional iff it is totally disconnected, i.e. the only connected sets are singletons. In the proof of the previous result we saw that S-T was compact. Thus the result can be restated to say that a Hausdorff compactification S of a locally compact space T is of Wallman type whenever S-T is totally disconnected.

We mention previously that the case of a one-point compactification of a locally compact Hausdorff space T is handled by the previous result. Indeed, as T is locally compact, S is Hausdorff and S-T, being a singleton, is zero-dimensional so that $S=W(T,z(E))$ where z(E) is the collection of all zero sets of the continuously extendible functions (each such function is locally constant on S-T). Another situation which is subsumed by Theorem 3.5-2 is the case of zero-dimensional Hausdorff compactifications S of a locally compact zero-dimensional space T. The question of whether a zero-dimensional Hausdorff compactification of a not necessarily locally compact 0-dimensional Hausdorff space is of Wallman type has been answered in the affirmative by Alo and Shapiro (Exercise 3.7(d)).

<u>3.6 EQUIVALENT WALLMAN SPACES</u> In this section the notion of equivalent Wallman spaces is defined and then a sufficient condition that two lattices of subsets of T produce equivalent Wallman spaces is given in Theorem 3.6-1. Theorem 3.6-1 is used in proving Theorems 5.4-1 and 2, concerning characterization of the space of maximal ideals of a topological algebra as a Wallman compactification of the space of closed maximum ideals.

<u>Definition 3.6-1</u> <u>EQUIVALENT WALLMAN SPACES</u> Let $\mathcal{L}$ and $\mathcal{H}$ be α-lattices of subsets of a set T, let φ and ψ be the canonical maps of T into $w(T,\mathcal{L})$ and $w(T,\mathcal{H})$:

$$\psi: T \to w(T,\mathcal{H}), \quad t \to \mathcal{G}_t = \{A \in \mathcal{H} \mid t \in A\}, \qquad \varphi: T \to w(T,\mathcal{L}), \quad t \to \mathcal{F}_t = \{A \in \mathcal{L} \mid t \in A\}$$

$w(T,\mathcal{L})$ and $w(T,\mathcal{H})$ are equivalent if there exists a homeomorphism σ

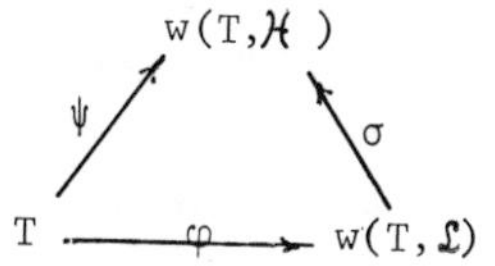

such that $\sigma \cdot \varphi = \psi$. We also say that the lattices $\mathcal{L}$ and $\mathcal{H}$ are <u>equivalent</u>.

$\mathcal{L}$ and $\mathcal{H}$ are required to be α-lattices in the above definition, so that the mapping φ and ψ can be defined ((3.3-1)(a)). It is not required that φ and ψ be 1-1, i.e. that $\mathcal{L}$ and $\mathcal{H}$ be β-lattices (cf. (3.3-1)(b)). We also note that T needn't be a topological space in Definition 3.6-1, so $w(T,\mathcal{L})$ and $w(T,\mathcal{H})$ certainly needn't be compactifications of T. Should it be the case that T is a topological space, $w(T,\mathcal{L})$ and $w(T,\mathcal{H})$ are compactifications of T, φ and ψ are homeomorphisms of T into $w(T,\mathcal{L})$ and $w(T,\mathcal{H})$ respectively, as in (3.3-2), and Definition 3.6-1 reduces to the fact that $w(T,\mathcal{L})$ and $w(T,\mathcal{H})$ are equivalent compactifications of T.

Below we state and prove a pair of necessary and sufficient conditions for the Wallman spaces $w(T,\mathcal{L})$ and $w(T,\mathcal{H})$ to be equivalent under the assumption that $\mathcal{L} \subset \mathcal{H}$ (cf. Theorem 1.3-2).

<u>Theorem 3.6-1</u> <u>EQUIVALENT WALLMAN SPACES</u> Let $\mathcal{L}$ and $\mathcal{H}$ be α-lattices of

subsets of T with φ and ψ as in Definition 3.6-1. If $\mathcal{L} \subset \mathcal{H}$, then the following are equivalent:

(1) If F, H $\in \mathcal{H}$, then $F \cap H = \emptyset$ iff $cl_{\mathcal{L}}\varphi(F) \cap cl_{\mathcal{L}}\varphi(H) = \emptyset$ where $cl_{\mathcal{L}}$ denotes closure in $w(T,\mathcal{L})$.

(2) If, F, H $\in \mathcal{H}$, then $cl_{\mathcal{L}}\varphi(F \cap H) = cl_{\mathcal{L}}\varphi(F) \cap cl_{\mathcal{L}}\varphi(H)$.

(3) $w(T,\mathcal{L})$ and $w(T,\mathcal{H})$ are equivalent.

<u>Proof</u> (1) → (2): We consider two cases.

<u>Case (a)</u>: Suppose $F \in \mathcal{L}$ and $H \in \mathcal{H}$. Certainly $cl_{\mathcal{L}}\varphi(F \cap H) \subset cl_{\mathcal{L}}\varphi(F) \cap cl_{\mathcal{L}}\varphi(H)$. Suppose $\mathcal{F} \notin cl_{\mathcal{L}}\varphi(F \cap H)$. By (3.2-1)(1), $\{\mathcal{F}\} = \bigcap_{K \in \mathcal{F}} \mathcal{B}_K$. Now $U = w(T,\mathcal{L}) - cl_{\mathcal{L}}(F \cap H)$ is an open subset of the compact space $w(T,\mathcal{L})$ containing the point $\mathcal{F}$. Thus $CU \cap (\bigcap_{K \in \mathcal{F}} \mathcal{B}_K) = \emptyset$ so there exist $K_1, \ldots, K_n \in \mathcal{F}$ such that $CU \cap (\bigcap_i \mathcal{B}_{K_i}) = \emptyset$. Since $\mathcal{B}_{\bigcap_i K_i} = \bigcap_i \mathcal{B}_{K_i}$, by (3.2-1)(2), and $K = \bigcap_i K_i \in \mathcal{F}$, it follows that $\mathcal{B}_K \subset U$. Thus $\mathcal{B}_K \cap cl_{\mathcal{L}}\varphi(F \cap H) = \emptyset$ and therefore $K \cap F \cap H = \emptyset$. Hence, by (3.2-1)(2), (3.3-3)(a), and (1), $\mathcal{B}_K \cap \mathcal{B}_F \cap cl_{\mathcal{L}}\varphi(H) = \mathcal{B}_{K \cap F} \cap cl_{\mathcal{L}}\varphi(H) = cl_{\mathcal{L}}\varphi(K \cap F) \cap cl_{\mathcal{L}}\varphi(H) = \emptyset$. Since $K \in \mathcal{F}$, then $\mathcal{F} \in \mathcal{B}_K$; as $\mathcal{B}_K \cap \mathcal{B}_F \cap cl_{\mathcal{L}}\varphi(H) = \emptyset$, $\mathcal{F} \notin \mathcal{B}_F \cap cl_{\mathcal{L}}\varphi(H) = cl_{\mathcal{L}}\varphi(F) \cap cl_{\mathcal{L}}\varphi(H)$. Thus $cl_{\mathcal{L}}\varphi(F) \cap cl_{\mathcal{L}}\varphi(H) \subset cl_{\mathcal{L}}\varphi(F \cap H)$ and the result for case (a) is proved.

<u>Case (b)</u>: Suppose F, H $\in \mathcal{H}$, and again choose $\mathcal{F} \notin cl_{\mathcal{L}}\varphi(F \cap H)$. Since $\mathcal{F} = \bigcap_{K \in \mathcal{F}} \mathcal{B}_K$, once again there exists $K \in \mathcal{F}$ such that $\mathcal{B}_K \subset w(T,\mathcal{L}) - cl_{\mathcal{L}}\varphi(F \cap H) = U$ and therefore $K \cap F \cap H = (K \cap F) \cap (K \cap H) = \emptyset$.

By (1), $cl_{\mathcal{L}}\varphi(K \cap F) \cap cl_{\mathcal{L}}\varphi(K \cap H) = \emptyset$. Since $K \in \mathcal{L}$, from part (a) it follows that $cl_{\mathcal{L}}\varphi(K) \cap cl_{\mathcal{L}}\varphi(F) \cap cl_{\mathcal{L}}\varphi(K) \cap cl_{\mathcal{L}}\varphi(H) = \emptyset$, or more simply, $cl_{\mathcal{L}}\varphi(K) \cap cl_{\mathcal{L}}\varphi(F) \cap cl_{\mathcal{L}}\varphi(H) = \emptyset$. As $cl_{\mathcal{L}}\varphi(K) = \mathcal{B}_K$ and $K \in \mathcal{F}$, $\mathcal{F} \in \mathcal{B}_K$ and therefore once again $\mathcal{F} \notin cl_{\mathcal{L}}\varphi(F) \cap cl_{\mathcal{L}}\varphi(H)$. Thus $cl_{\mathcal{L}}\varphi(F) \cap cl_{\mathcal{L}}\varphi(H) \subset cl_{\mathcal{L}}\varphi(F \cap H)$.

(2) → (3): We must construct a homeomorphism σ between $w(T,\mathcal{L})$ and $w(T,\mathcal{H})$ such that $\sigma \cdot \varphi = \psi$. Let $\mathcal{F} \in w(T,\mathcal{L})$ and define

$$\sigma(\mathcal{F}) = \{B \in \mathcal{H} \mid \mathcal{F} \in cl_{\mathcal{L}}\varphi(B)\}$$

Note that $\mathcal{F} \subset \sigma(\mathcal{F})$ for if $F \in \mathcal{F}$, $\mathcal{F} \in cl_{\mathcal{L}}\varphi(F) = \mathcal{B}_F$. We prove:

(a) $\sigma(\mathcal{F}) \in w(T,\mathcal{H})$ and therefore $\sigma(w(T,\mathcal{L})) \subset w(T,\mathcal{H})$.

By (2) it follows that $\sigma(\mathcal{F})$ is a filter. Suppose $A \in \mathcal{H}$ and $A \notin \sigma(\mathcal{F})$. Then $\mathcal{F} \notin cl_{\mathcal{L}}\varphi(A)$. As $cl_{\mathcal{L}}\varphi(A)$ is the intersection of basic closed subsets $\mathcal{B}_K$ there exists $K \in \mathcal{L}$ such that $cl_{\mathcal{L}}\varphi(A) \subset \mathcal{B}_K$ and $\mathcal{F} \notin \mathcal{B}_K$. Thus $K \notin \mathcal{F}$ and, since $\mathcal{F}$ is an $\mathcal{L}$-ultrafilter, there exists $F \in \mathcal{F}$ such that $F \subset CK$. Thus since $\varphi(A) \subset \mathcal{B}_K$ then $A \subset K$ and therefore $F \cap A = \emptyset$. Since $F \in \mathcal{F} \subset \sigma(\mathcal{F})$ we have found a set in the filter $\sigma(\mathcal{F})$ disjoint from A and therefore $\mathcal{F}$ is an ultrafilter (Theorem 3.1-1).

Next we claim that σ is onto. Suppose that $\mathcal{G} \in w(T,\mathcal{H})$. Then as the sets in $\mathcal{G}$ satisfy the finite intersection property, the family of closed sets $\{cl_{\mathcal{L}}\varphi(D) \mid D\in\mathcal{G}\}$ satisfies the finite intersection property and since $w(T,\mathcal{L})$ is a compact space, $\bigcap_{D\in\mathcal{G}} cl_{\mathcal{L}}\varphi(D)\neq\emptyset$. Let $\mathcal{F}\in\bigcap_{D\in\mathcal{G}} cl_{\mathcal{L}}(D)$; we contend that $\bigcap_{D\in\mathcal{G}} cl_{\mathcal{L}}\varphi(D)=\{\mathcal{F}\}$ and $\sigma(\mathcal{F})=\mathcal{G}$. Suppose $\mathcal{F}_1, \mathcal{F}_2\in\bigcap_{D\in\mathcal{G}} cl_{\mathcal{L}}\varphi(D)$ and $\mathcal{F}_1\neq\mathcal{F}_2$. Then by Theorem 3.3-1 there exist $F_1\in\mathcal{F}_1$ and $F_2\in\mathcal{F}_2$ such that $F_1\cap F_2=\emptyset$. We shall now show that both $F_1, F_2\in\mathcal{G}$ - a conclusion which is at variance with the fact that $\mathcal{G}$ is a filter. To prove this, note that $\mathcal{F}_1\in cl_{\mathcal{L}}D$ for all $D\in\mathcal{G}$, and $\mathcal{F}_1\in\mathcal{B}_{F_1}=cl_{\mathcal{L}}\varphi(F_1)$. If for some $D\in\mathcal{G}$, $F_1\cap D=\emptyset$, then, by (2), $cl_{\mathcal{L}}\varphi(F_1)\cap cl_{\mathcal{L}}D=\emptyset$ and, as $\mathcal{F}_1\in cl_{\mathcal{L}}\varphi(F_1)\cap cl_{\mathcal{L}}D$, this is a contradiction. Thus $F_1\cap D\neq\emptyset$ for all $D\in\mathcal{G}$ and, since $F_1\in\mathcal{L}\subset\mathcal{H}$ and $\mathcal{G}$ is an ultrafilter, $F_1\in\mathcal{G}$. Similarly $F_2\in\mathcal{G}$ and $\bigcap_{D\in\mathcal{G}} cl_{\mathcal{L}}\varphi(D)=\{\mathcal{F}\}$. As $\mathcal{F}\in cl_{\mathcal{L}}\varphi(D)$ for all $D\in\mathcal{G}$, it follows by the definition of $\sigma(\mathcal{F})$ that $\mathcal{G}\subset\sigma(\mathcal{F})$. Thus $\mathcal{G}=\sigma(\mathcal{F})$ by the maximality of $\mathcal{G}$. Our next contention is that :

(c) σ is 1-1.

Recall the definition of $\sigma(\mathcal{F})$ and the observation that $\mathcal{F}\subset\sigma(\mathcal{F})$. If $\mathcal{F}_1\neq\mathcal{F}_2$, there exists $F_1\in\mathcal{F}_1$, $F_2\in\mathcal{F}_2$ such that $F_1\cap F_2=\emptyset$. But then if $\sigma(\mathcal{F}_1)=\sigma(\mathcal{F}_2)$, $\mathcal{F}_1, \mathcal{F}_2\subset\sigma(\mathcal{F}_1)$ so that the absurd conclusion that $\emptyset=F_1\cap F_2\in\sigma(\mathcal{F}_1)$ follows.

We prove that:

(d) $\sigma\cdot\varphi = \psi$.

Let $t\in T$ and $\mathcal{F}_t=\{A\in\mathcal{L} \mid t\in A\}$. Suppose $B\in\sigma(\mathcal{F}_t)$. Then $\mathcal{F}_t\in cl_{\mathcal{L}}\varphi(B)$. If $t\notin B$ then, as $\mathcal{H}$ is an α-lattice, there exists $C\in\mathcal{H}$ such that $t\in C$ and $C\cap B=\emptyset$. But $t\in C$ so that $\mathcal{F}_t\in cl_{\mathcal{L}}\varphi(C)$. Hence $\mathcal{F}_t\in cl_{\mathcal{L}}\varphi(C)\cap cl_{\mathcal{L}}\varphi(B)$. By (2), $\mathcal{F}_t\in cl_{\mathcal{L}}\varphi(C)\cap cl_{\mathcal{L}}\varphi(B)=cl_{\mathcal{L}}\varphi(B\cap C)=\emptyset$ and we have a contradiction. Thus $t\in B$ for each $B\in\sigma(\mathcal{F}_t)$ and therefore $\sigma(\mathcal{F}_t)\subset\mathcal{G}_t=\{B\in\mathcal{H} \mid t\in B\}$. But both $\sigma(\mathcal{F}_t)$ and $\mathcal{G}_t$ are ultrafilters; hence $\sigma(\mathcal{F}_t)=\mathcal{G}_t$ and $\sigma\cdot\varphi=\psi$. To finish the proof that (2) → (3) it only remains to show that:

(e) σ is bicontinuous.

To see that σ is continuous, let $B\in\mathcal{H}$ and consider the basic closed set

$$\mathcal{A}_B = \{\mathcal{G}\in w(T,\mathcal{H}) \mid B\in\mathcal{G}\} \text{ of } w(T,\mathcal{H}).$$

Then

$$\sigma^{-1}(\mathcal{A}_B) = \sigma^{-1}(\{\mathcal{G}\in w(T,\mathcal{H}) \mid B\in\mathcal{G}\}) = \{\mathcal{F}\in w(T,\mathcal{L}) \mid \sigma(\mathcal{F})\in\mathcal{A}_B\}$$
$$= \{\mathcal{F} \mid B\in\sigma(\mathcal{F})\} = \{\mathcal{F} \mid \mathcal{F}\in cl_{\mathcal{L}}\varphi(B)\} = cl_{\mathcal{L}}\varphi(B)$$

and $\sigma^{-1}(\mathcal{A}_B)$ is a closed set. Therefore σ is continuous.

To show that σ^{-1} is continuous, let $A \in \mathcal{L} \subset \mathcal{H}$. We recall that $\mathcal{F} \subset \sigma(\mathcal{F})$ for each $\mathcal{F} \in w(T,\mathcal{L})$. Thus $\sigma(\mathcal{B}_A) = \sigma(\{\mathcal{F} | A \in \mathcal{F}\}) \subset \mathcal{A}_A$. However, if $\mathcal{G} \in \mathcal{A}_A$ then $\mathcal{G} = \sigma(\mathcal{F})$ for some $\mathcal{F} \in w(T,\mathcal{L})$ and, as $\mathcal{F} \subset \sigma(\mathcal{F}) = \mathcal{G}$, A meets each element of $\mathcal{F}$. Thus A must actually belong to the ultrafilter $\mathcal{F}$. So that $\mathcal{F} \in \mathcal{B}_A$ and $\mathcal{A}_A \subset \sigma(\mathcal{B}_A)$. Thus $\mathcal{A}_A = \sigma(\mathcal{B}_A)$ and σ^{-1} is continuous.

<u>(3) → (1)</u>: Let F, $H \in \mathcal{H}$ be such that $F \cap H = \emptyset$. If $t \in F$ then $\sigma(\mathcal{F}_t) = \mathcal{G}_t$ and, as $\mathcal{G}_t \in \mathcal{A}_F$, $\mathcal{F}_t \in \sigma^{-1}(\mathcal{A}_F)$. Consequently $\varphi(F) \subset \sigma^{-1}(\mathcal{A}_F)$. As σ is a homeomorphism, $\sigma^{-1}(\mathcal{A}_F)$ is closed in $w(T,\mathcal{L})$ and $cl_{\mathcal{L}}\varphi(F) \subset \sigma^{-1}(\mathcal{A}_F)$. Similarly $cl_{\mathcal{L}}\varphi(H) \subset \sigma^{-1}(\mathcal{A}_H)$. As $F \cap H = \emptyset$, $\mathcal{A}_F \cap \mathcal{A}_H = \emptyset$ and, therefore, $cl_{\mathcal{L}}\varphi(F) \cap cl_{\mathcal{L}}\varphi(H) = \emptyset$. ▽

Exercises 3

3.1 SEPARATION IN LATTICES Let L be a distributive lattice with 0 and 1. L is a disjointed lattice if a, b$\in$L, a$\neq$0, b$\neq$0, then there exists c$\in$L such that ac$\neq$0 and bc=0 or bc$\neq$0 and ac=0.

(a) If L is a disjointed lattice then $a\in\mathcal{F}$ for all $\mathcal{F}\in w$ L (Sec. 3.2) iff a=1.

(b) The lattice of all closed subsets of a topological T_o space T is a disjointed lattice iff T is a T_1 space.

(c) If T is a topological space and L the lattice of all closed subsets of T, then if L is an α-lattice, L is a disjointed lattice. Conversely if L is a disjointed β-lattice, then L is an α-lattice.

3.2 BOOLEAN LATTICES; THE STONE REPRESENTATION THEOREM In this exercise some results are presented about Boolean lattices (complemented distributive lattices with 0 and 1) including the Stone representation theorem (f) which can be generalized to distributive lattices (Stone, 1936), as can a number of the parts of this exercise.

All lattices are assumed to contain 0 and 1; a' denotes 1-a. A non-zero element a of a lattice L is an atom if when b$\in$L and b $\leq$ a then b=0 or b=a.

(a) If L is a finite lattice then for every c$\in$L there exists an atom a$\in$L such that a $\leq$ c.

(b) If L is a lattice and a an atom in L, then for every b$\in$L, ab=0 or ab=a.

(c) Let a$\in$L where L is a lattice and let R(a) be the set of all atoms that are $\leq$ a. Then R(ab)=R(a)$\cap$R(b), R(a')=R(1)-R(a). If L is finite, then R(a)=R(b) if and only if a=b. Let $a_1,\ldots,a_k$ be atoms in a Boolean algebra L. Then $R(\max_i a_i)=\{a_1,\ldots,a_n\}$.

Two Boolean algebras L_1 and L_2 are said to be isomorphic if there exists a 1-1 correspondence f between them such that f(a')=f(a)' for all a$\in L_1$ and $f(a_1\wedge a_2)=f(a_1)\wedge f(a_2)$ for all a_1, $a_2\in L_1$.

(d) If L is a finite Boolean lattice, then L is isomorphic to the Boolean lattice of all subsets of the set of atoms of L.

Hint: Let $X=\{a_1,\ldots,a_n\}$ be the atoms of L and consider

$$f:L \longrightarrow \mathcal{P}(X)$$
$$a \longrightarrow R(a)$$

(e) Two finite Boolean lattices with the same number of elements are isomorphic.

Statement (d) is true for infinite Boolean lattices as well. However, in proving this, use is made of Zorn's lemma. As is often the case, the lemma is used in showing that every nonzero $a \in L$ can be embedded in an ultrafilter.

(f) Every Boolean lattice L is isomorphic to a Boolean lattice of subsets of a set X.

Hint: Let X be the set of ultrafilters of L. Let

$$T: L \longrightarrow \mathcal{P}(X)$$
$$a \longrightarrow \{\mathcal{F} \in X \mid a \in \mathcal{F}\} \quad \text{for } a \neq 0$$

and $T(0)=\emptyset$.

A Boolean lattice is called complete if for every $W \subset L$, sup W exists. L is called atomic if every element in L is $\geq$ some atom.

(g) A Boolean lattice L is isomorphic to $\mathcal{P}(X)$ under the mapping T of (f) if and only if L is complete and atomic.

3.3 ZERO-DIMENSIONALITY OF $w(T,\mathcal{C})$ AND ULTRANORMALITY In this exercise T is a T_1 space. It is shown in (a) that $w(T,\mathcal{C})$ is zero-dimensional iff T is ultranormal.

(a) Let $\mathcal{C}$ be the lattice of all closed subsets of T. Show that $w(T,\mathcal{C})$ is zero-dimensional iff T is ultranormal.

Hint: If $w(T,\mathcal{C})$ is ultraregular then, being compact, $w(T,\mathcal{C})$ is ultranormal. Hence if $\mathcal{B}_F$ and $\mathcal{B}_K$ are disjoint basic closed sets in $w(T,\mathcal{C})$, there exists (see Example 3.2-1) a clopen set $\bigcup_{i\in I} V_{U_i}$ (where $T-U_i \in \mathcal{C}$ for $i\in I$) such that $\mathcal{B}_F \subset \bigcup_{i\in I} V_{U_i}$ and $\mathcal{B}_K \cap \bigcup_{i\in I} V_{U_i} = \emptyset$. Thus with φ as in Sec. 3.3, $F = \mathcal{B}_F \cap \varphi(T) \subset \bigcup_{i\in I} U_i$, $K \cap \bigcup_{i\in I} U_i = \mathcal{B}_K \cap \varphi(T) \cap \bigcup_{i\in I} V_{U_i} = \emptyset$, and $\bigcup_{i\in I} U_i$ is clopen.

Conversely, if T is ultranormal, the clopen subsets of T separate the closed sets. Letting $\mathcal{G}$ denote the clopen subsets of T, by Example 3.2-1(b), $\{V_U \mid U \in \mathcal{G}\}$ is a base for the topology of $w(T,\mathcal{G})$. As each $U \in \mathcal{G}$ is clopen, it follows that V_U for each $U \in \mathcal{G}$ is clopen.

(b) A normal T_1 space T is ultranormal iff βT is ultraregular and in this case $\beta T = w(T,\mathcal{C})$.

Hint: If βT is ultraregular then show that βT is ultranormal. Utilizing normality of T, show that the disjoint closed sets in T have disjoint closures in βT.

Conversely, suppose that T is ultranormal and $s\in\beta T$. Let $s\in U$ where U is open in βT and take $x^\beta\in C(\beta T,\underset{\sim}{R})$ such that $z(x^\beta)\subset U$ where $z(x^\beta)$ is a neighborhood of s. As T is ultranormal, let V be clopen in T such that $z(x)=z(x^\beta)\cap T\subset V\subset U\cap T$. Observing that V and CV are zero sets in T show that $cl_\beta V$ and $cl_\beta CV$ are disjoint clopen sets in βT with $s\in cl_\beta V\subset U$.

(c) Let T be an ultraregular space which is not ultranormal. Show that $\{\mathcal{V}_U \mid U\in\mathcal{G}\}$ is not a base for the topology of $w(T,\mathcal{G})$.

Hint: The sets $\mathcal{V}_U$ are clopen in $w(T,\mathcal{G})$ and $w(T,\mathcal{G})$ is not ultraregular.

(d) AN ULTRAREGULAR SPACE WHICH IS NOT ULTRANORMAL Let Ω be the first uncountable ordinal and ω the first infinite cardinal. Let $S=[0,\omega]\times[0,\Omega]$ and $T=S-\{(\omega,\Omega)\}$. Show that S and T (carrying the product of the order topology on the ordinal numbers) are both ultraregular but that T is not ultranormal.

Hint: T is not normal as the closed sets $K=\{(\alpha,\omega)\mid 0\leq\alpha<\Omega\}$ and $F=\{(n,\Omega)\mid 0\leq n<\omega\}$ cannot be separated by disjoint open sets (Dugundjii 1966, p. 145).

Given a T_1 space T and a lattice $\mathcal{L}$ of closed subsets of T, it was found that under certain conditions the collection $w(T,\mathcal{L})$ of all ultrafilters can be topologized to obtain a T_1-compactification of T ((3.3-2)). In Exercises 3.4 - 3.6 we consider a collection $J(T,\mathcal{L})$ of prime filters containing $w(T,\mathcal{L})$, topologized in such a way as to make $J(T,\mathcal{L})$ a T_o-compactification of T.

3.4 VERY DENSE SPACES Let Y be a topological space. A subspace X is said to be very dense in Y if, when F is any closed set in Y, then $cl_Y(F\cap X)=F$, i.e. X is "dense" in any closed subset.

(a) X is very dense in Y iff for any $y\in Y$, $y\in cl_Y(cl_Y\{y\}\cap X)$.

(b) If X is very dense in Y then for each relatively open subset G of X there exists a unique open set $G'\subset Y$ such that $G'\cap X=G$. The set G is compact iff G' is compact.

Hint: Show that a closed subset F of X meets G iff $cl_Y F$ meets G'.

(c) If X is very dense in Y and (B_α) is a base for the closed sets of X, then $(cl_Y B_\alpha)$ is a base for the closed sets of Y.

(d) If X is very dense in Y, then if (F_α) is a family of closed subsets of X, then $cl_Y(\bigcap_\alpha F_\alpha)=\bigcap_\alpha cl_Y F_\alpha$.

(e) If X is very dense in Y, then X is normal iff Y is normal. X is a T_o-space iff Y is a T_o-space.

3.5 VERY DENSE SPACES AND IRREDUCIBLE CLOSED SETS A set F in a topological space T is irreducible if when $F=F_1\cup F_2$ with F_1 and F_2 closed sets, then $F=F_1$ or $F=F_2$.

(a) Let X be very dense in Y. Then a closed subset F of X is irreducible in X iff $cl_Y F$ is irreducible in Y. If $y \in Y$, then $cl_Y\{y\}$ is irreducible in Y and $cl_Y\{y\} \cap X = F_y$ is irreducible in X.

(b) Let X be very dense in a T_o-space Y and Irr X denote the collection of all irreducible closed subsets of X. Then

$$I: Y \longrightarrow \text{Irr } X$$
$$y \longrightarrow cl_Y\{y\} \cap X$$

is a 1-1 mapping.

If the mapping I is onto, then Y is referred to as a <u>Jacobson completion</u> of X. If X=Y, X is said to be <u>Jacobson complete</u>. In the event that X (and therefore Y) is compact, with the compact open subsets of X forming a base for the topology closed under the formation of finite intersections, then Y is called a <u>spectral completion</u> of X and if X=Y, then X is called <u>spectrally complete</u>.

If F is an irreducible closed subset of a space X, then $x \in X$ is a <u>generic point</u> of F if $cl_X\{x\}=F$. If X is very dense in Y, then we say that $y \in Y$ is a <u>generic point of F in Y</u> if $cl_Y\{y\} \cap X = F$.

(c) An ultraregular space is spectrally complete. Part (d) is included to make the more general case of (e) clearer.

(d) Let F be a closed subset of a T_1 space X with no generic point. Then a T_o space $Y = X \cup \{y\}$ can be formed in which X is very dense and y is a generic point of F in Y.

<u>Hint</u>: Let $\mathcal{B} = \{G \mid G$ is an open neighborhood of some $x \in F\}$. As F is irreducible, $\mathcal{B}$ is a filter among the open sets of X. For $y \notin X$, define neighborhoods in $Y = X \cup \{y\}$ as follows.

(1) If $y \in N \subset Y$ and $N \supset G$ for some $G \in \mathcal{B}$, N is a neighborhood of y.

(2) If $x \in N \subset X$ and $x \in G \subset N$ for some $G \in \mathcal{B}$, then $N \cup \{y\}$ is a neighborhood of x.

(3) If N is a neighborhood of $x \in X$ and the condition of (2) does not hold, then N is a neighborhood of x.

(e) If X is a T_1-space, a Jacobson completion of X exists.

<u>Hint</u>: Let W be a set whose cardinality is strictly bigger than the cardinality of Irr X. Let $Z = X \cup W$ and $\mathcal{A} = \{S \subset Z \mid S$ is a T_o-space with respect to some topology $\mathcal{J}$ and X is very dense in S$\}$. Order the objects $(S, \mathcal{J})$ with $(S_1, \mathcal{J}_1) \leq (S_2, \mathcal{J}_2)$ iff $S_1 \subset S_2$ and $\mathcal{J}_2 \cap S_1 = \mathcal{J}_1$. By Zorn's lemma there exists a maximal element $(S_M, \mathcal{J})$. Show that S_M is a Jacobson completion of X

according to the following scheme: If $F \subset X$ and F is irreducible with no generic point in S_M, let $y \in Z - S_M$ and define a topology on $S_M \cup \{y\}$ as follows. If $\mathcal{B}$ is the filter among the open subsets of X associated with F as in the hint to (d), let $\hat{\mathcal{B}}$ be the open subsets of S_M associated with the open subsets in $\mathcal{B}$ by Exercise 3.4(b). Letting N denote a subset of S_M,

(1) If there exists $G' \in \hat{\mathcal{B}}$ such that $G' \subset N$, then $N \cup \{y\}$ is a neighborhood of y;

(2) If $s \in S_M$ and there exists $G \subset N$ such that $s \in G \subset N$, then $N \cup \{y\}$ is a neighborhood of s;

(3) If $s \in S_M$ and N is a neighborhood of s which fails to satisfy the condition of (2), then N is a neighborhood of s.

Show that $S_M \cup \{y\}$ violates the maximality of S_M.

(f) Let X be a compact T_1-space whose compact open sets form a base for the topology which is closed under the formation of finite intersections. Then X is spectrally complete iff for any filterbase $\mathcal{B}$ of compact open subsets of X, $\cap \mathcal{B} \neq \emptyset$.

Hint: If there exists $\mathcal{B}$ such that $\cap \mathcal{B} = \emptyset$, then assume that $\mathcal{B}$ is an ultrafilter in the class of compact open sets of X. A compact open set belongs to $\mathcal{B}$ iff it meets every set in $\mathcal{B}$. Define neighborhoods in $Y = X \cup \{y\}$ for $y \notin X$ as follows:

(1) If $N \subset X$ and there exists $B \in \mathcal{B}$ with $B \subset N$, then $N \cup \{y\}$ is a neighborhood of y.

(2) If $x \in X$ and $N \subset X$ is a neighborhood of x in X with $x \in B \subset N$ for some $B \in \mathcal{B}$, then $N \cup \{y\}$ is a neighborhood of x.

(3) If $x \in X$ and $N \subset X$ is a neighborhood of x and the condition of (2) is not satisfied, then N is a neighborhood of x in Y.

Y is a T_0-space and X is very dense in Y. Note that $F_y = cl_Y\{y\} \cap X = \{x \in X \mid \text{all neighborhoods of } x \text{ are of type (2)}\}$. Show that $y \in cl_Y F_y$.

3.6 JACOBSON FILTERS IN A LATTICE AND JACOBSON COMPLETIONS In this exercise L is a distributive lattice with 0 and 1. A prime filter $\mathcal{P}$ in L is a filter such that if $a, b \in L$, and $a+b \in \mathcal{P}$, then a or $b \in \mathcal{P}$. A Jacobson filter is a prime filter which is equal to the intersection of all the ultrafilters containing it. The set of all ultrafilters, Jacobson filters, and prime filters will be denoted by W(L), J(L), and P(L) respectively, If L is a lattice of subsets of a set T, these sets will be denoted by W(T,L), J(T,L), and P(T,L) respectively. We topologize P(L) by taking as a base for the closed sets the sets

$$\mathcal{B}_a = \{\mathcal{P}\in P(L)\,|\,a\in\mathcal{P}\}$$

with $a\in L$.

(a) $cl_{P(L)}\{\mathcal{P}\}=\{\mathcal{P}'\in P(L)\,|\,\mathcal{P}\subset\mathcal{P}'\}$ and P(L) is a T_0-space.

(b) Show that $C\mathcal{B}_a\cap W(L)=\{\mathcal{F}\in W(L)\,|$ there exists $b\in\mathcal{F}$ with $ab=0\}$ but that in P(L) this relationship does not hold.

(c) Show that if $W(L)\subset S$, then w(L) is very dense in S iff $S\subset J(L)$.

(d) Show that for every S such that $W(L)\subset S\subset J(L)$, S is a normal space iff L is a normal lattice.

For the remaining parts of this exercise we assume that T is a T_1-space and $\mathcal{C}$ the lattice of all closed subsets of T.

(e) If $F\in\mathcal{C}$, then the filter $\mathcal{P}_F=\{K\in\mathcal{C}\,|\,F\subset K\}$ is a prime filter iff $F\in$ IrrT.

(f) If T is compact, a filter $\mathcal{F}$ is an ultrafilter iff $\mathcal{F}=\mathcal{F}_t=\{K\in\mathcal{C}\,|\,t\in K\}$ for some $t\in T$.

(g) When T is compact, a filter $\mathcal{H}$ is a Jacobson filter iff $\mathcal{H}=\mathcal{P}_F$ for some $F\in$ IrrT.

(h) Let T be compact and very dense in the T_0-space S. Then letting $F_s=cl_S\{s\}\cap T$, the mapping

$$\sigma: S\to J(T,\mathcal{C})$$
$$s\to\mathcal{P}_{F_s}$$

is a homeomorphism such that $\sigma(T)=w(T,\mathcal{C})$ and σ is onto $J(T,\mathcal{C})$ iff S is a Jacobson completion of T.

<u>Note</u>: It now follows that a Jacobson completions of a compact T_1-space is unique and topologically equivalent to $J(T,\mathcal{C})$.

<u>3.7 WHEN COMPACTIFICATIONS ARE OF WALLMAN TYPE</u> (Alo and Shapiro 1968a and 1968b, Brooks 1967a; cf. also Sec. 3.5). In this exercise Y denotes a Hausdorff compactification of the completely regular Hausdorff space T; otherwise the notation of Secs. 3.2 - 3.4 is assumed to be in force. In (b), (c), and (d), conditions are obtained which make Y a Wallman type compactification of T.

(a) Let $\mathcal{L}$ be a normal base (Definition 3.4-1 and (3.4-1)) for T. If $\mathcal{F}\in w(T,\mathcal{L})$ and G is open in $w(T,\mathcal{L})$ with $\mathcal{F}\in G$, then there exists $A\in\mathcal{L}$ such that $\mathcal{F}\in cl_w\varphi(A)=\mathcal{B}_A\subset G$ and $\mathcal{B}_A$ is a neighborhood of $\mathcal{F}$.

<u>Hint</u>: Since $w(T,\mathcal{L})$ is a compact Hausdorff space there exists $A\in\mathcal{L}$ such that $CG\subset\hat{\mathcal{B}}_{CA}$ while $\hat{\mathcal{B}}_{CA}$ is disjoint from some open neighborhood U of $\mathcal{F}$. Hence $\mathcal{F}\in U\subset\hat{\mathcal{B}}_{CA}\subset G$.

(b) If $\mathcal{L}$ is a normal base for T then Y and $w(T,\mathcal{L})$ are equivalent compactifications of T iff the class of sets $\{cl_Y A \mid A\in\mathcal{L}\}$ contains a neighborhood base for Y.

Hint: Let $V(\mathcal{F})=\{\hat{\mathcal{B}}_{CA} \mid \mathcal{F}\in\hat{\mathcal{B}}_{CA}\}$ and $B(\mathcal{F})=\{cl_Y CA \mid \hat{\mathcal{B}}_{CA}\in V(\mathcal{F})\}$. The sets of $B(\mathcal{F})$ satisfy the finite intersection property and the condition of (b) yields $\bigcap B(\mathcal{F})=\{y\}$. Define $y=f(\mathcal{F})$. Clearly if $\mathcal{F}=\varphi(t)$ for some $t\in T$, then $f(\mathcal{F})=f(\varphi(t))=t$. Also $B(\mathcal{F})\to y$; hence if V is a neighborhood of y, then there exists $\hat{\mathcal{B}}_{CA}\in V(\mathcal{F})$ such that $cl_Y CA\subset V$. Thus if $\mathcal{G}\in\hat{\mathcal{B}}_{CA}$, $f(\mathcal{G})\in cl_Y CA\subset V$, $f(\hat{\mathcal{B}}_{CA})\subset V$ and f is continuous. The fact that both spaces $w(T,\mathcal{L})$ and Y are compact leads to the completion of the proof.

(c) (cf. Theorem 3.5-1). Y is equivalent to $w(T,\mathcal{L})$ for some lattice $\mathcal{L}$ of closed subsets of T iff $\mathcal{L}$ satisfies the following two conditions:

(1) $cl_Y(A\cap B)=cl_Y A\cap cl_Y B$ for all $A, B\in\mathcal{L}$.

(2) If V is a neighborhood of y in Y, there exists $A\in\mathcal{L}$ such that $y\in cl_Y A\subset V$.

Hint: Note first that statement (2) is weaker than the condition of (b) in that the sets $cl_Y A$ need not be neighborhoods of y. Let $y\in Y$ and $\mathcal{F}_y=\{A\in\mathcal{L} \mid y\in cl_Y A\}$. Then by (1) and (2) $\mathcal{F}_y$ is an ultrafilter. Since Y is a compact Hausdorff space, as $\mathcal{F}$ is an ultrafilter there exists $y\in\bigcap_{A\in\mathcal{F}} cl_Y A$ and $\mathcal{F}\subset\mathcal{F}_y$. Hence $\mathcal{F}=\mathcal{F}_y$ and the mapping

$$g: Y\to w(T,\mathcal{L})$$
$$y\to\mathcal{F}_y$$

is 1-1 and onto. Let $g(y)=\mathcal{F}_y\in\hat{\mathcal{B}}_{CA}$. Then there exists $B\in\mathcal{F}_y$ such that $B\subset CA$. If $y\in cl_Y A$, then as $B\in\mathcal{F}_y$, $y\in cl_Y A\cap cl_Y B=cl_Y(A\cap B)$ and we have a contradiction. Hence $y\in C(cl_Y A)$. If $w\in C(cl_Y A)$, then obviously $A\notin\mathcal{F}_w$ and $\mathcal{F}_w\in\hat{\mathcal{B}}_{CA}$. Hence $g(C(cl_Y A))\subset\hat{\mathcal{B}}_{CA}$ and g is continuous.

(d) Show that any ultraregular compactification Y of (the necessarily ultraregular space) T is equivalent to $w(T,\mathcal{L})$ for some lattice $\mathcal{L}$ of closed subsets of T.

Hint: Let $\mathcal{L}=\{G\cap T \mid G$ is clopen in $Y\}$.

3.8 WALLMAN COMPACTIFICATIONS AND E-COMPACTIFICATIONS Let T be a T_1-space and E be a Hausdorff space.

(a) E-closed sets A subset A of T is E-closed if for some $n\in\mathbb{N}$ and $x\in C(T,E^n)$, $A=x^{-1}(F)$ for some closed subset F of the product E^n.

Finite unions and finite intersections of E-closed sets are E-closed so that the E-closed sets form a ring of sets. T is E-completely regular (Exercise 1.11) iff for each closed subset F of T and each t not in F there are disjoint E-closed subsets A and B of T such $t \in A$ and $F \subset B$.

(b) Separating families (Steiner 1968). A family $\mathcal{F}$ of closed subsets of T is separating if for each closed subset F of T and each t not in F there are disjoint sets A and B in $\mathcal{F}$ such that $t \in A$ and $F \subset B$. If $\mathcal{F}$ is a separating family then the Wallman space $w(T,\mathcal{F})$ is a compactification of T. Conversely, for any topological space T, if $w(T,\mathcal{F})$ is a compactification, then T must be T_1 and the ring of sets generated by $\mathcal{F}$ is separating.

(c) E-closed sets and Wallman compactifications (Piacun and Su 1973). By Exercise 1.11(a), if T is E-completely regular there is a map h which homeomorphically embeds T in the product E^B for some index set B. If $\mathcal{S}$ denotes the family of all E-closed subsets of E^B and $\mathcal{F} = \{F \subset T \mid F = h^{-1}(F')$ for some $F' \in \mathcal{S}\}$ then $w(T,\mathcal{F})$ is a compactification of T. (Show that $\mathcal{F}$ is a separating family and use (b).) If $\mathcal{F}$ is the family of all E-closed subsets of T then $w(T,\mathcal{F})$ is a compactification of T iff T is E-completely regular.

If the closed interval $[0,1]$ in its usual topology and T are each E-completely regular and $\mathcal{F}$ denotes the collection of all E-closed subsets of T, then $w(T,\mathcal{F})$ is an E-compactification (Exercise 1.11) of T.

(d) (Piacun and Su 1973) If the zero-dimensional and normal space T has more than one point, each closed subset of T is E-closed, and $\mathcal{F}$ denotes the ring of all closed subsets of T, then $w(T,\mathcal{F})$ is an E-compactification of T.

If $\mathcal{F}$ is the family of all closed subsets of the discrete space T then $w(T,\mathcal{F})$ is an E-compactification of T.

3.9 (Brooks 1967a) HAUSDORFF COMPACTIFICATIONS ARE "WALLMAN" QUOTIENTS

It is well known that each Hausdorff compactification S of a completely regular Hausdorff space T is a quotient of the Stone-Cech compactification βT (Dugundji 1966, XI, 8.2). In this exercise it is established that each such compactification S is also realizable as a quotient of the Wallman-type compactification $W(T,\mathcal{Z}_e)$ obtained by utilizing the lattice $\mathcal{Z}_e$ of zero sets of continuous real-valued functions which are continuously extendible to S.

(a) $W(T,\mathcal{Z}_e)$ is a T_1-compactification of T: $\mathcal{Z}_e$ is an $\alpha\beta$-lattice, (see Definition 3.3-1 and Example 3.3-1) of subsets of T that serves as a base of closed subsets of T. Thus, by (3.3-1) and (3.3-2), $w(T,\mathcal{Z}_e)$ is a T_1-compactification of T.

Before proceeding any further we settle the question of whether S and $w(T,\mathcal{Z}_e)$ are equivalent compactifications (a conjecture motivated perhaps by the fact that βT and $w(T,\mathcal{Z})$ are always equivalent ((3.4-3))).

(b) S and $w(T,\mathcal{Z}_e)$ are not always equivalent compactifications

Hint: Let $T=\underset{\sim}{N}$ with the discrete topology and S be the one-point compactification of $\underset{\sim}{N}$. In this case $\mathcal{Z}_e=\mathcal{Z}=\mathcal{P}(T)$ so that $w(T,\mathcal{Z}_e)=w(T,\mathcal{Z})=\beta T$.

Let $C_s(T,\underset{\sim}{R})$ denote the continuous real-valued functions of T which are continuously extendible to S and $C_w(T,\underset{\sim}{R})$ those continuously extended to $w(T,\mathcal{Z}_e)$.

(c) $C_s(T,\underset{\sim}{R})\subset C_w(T,\underset{\sim}{R})$ Each continuous real-valued function of T which is continuously extendible to S is $\mathcal{Z}_e$-uniformly continuous (Definition 3.4-2) and is therefore continuously extendible to $w(T,\mathcal{Z}_e)$ by Theorem 3.4-1. (Note that this theorem is valid even if the requirement that the lattice be normal is dropped).

Hint: Let $x\in C_s(T,\underset{\sim}{R})$ and choose $a, b\in\underset{\sim}{R}$ such that $x(T)\subset(a,b)$. Given $\epsilon > 0$ choose a partition $a=t_o \leq t_1 \leq \cdots \leq t_n=b$ of $[a,b]$ such that $t_i-t_{i-1} \leq \epsilon/4$ for $i=1,\ldots,n$. For each $1 \leq i \leq n-2$ set $y_i=(t_{i+2} - \min(t_{i+2},x))(\max(t_i,x)-t_i)$. Clearly each $y_i\in C_s(T,\underset{\sim}{R})$ and $z(y_i)=x^{-1}(-\infty,t_i]\cup x^{-1}[t_{i+2},\infty)=T-x^{-1}(t_i,t_{i+2})$. Hence $\bigcap_i z(y_i)=\emptyset$ or, equivalently, $T=\bigcup_i Cz(y_i)$. Furthermore, since $Cz(y_i)=x^{-1}(t_i,t_{i+2})$, $0(x,Cz(x_i))=\sup\{|x(t)-x(s)|\,|\,s,t\in Cz(y_i)\} < \epsilon$ and x is $\mathcal{Z}_e$-uniformly continuous.

Thus each $x\in C_s(T,\underset{\sim}{R})$ can be extended uniquely to a continuous function $x^e: w(T,\mathcal{Z}_e) \to \underset{\sim}{R}$ such that $x^e(\mathcal{F}_t)=x(t)$ for all $t\in T$. Then $\mathcal{R}=\{(\mathcal{F},\mathcal{G})\in w(T,\mathcal{Z}_e)\times w(T,\mathcal{Z}_e)\;|x^e(\mathcal{F})=x^e(\mathcal{G})$ for all $x\in C_s(T,\underset{\sim}{R})\}$ is an equivalence relation and we consider $w(T,\mathcal{Z}_e)/\mathcal{R}$ with the quotient topology.

(d) $w(T,\mathcal{Z}_e)/\mathcal{R}$ and S are equivalent compactifications of T

Hint: Certainly the canonical map $K: w(T,\mathcal{Z}_e) \to w(T,\mathcal{Z}_e)/\mathcal{R}$ is continuous so $w(T,\mathcal{Z}_e)/\mathcal{R}$ is compact. As each x^e is constant on the elements of $K\mathcal{F}$, $x'(K\mathcal{F})=x^e(\mathcal{F})$ is a well-defined continuous real-valued function on $w(T,\mathcal{Z}_e)/\mathcal{R}$ (see Dugundji, 1966, pp. 123-124). The family $\mathcal{U}$ of all such functions x' is a uniformly closed subalgebra of $C(w(T,\mathcal{Z}_e)/\mathcal{R},\underset{\sim}{R},c)$ which contains the constants and separates the points of $w(T,\mathcal{Z}_e)/\mathcal{R}$. Thus $w(T,\mathcal{Z}_e)/\mathcal{R}$ is Hausdorff and by the Stone-Weierstrass Theorem (Dunford and Schwartz 1958, Vol. 1, IV 6.16, p. 272), $\mathcal{U}=C(w(T,\mathcal{Z}_e)/\mathcal{R},\underset{\sim}{R})$. Consider the mapping

$$f: C(S,\underset{\sim}{R}) \to C_s(T,\underset{\sim}{R}) \to C(w(T,\mathcal{Z}_e)/\mathcal{R},\underset{\sim}{R})$$
$$y \to y|_T \to (y|_T)'$$

is a surjective algebraic isomorphism. Thus it follows that there exists a homeomorphism h taking $w(T,\mathcal{Z}_e)/\mathcal{R}$ onto S such that $(fy)(K\mathcal{F})=y(h(K\mathcal{F}))$ for each $y\in C(S,\underset{\sim}{R})$ and $K\mathcal{F}\in w(T,\mathcal{Z}_e)/\mathcal{R}$ (see (1.6-3)). Finally, since

$K\mathcal{F}_t = \{\mathcal{F}_t\}$ for each $t \in T$

$$y(t) = y^e(\mathcal{F}_t) = (y|_T)'(K\mathcal{F}_t) = (fy)(K\mathcal{F}_t) = y(h\{\mathcal{F}_t\})$$

for each $y \in C(S, \underset{\sim}{R})$ and $t \in T$. Thus $t = h\{\mathcal{F}_t\}$ for each $t \in T$ and $w(T, \mathcal{Z}_e)/\mathcal{R}$ and S are equivalent compactifications.

FOUR

Cmmutative Topological Algebras

The Theory of Banach algebras began in 1939 with the publication of Gelfand's striking paper "On normed rings" followed by his "Normierte Ringe" in 1941. From then on it gained momentum and tho its development abated somewhat in the mid-fifties, it remains an active area. Considering the development of the theories of topological groups and topological vector spaces throughout the forties, the investigation of topological algebras was ineluctable. For the study to begin in earnest it was desirable that the theory of Banach algebras mature further so that its most prominent distinguishing features showed more clearly. But still another factor had to go into the mix: there had to be some reason to examine topological algebras *as algebras* and not merely topological rings, that is, some way to get the scalars to play a more significant role.

During the forties the theories of topological algebras and topological rings began to undergo simultaneous development. Topological rings were extensively treated in Kaplansky 1947b and 1948 and topological algebras received the feature to distinguish them from topological rings in Arens 1946b with the introduction of local *m-convexity*, a subset U of a topological algebra being called *multiplicatively convex*, or *m-convex* for short, if it was convex and UU was again a subset of U. A topological algebra is locally *m-convex* if it possesses a base at 0 of balanced m-convex sets. The next major development in the theory of topological algebras, or more accurately locally m-convex algebras, came with the publication of Michael's 1952 Memoir on the subject.

We begin the study of commutative topological algebras in this chapter. The algebras $C(T,\underset{\sim}{F})$ of continuous functions of Chapter 1 are endowed with the compact-open topology and drawn upon to motivate certain results as well as to serve as examples and counterexamples. Banach algebras too of course serve in a similar way. And altho we consider it unlikely that a reader of this book would be completely unfamiliar with Banach algebras, no formal knowledge of them to speak of is required. In any event the standard references of Rickart 1960 and Naimark 1960 contain whatever background material one might want for the sake of comparison of Banach algebras with topological algebras. Chapters 5 and 6 continue what has been begun here. In Chapter 5 the primary thrust is devoted to the space of maximal ideals of a locally m-convex algebra, especially in regard to its relationship to the space of closed maximal ideals as a compactification of it. In Chapter 6 a special type of topological algebra--essentially inductive limits of Banach algebras--is investigated.

ALL OUR ALGEBRAS ARE ASSUMED TO BE COMMUTATIVE AND TO POSSESS AN IDENTITY. THE UNDERLYING FIELD IS $\underset{\sim}{R}$ OR $\underset{\sim}{C}$, DENOTED GENERICALLY BY $\underset{\sim}{F}$. For some results the underlying field must be $\underset{\sim}{C}$. When this is necessary the more special hypothesis is explicitly indicated.

4.1 Topological Algebras

Definition 4.1-1 BASIC NOTIONS A linear space X equipped

with an additional binary operation, called <u>vector multiplica-</u><u>tion</u> and denoted by xy for x,y ∈ X, is called an <u>algebra</u> if X is a ring with respect to vector addition and vector multiplication, and α(xy) = (αx)y = x(αy) for all scalars α and all x,y ∈ X. A linear map A:X→Y, where X and Y are algebras is called an <u>(algebra)</u> <u>homomorphism</u> if A(xy) = (Ax)(Ay) for all x,y ∈ X. If A is also 1-1 it is referred to as an <u>(algebra)</u> <u>isomorphism</u> or an <u>embedding</u>. The null space of an algebra homomorphism A, i.e. $A^{-1}(0)$, is renamed the <u>kernel of A</u> and is denoted by <u>ker A</u>. "Homomorphism of X" unmodified signifies a homomorphism whose range is in X's underlying field which (as is mentioned below) is always $\underset{\sim}{F} = \underset{\sim}{R}$ or $\underset{\sim}{C}$; such homomorphisms are also called <u>real</u> or <u>complex homomor-</u><u>phisms</u>, as the case may be.

A <u>topological algebra</u> is an algebra endowed with a nontrivial topology $\mathcal{J}$ which is compatible with X's linear structure and such that the map X×X→X , (x,y)→xy is continuous. The algebra X and the topology $\mathcal{J}$ are <u>compatible</u> or $\mathcal{J}$ is <u>compatible with the algebraic structure of X</u> when (X,$\mathcal{J}$) is a topological algebra. <u>Topological isomorphism</u> in the context of topological algebras means an algebra isomorphism which is also a homeomorphism.

For the sake of simplicity, <u>all algebras considered are</u> <u>assumed to be commutative,</u> <u>and to possess an identity</u>. But how much is lost by specializing to commutative algebras with identity? As regards the presence of an identity: not much; as regards commutativity: a good deal. Little is lost through the identity requirement because any algebra X may be embedded (via a topological isomorphism) in an algebra X_1 which has an identity* so that X may be viewed as a subalgebra of an algebra with identity. Results proved for X_1 are then "restricted" to X. Commutativity, however, is a different matter: there is just no way to identify a noncommutative algebra with a subalgebra of a commutative algebra and generally the theory of commutative algebras is quite different from that of noncommutative algebras.

An immediate consequence of the definition of a topological algebra is that the map $x \to wx$ is continuous for any $w \in X$; it is a homeomorphism if w is a unit. In fact continuity of the map $(x,y) \to xy$ is equivalent (using the fact that X is a TVS) to the continuity of each of the maps $x \to wx$ and continuity of $(x,y) \to xy$ at 0. This is easily verified using the identity $xy - x_0y_0 = (x-x_0)(y-y_0) + (x-x_0)y_0 + x_0(y-y_0)$.

Theorem 4.1-1 BASES FOR COMPATIBLE TOPOLOGIES The filterbase $\mathcal{B}$ in the algebra X determines a base at 0 for a compatible topology for X iff

(a) $\mathcal{B}$ is a neighborhood base at 0 for a topology which is compatible with X's linear structure

(b) For each $V \in \mathcal{B}$ there exists a $B \in \mathcal{B}$ such that $BB \subset V$.

Proof. We prove only the sufficiency of the conditions. Let $V(0)$ denote the filter of neighborhoods of 0 determined by $\mathcal{B}$.

---

* To embed a topological algebra X without identity in a topological algebra X_1 with identity, consider $F \times X$ with the product topology and pointwise operations. The element $(1,0)$ then serves as an identity for X_1.

To prove that $(x,y) \to xy$ is continuous, consider a neighborhood of $xy, xy + U$ where $U \in V(0)$, and choose a balanced neighborhood V of 0 such that $V+V+V \subset U$. By (b) there exists a $B \in \mathcal{B}$ such that $BB \subset V$. Now choose a real number a, $0 < a < 1$, such that $ax \in B$ and $ay \in B$. Since

$$\begin{aligned}(x+aB)(y+aB) &= xy + axB + ayB + a^2BB \\ &\subset xy + BB + BB + a^2V \\ &\subset xy + V + V + V \\ &\subset xy + U,\end{aligned}$$

it follows that X is a topological algebra.

Corollary 4.1-1 SUBBASES If $\mathcal{S}$ is a collection of subsets of the algebra X which is a neighborhood subbase at 0 for a topology which is compatible with X's linear structure and such that for each $S \in \mathcal{S}$ there exists a $V \in \mathcal{S}$ such that $VV \subset S$, then the filterbase generated by $\mathcal{S}$ is a base at 0 for a compatible topology for the algebra X.

The prototype topological algebra is the normed algebra: An algebra X is a normed algebra if it is a normed space and $\|xy\| \le \|x\|\|y\|$. A normed algebra which is also a Banach space is a Banach algebra. The space of n-tuples $\underset{\sim}{F}^n$ with sup norm and pointwise operations is a Banach algebra, as is the space $C(T,\underset{\sim}{F},c)$ of Example 0.1-2 of scalar-valued functions of the topological space T with pointwise oeprations when T is compact (i.e. $(xy)(t) = x(t)y(t)$ and sup norm).

If T is not necessarily compact, the compact-open and point-open topologies (Example 0.1-2) are each compatible with the algebraic structure but we defer proving this until Section 4.3. Other examples which include such things as products and quotients are given throughout the chapter.

As an immediate application of the basis theorem we have:

(4.1-1) INITIAL TOPOLOGIES Suppose that X is an algebra, Y is a topological algebra with neighborhood filter at 0 denoted by $V(0)$, and $A:X\to Y$ a homomorphism. It is easy to verify that the filter $A^{-1}(V(0))$ determines a topology which is compatible with X's linear structure. To see that it is compatible with the algebraic structure as well, we first note that for any $V \in V(0)$, there is a $B \in V(0)$ such that $BB\subset V$. Hence $A^{-1}(B)A^{-1}(B) \subset A^{-1}(BB) \subset A^{-1}(V)$. The topology determined by $A^{-1}(V(0))$ is called the initial (inverse image, weak) topology induced by the homomorphism A. One of the consequences of this result is that any subalagebra M of a topological algebra X is a topological algebra in its relative topology as its relative topology is that induced by the homomorphism $x\to x$ of M into X.

(4.1-2) FINAL TOPOLOGIES Suppose X is a topological algebra with filter of neighborhoods of 0 denoted by $V(0)$, Y an algebra, and $A:X\to Y$ a homomorphism. It is easy to see that the collection $\mathcal{B}$ of subsets U of Y such that $A^{-1}(U)\in V(0)$ forms a base at 0 for a topology compatible with X's linear structure. For any $U \in \mathcal{B}$, we may select $B \in V(0)$ such that $BB\subset A^{-1}(U)$. Thus $A(B)A(B) = A(BB) \subset A(A^{-1}(U)) \subset U$. Since $A^{-1}(A(B)) \supset B \in V(0)$, it follows that $A^{-1}(A(B)) \in V(0)$, i.e. that $A(B) \in \mathcal{B}$, so that $\mathcal{B}$ is a base at 0 for a topology which is compatible with Y's algebraic structure. The topology generated by $\mathcal{B}$ is called the final topology for Y determined by the homomorphism A.

4.2 Multiplicative sets and miltiplicative seminorms

For most of our subsequent results, we shall specialize our investigation to a particular type of topological algebra--"locally m-convex algebras." We prepare the ground for that discussion now.

Definition 4.2-1 MULTIPLICATIVE CONVEXITY. A subset U of an algebra X is called multiplicative (idempotent) if $U^2 = UU \subset U$. It is called multiplicatively-convex or m-convex if it is convex and multiplicative, absolutely m-convex if it is balanced and m-convex.

An immediate example of multiplicative sets is afforded by the spheres-open or closed-of radius 1/n, $n \in \underset{\sim}{N}$, about 0 in any normed algebra. As is apparent, each such sphere is absolutely m-convex as well.

Preservation of multiplicativity is the subject of our next result.

(4.2-1) PROPERTIES OF MULTIPLICATIVE SETS. Let X be an algebra. If $U \subset X$ is multiplicative, then so is

(a) its convex hull U_c;

(b) λU if U is balanced and $|\lambda| \leq 1$;

(c) its balanced hull U_b;

(d) its balanced convex hull U_{bc};

(e) any direct or inverse homomorphic image;

(f) its closure cl(U) if X is a topological algebra.

Proof. (a) Consider any two elements of U_c: $x = \Sigma a_n x_n$ and $y = \Sigma b_m y_m (a_n, b_m \geq 0,\ \Sigma a_n = \Sigma b_m = 1)$ and their product $xy = \Sigma_{n,m} a_n b_m x_n y_m$. Since, for all n and m, $a_n b_m \geq 0$, $\Sigma_{n,m} a_n b_m = 1$, and $x_n y_m \in U$, it follows that $(U_c)^2 \subset U_c$.

(b) Clear.

(c) If λx, $\mu y \in U_b = C_1(0)U$, then $\lambda x \cdot \mu y \in C_1(0)U$.

(d) This follows from (a) and (c), since $U_{bc} = (U_b)_c$.

(e) For any homomorphism A, $A^{-1}(U)A^{-1}(U) \subset A^{-1}(UU) \subset A^{-1}(U)$ and $A(U)A(U) = A(UU) \subset A(U)$.

(f) Similar to what was done in (3.1-11), it follows that $cl(U)cl(U) \subset cl(U^2) \subset cl(U)$ by continuity of multiplication.∇

By well-known results on convexity parts (e) and (f) remain true if you substitute "m-convex" for "multiplicative."

In making the transition from normed spaces to locally convex spaces, one goes from topologies determined* by a norm to topologies determined by families of seminorms. The analog for normed algebras is from a topology determined by a single norm to that determined by a family of <u>multiplicative</u> seminorms.

<u>Definition 4.2-2 MULTIPLICATIVE SEMINORMS</u>. A seminorm p on an algebra X is <u>multiplicative</u> if $p(xy) \leq p(x)p(y)$ for all $x, y \in X$.

We note that for a multiplicative seminorm p to be nontrivial on an algebra X, it is necessary and sufficient that p(e) be non-zero. The trivial seminorm (i.e. identically zero) is multiplicative and generates the trivial topology.

<u>(4.2-2) MAXIMA OF MULTIPLICATIVE SEMINORMS</u>. If $p_1, \ldots, p_n$ are multiplicative then $\max a_j p_j$ is a multiplicative seminorm for any collection of non-negative numbers $a_1, \ldots, a_n$.

---

*"Determined" in the sense that the "norm topology" is the weakest topology with respect to which the norm is a continuous map; a similar meaning is attached to "determined by a family of seminorms."

Proof. Since $\max_j a_j p_j$ is clearly a seminorm we show that $\max a_j p_j$ is multiplicative. To do this it suffices to consider two multiplicative seminorms q and r, and show that $p = \max(q,r)$ is multiplicative. To this end consider $p(xy) = \max(q(xy), r(xy))$. Without loss of generality we may suppose that $p(xy) = q(xy)$. Thus $p(xy) = q(xy) \leq q(x)q(y) \leq \max(q(x), r(x)) \max(q(y), r(y)) = p(x)p(y)$. ∇

It is straightforward to verify that a base at 0 for the initial topology generated by a family P of seminorms is given by positive multiples of finite intersections of sets of the form $V_p = \{x|p(x) < 1\}$, $p \in P$. To enable us to be rid of finite intersections and be able to use simply positive multiples of the V_p, the notion of a "saturated" family of multiplicative seminorms is introduced.

Definition 4.2-3 SATURATED FAMILIES. A family P of seminorms (multiplicative or not) is saturated if $\max_j p_j \in P$ for any finite subset $\{p_1, \ldots, p_n\}$ of P.

It follows immediately from (4.2-2) that there is never any loss in generality in assuming a family of seminorms to be saturated. As a particular example of a saturated family of multiplicative seminorms, consider the algebra $C(\underset{\sim}{R}, \underset{\sim}{R})$ of continuous real-valued functions endowed with the topology generated by the multiplicative seminorms $p_n, n \in \underset{\sim}{N}$, defined at each $x \in C(\underset{\sim}{R}, \underset{\sim}{R})$ by: $p_n(x) = \sup_{t \in [-n,n]} |x(t)|$. These seminorms generated the compact-open topology for $C(\underset{\sim}{R}, \underset{\sim}{R})$ (cf. Example 4.3-1).

If X and Y are algebras and $A: X \to Y$ is a homomorphism then $p \cdot A$ determines a multiplicative seminorm on X for any multiplicative seminorm p on Y. Any normed algebra norm is a multiplicative seminorm. Further examples of multiplicative seminorms are given in the examples in Section 4.3.

(4.2-3) MULTIPLICATIVE SEMINORMS AND GAUGES. (a) If p is a multiplicative seminorm, then $V_p = \{x | p(x) < 1\}$ is absolutely m-convex and absorbent. (b) If U is absolutely m-convex and absorbent then its gauge p_U, $p_U(x) = \inf\{a>0 | x \in aU\}$, is a multiplicative seminorm.

Proof. (a) If $x,y \in V_p$, then $p(xy) \leq p(x)p(y) < 1$, which proves that V_p is multiplicative. (b) It is well-known that the gauge of an absolutely convex absorbent set is a seminorm, so it only remains to show that p_U is multiplicative. To this end, suppose $x,y \in X$ and $a,b > 0$ are such that $x \in aU$ and $y \in bU$. Then $xy \in abU^2 \subset abU$ so $p_U(xy) \leq ab$. The result now follows from the arbitrary nature of a and b.∇

4.3 Locally m-convex algebras

A topological algebra $(X,\mathcal{T})$ is a locally m-convex algebra (LMC algebra) if there is a base of m-convex sets for V(0). We also say that $\mathcal{T}$ is locally m-convex or is an LMC-topology. X is a locally convex algebra if X is a topological algebra which carries a locally convex linear space structure.* If, in addition to being locally m-convex, $\mathcal{T}$ is Hausdorff, we say that X is an LMCH algebra, and $\mathcal{T}$ to be LMCH. An LMC algebra which is a complete metrizable topological space is a Frechet algebra.

Clearly each normed algebra is an LMC algebra and each

* Clearly any LMC algebra is a locally convex algebra. The converse is not true, however, and a counter-example may be found in Exercise 4.7 (c).

Banach algebra is a Frechet algebra.

Our first result about LMC algebras shows that we can do a little better than just say that there is a base of m-convex sets at 0. Our second, (4.3-2), shows LMC topologies to inevitably be generated by families of multiplicative seminorms.

(4.3-1) BASES AT 0 FOR LMC TOPOLOGIES. The following conditions are equivalent on any algebra X:

(a) X is an LMC algebra;

(b) X is a locally convex TVS and there exists a base of multiplicative sets at 0;

(c) X is a TVS and there exists a base of absolutely m-convex sets at 0;

(d) X is a TVS and there exists a base of closed absolutely m-convex sets at 0.

Proof. (a)$\Rightarrow$(b): X is clearly a locally convex TVS if it is an LMC algebra and it possesses a base at 0 of m-convex sets.

(b)$\Rightarrow$(c): Since X is a locally convex TVS it has a base $\mathcal{B}'$ at 0 of balanced convex sets. It also has a base $\mathcal{B}$ at 0 of multiplicative sets. Thus each $B' \in \mathcal{B}'$ contains a multiplicative set $B \in \mathcal{B}$ and the collection of absolutely m-convex sets B_{bc} also serves as a base at 0.

(c)$\Rightarrow$(a): Let $\mathcal{B}$ be a base of absolutely m-convex sets at 0. $\mathcal{B}$ clearly satisfies the conditions of the basis theorem for topological algebras, Theorem 4.1-1, from which (a) follows.

(c)$\Rightarrow$(d): By (c) X is a TVS and has a base at 0, $\mathcal{B}$, of absolutely m-convex sets. Since X is a topological algebra (since (c)$\Rightarrow$(a)), the closure of a multiplicative set is also multiplicative [(4.2-1)(f)] while absolute convexity is preserved by closure in any TVS.

Now for any $V \in \mathcal{V}(0)$ there exists $U \in \mathcal{V}(0)$ such that $\mathrm{cl}(U) \subset V$, since any TVS is regular; $\mathrm{cl}(U)$, in turn, contains some $B \in \mathcal{B}$. It follows that V contains the closed absolutely m-convex set $\mathrm{cl}(B)$.

(d)$\Longrightarrow$(c): Clear.∇

(4.3-2) LMC ALGEBRAS AND SEMINORMS. A topological algebra X is locally m-convex iff its topology is generated by a family of multiplicative seminorms.

Proof. If X is locally m-convex, it has a base $\mathcal{B}$ at 0 of absolutely m-convex neighborhoods by (4.3-1). The gauges of the sets $B \in \mathcal{B}$ are multiplicative seminorms by (4.2-3) which obviously generate the topology. The converse follows immediately from (4.2-3) (a).∇

Our assumption that a topological algebra not carry the trivial topology implies that if the topology is generated by a family P of multiplicative seminorms, then there exists some $p \in P$ which is not trivial.

Some examples of LMC algebras follow.

Example 4.3-1 COMPACT-OPEN AND POINT-OPEN TOPOLOGIES. Consider the linear space $C(T,\underset{\sim}{F})$ of all scalar-valued continuous functions on the topological space T with pointwise linear operations. For $x,y \in C(T,\underset{\sim}{F})$ define the pointwise product of x and y, xy, at any $t \in T$ as $(xy)(t) = x(y)y(t)$. With these operations, $C(T,\underset{\sim}{F})$ becomes an algebra whose identity is the map which takes each $t \in T$ into 1. The compact-open topology for $C(T,\underset{\sim}{F})$ ((Example 0.1-2) was generated by the maps p_G where

G is a compact subset of T and $p_G(x) = \sup|x(G)|$. Since each such seminorm is clearly multiplicative with respect to pointwise multiplication, it follows that $C(T,\underset{\sim}{F},c)$, $C(T,\underset{\sim}{F})$ with compact-open topology, is locally m-convex.

With $C(T,\underset{\sim}{F})$ as above then $C(T,\underset{\sim}{F},p)$ - $C(T,\underset{\sim}{F})$ endowed with the point-open topology (Example 0.1-2)-is generated by the seminorms $\{p_t | t \in T\}$ where $p_t(x) = |x(t)|$. Each p_t is multiplicative and so $C(T,\underset{\sim}{F},p)$ is a locally m-convex algebra.

Example 4.3-2 INFINITELY DIFFERENTIABLE FUNCTIONS. Consider the algebra $\mathcal{D}$ of infinitely differentiable functions on $[0,1]$ (or $[a,b]$) with pointwise operations. The seminorms $p_n(x) = \sup_{t \in [0,1]}|x^{(n)}(t)|$, $(n=0,1,\ldots)$ determine a first-countable T_2-hence metrizable-topology for $\mathcal{D}$. These seminorms are not multiplicative, however, [if $x(t) = t$, then $p_1(x^2) = 2$ while $p_1(x) = 1$ so that $p_1(x^2) \not\leq p_1(x)\cdot p_1(x)$] but the topology is locally m-convex nevertheless. To see this, note that the same topology is generated by the seminorms $r_n = \max_{1 \leq j \leq n} p_j$. A straightforward computation using Leibniz's rule for computing the j-th derivative shows that for each n $r_n(xy) \leq 2^n r_n(x) r_n(y)$. Thus the seminorms $q_n = 2^n r_n$ are multiplicative as $q_n(xy) = 2^n r_n(xy) \leq 2^n(2^n r_n(x) r_n(y)) = q_n(x) q_n(y)$ and (q_n) also generates the original topology. Thus $\mathcal{D}$ is a metrizable LMCH algebra. Actually $\mathcal{D}$ is a Frechet algebra, i.e. $\mathcal{D}$ is complete. To see this, suppose (x_n) is a Cauchy sequence in $\mathcal{D}$. It follows that (x_n) is a Cauchy sequence for each $p_j, j=0,1,\ldots$. In particular for $j=0$, we see that (x_n) is a uniform Cauchy sequence, so there is some x such that $x_n \to x$ uniformly in the usual function-theoretic sense on

[0,1]. Is x infinitely differentiable? Since (x_n) is Cauchy with respect to p_1, then the sequence of derivatives (x_n') is also a uniform Cauchy sequence on [0,1]. Hence x is differentiable and x' is the uniform limit of the x_n'. Continuing in this fashion, x is seen to be infinitely-differentiable and $x_n^{(k)} \to x^{(k)}$ uniformly on [0,1] for each k=0,1,... . In turn this is the same as saying that $p_k(x_n - x) \to 0$ for each k=0,1,... and therefore $x_n \to x$ in $\mathcal{D}$'s topology.

$\mathcal{D}$ is not normable, however, as it cannot contain a bounded neighborhood of 0: Any neighborhood of 0 will contain sequences (x_n) whose subsequently derived sequences will achieve arbitrarily large values, thus violating the condition that a set B is bounded iff for each seminorm p generating the topology, $(p(x_n))$ is a bounded set, for each sequence (x_n) from B. The sequence determined by taking $x_n(t) = 10^{-10} \sin 10^n t$ serves as a fairly dramatic example of this phenomenon for V_{p_o}.

Example 4.3-3 INITIAL TOPOLOGIES; INITIAL LMC TOPOLOGIES. Let $(x_\mu)_{\mu \in M}$ be a family of topological algebras, let X be an algebra, and suppose for each $\mu \in M$ there is a homomorphism $A_\mu : X \to X_\mu$. The initial topology $\mathcal{T}$ determined by the family $(A_\mu)_{\mu \in M}$ for X ((0.1-1)) has a neighborhood subbase at 0 of sets of the form $A_\mu^{-1}(B_\mu)$, where B_μ is a basic neighborhood of 0 in X_μ. That $\mathcal{T}$ is compatible with X's algebraic structure is evident by virtue of the results of (4.1-1)(on initial topologies) and the subbase theorem, Corollary 4.1-1.

Moreover, since the inverse homomorphic image of a multiplicative set is multiplicative [(4.2-1)(e)] and finite intersections of multiplicative sets are multiplicative, it follows that $\mathcal{T}$ is LMC if each X_μ is LMC.

If (X_μ) is a family of topological algebras, then $\Pi X_\mu = X$ with the product topology and pointwise operations ($(x_\mu)(y_\mu) = (x_\mu y_\mu)$, etc.) is also a topological algebra: The projection maps pr_μ are homomorphisms of X into X_μ and the product topology is the initial topology determined by the family (pr_μ). Thus by Example 4.3-3 the product topology is compatible with the algebraic structure of X. Moreover by the same example:

(4.3-3) PRODUCTS OF LMC ALGEBRAS. A product of topological algebras is LMC iff each component space is LMC.

4.4 Final topologies and quotients. In (4.1-1) we observed that any subalgebra of a topological algebra is a topological algebra in its relative topology. To deal with quotients of topological algebras, we use (4.1-2). The linear subspace I of an algebra X is an ideal if $xI \subset I$ for each $x \in X$. If X is a topological algebra and X/I carries the final topology induced by the canonical homomorphism $x \to x+I$ of X onto X/I, then X/I is a topological algebra by (4.1-2). We denominate this topology for X/I the quotient topology. (The term "factor topology" will be reserved for a different object, considered in the next section.) Moreover, since homomorphisms preserve multiplicity, X/I is seen to be LMC if X is.

The topology of the following example facilitates the passage from consideration of final topologies generated by a single map to final topologies determined by a family of maps.

Example 4.4-1 SUPREMUM TOPOLOGY. Let X be an algebra and let $(\mathcal{T}_\mu)_{\mu \in M}$ be a family of compatible topologies for X. By the supremum (sup) topology $\mathcal{T}$ for X we mean the topology generated by the sets $\bigcup_{\mu \varepsilon M} \mathcal{J}_\mu$; we denote $\mathcal{J}$ by $\bigvee_\mu \mathcal{J}_\mu$.

By letting X_μ denote X topologized by $\mathcal{J}_\mu$ and letting i_μ denote the canonical injection of X into X_μ, $\mathcal{J}$ is readily identified as the weakest topology for X with respect to which each i_μ is continuous. It follows from Example 4.3-3 that $\mathcal{J}$ is compatible with the algebraic structure of X. By the same reasoning it follows that $\mathcal{J}$ is locally m-convex if $\mathcal{J}_\mu$ is.

A neighborhood base at 0 for $\mathcal{J}$ is given by finite intersections of the form $\bigcap_{i=1}^n V_i$ where each V_i is a $\mathcal{J}_{\mu_i}$-neighborhood of 0, i=1,...,n.

We define the <u>final topology</u> on an algebra X determined <u>by a family</u> (A_μ) of homomorphisms $A_\mu : X_\mu \to X$, where each X_μ is a topological algebra, as the supremum of all compatible topologies for X with respect to which each A_μ is continuous. The trivial topology is one such topology for X, so the family of all such topologies is not empty. Similarly we obtain the <u>final LMC</u> topology for X determined by the (A_μ) as the supremum of all LMC topologies for X with respect to which each A_μ is continuous. This topology has a neighborhood base at 0 given by the collection of all absorbent absolutely m-convex subsets V of X for which $A_\mu^{-1}(V)$ is a neighborhood of 0 in X_μ for each index μ. It immediately follows that any homomorphism A taking X equipped with the final LMC topology determined by (A_μ) into an LMC algebra Y is continuous iff $A \cdot A_\mu$ is continuous for each index μ.

As the final LMC topology discussed above is clearly locally convex, it is coarser than the final locally convex topology for X determined by the A_μ, the supremum of all locally convex topologies making each A_μ continuous. These two topologies do not always coincide (see Exercise 4.7). In the discussion to follow we present two instances in which the final locally convex and final LMC topologies coincide.

Example 4.4-2. If X is a normed algebra and (X_μ) is a collection of ideals of X such that $X = \cup X_\mu$ then the final locally convex and final LMC topologies determined by the injection maps $i_\mu : X_\mu \to X$ coincide.

Proof. It is only necessary to show that a typical neighborhood of 0 in the final locally convex topology contains a neighborhood of 0 in the final LMC topology. To this end let V be an absorbent absolutely convex subset of X such that $V \cap X_\mu = i_\mu^{-1}(V)$ is a neighborhood of 0 in X_μ for each μ. Then for each μ there is a positive number $\varepsilon_\mu \le 1$ such that $W_\mu = \{x \varepsilon X_\mu \mid \|x\| < \varepsilon_\mu\} \subset V \cap X_\mu$. Setting $W = \cup W_\mu$ it follows that W is absorbent. To see that W is multiplicative, let $x, y \varepsilon W$, so that $x \varepsilon W_\mu$ and $y \varepsilon W_\beta$ for some indices μ and β. X_μ is an ideal so $xy \varepsilon X_\mu$. Since $\|xy\| \le \|x\| \cdot \|y\| < \varepsilon_\mu 1 = \varepsilon_\mu$, $xy \varepsilon W_\mu \subset W$, and W is multiplicative. The balanced convex hull W_{bc} of W is an absorbent absolutely m-convex subset of V. Moreover since $i_\mu^{-1}(W_{bc}) = X_\mu \cap W_{bc}$ contains W_μ, a neighborhood of 0 in X_μ for each μ, W_{bc} is a neighborhood of the origin in the final LMC topology generated by the maps (i_μ) contained in V.

Example 4.4-3. If X is an LMC algebra containing an increasing sequence of ideals X_n, each carrying the subspace topology, such that $X = \cup X_n$, then the final locally convex and LMC topologies generated by the canonical injection maps i_n coincide.

Proof. Let P be a family of multiplicative seminorms generating the LMC topology of X and let V be an absorbing absolutely convex set such that, for every $n, i_n^{-1}(V) = V \cap X_n$ is a neighborhood of 0 in X_n. Then for each $k \in N$ there is a finite subset $P_k \subset P$ and a positive number $\varepsilon_k \leq 1$ such that $V_k = \{x \varepsilon X_k | p(x) < \varepsilon_k, p \varepsilon P_k\} \subset V \cap X_k$. For each $n \varepsilon N$, let $Q_n = \cup_{k=1}^{n} P_k$, $W_n = \{x \varepsilon X_n | p(x) < \min_{1 \leq k \leq n} \varepsilon_k, p \varepsilon Q_n\}$, and $W = \cup W_n$. As in the preceding example, once it has been shown that W is absorbing and multiplicative, it follows that W_{bc} is a neighborhood of 0 in the final LMC topology and $W_{bc} \subset V$. It is clear that W is absorbing; to see that it is multiplicative as well let $x, y \varepsilon W$. Then there are $n, m \varepsilon N$ such that $x \varepsilon W_n$ and $y \varepsilon W_m$. Assuming that $n \geq m$ and noting that $xy \varepsilon X_m$ since X_m is an ideal in X, it follows that for each $p \varepsilon Q_m \subset Q_n$, $p(xy) \leq p(x)p(y) < \min_{1 \leq k \leq n} \varepsilon_k \min_{1 \leq k \leq n} \varepsilon_k \leq \min_{1 \leq k \leq n} \varepsilon_k$. Thus $xy \varepsilon W_m \subset W$ and W is multiplicative.

4.5 The Factor Algebras. It is well-known that any locally convex Hausdorff space X is embedded in a product of Banach spaces. It is easily seen that the analogous statement for LMCH algebras also holds [(4.5-1)] and this fact has a number of ramifications such as Arens' invertibility criterion of the next section [Theorem 4.6-1 (e)]. Generally the effect of (4.5-1) is to produce theorems of the form: If each factor algebra (defined below) of the LMCH algebra X has property (*), then so does X.

If p is a multiplicative seminorm on an algebra X, then the null space or kernel of p, $N_p = p^{-1}(0)$, is an ideal since, for any $x \in N_p$ and $y \in X$, $p(xy) \leq p(x)p(y) = 0$. Let $\dot{p}$ denote the factor norm: $\dot{p}(x+N_p) = \inf p(x+N_p) = p(x)$. In view of the multiplicativity of p and the above relationship it is clear that X/N_p is a normed algebra with respect to $\dot{p}$. This normed algebra is the pre-factor algebra (associated with p); its completion is the <u>factor algebra</u> associated with p. The extension of $\dot{p}$ from X/N_p to the completion will still be denoted by $\dot{p}$.

<u>Definition 4.5-1</u>. When X is an LMC algebra and (p_μ) is a saturated (Definition 4.2-3) family of seminorms generating X's topology, the completions X_μ of the normed algebras X/N_{p_μ} are called a <u>set of factor algebras for X</u>.

<u>(4.5-1) AN LMCH ALGEBRA IS A SUBSPACE OF A PRODUCT OF BANACH ALGEBRAS</u>. Let P be a saturated family of multiplicative seminorms for the LMCH algebra X determining a set $(X_p)_{p \in P}$ of factor algebras for X. Then the map $x \to (x+N_p)_{p \in P}$ embeds X in the topological algebra $\Pi_{p \in P} X_p$.

<u>Proof</u>. It is readily seen that the given LMCH topology of X coincides with the initial topology ((4.1-1)) generated by the linear map $A: X \to \Pi_p X_p$, $x \to (x+N_p)$ so that A is continuous and relatively open (i.e. A maps open subsets of X into open subsets of A(X)). That A is 1-1 follows immediately from the fact that X is Hausdorff.∇

We note that the assumption that the generating family of seminorms be saturated involves no loss of generality because of (4.2-2), i.e. any family of seminorms generating the topology may be extended to a saturated family of seminorms generating the same topology.

We might also add that though we can't speak of the set of factor algebras for an LMC algebra, it doesn't make any real difference for our purpose: one set is as good as another for the applications that follow.

Another feature of the Banach algebras X_μ worth noting is that each of the canonical homomorphisms $\kappa_\mu : X \to X_\mu$, $x \to x + N_{p_\mu}$ is continuous as $\kappa_\mu^{-1}(\{x \in X_\mu | \dot{p}_\mu(x) < 1\}) = V_{p_\mu} = \{x | p_\mu(x) < 1\}$. In fact the original topology on X is the same as the initial topology for X induced by the family (κ_μ).

Some examples of factor algebras follow. In the first, there is a natural set of factor algebras, each of which is complete to begin with. Using some of the information about the factor algebras of these two algebras gleaned here some information about the spectrum of elements in these algebras is obtained in the next section.

Example 4.5-1. FACTOR ALGEBRAS OF $C(T,\underset{\sim}{F},c)$ ARE SUPREMUM NORMED ALGEBRAS. Let $C(T,\underset{\sim}{F},c)$ be the LMC algebra of Example 4.3-1: continuous scalar-valued functions on the topological space T with compact-open topology. Only this time assume that T is a completely regular Hausdorff space so that continuous functions on compact subsets may be continuously extended to the whole space*. Letting $\mathcal{G}$ denote the family of compact subsets of T, the topology on $C(T,\underset{\sim}{F},c)$ is that generated by the semi-norms $p_G(x) = \sup|x(G)|$ as G runs through $\mathcal{G}$. We denote the null space of p_G by simply N_G.

* Indeed, if G is compact in T it is closed in the compact Hausdorff space βT. Thus, as βT is normal, any continuous function x on G may be continuously extended to βT.

We observe that (p_G) is a saturated family of seminorms. We now show that each of the normed algebras $C(T,\underset{\sim}{F},c)/N_G$ is isometrically isomorphic to the Banach algebra $C(G,\underset{\sim}{F})$ of continuous complex-valued functions on the compact set G with supremum norm $(= C(G,\underset{\sim}{F},c))$. In particular consider the continuous linear map: $A_G : C(T,\underset{\sim}{F}) \to C(G,\underset{\sim}{F})$, $x \to x|_G$ $(G \in \mathcal{J})$. Certain things about A_G are evident: A_G is a multiplicative, onto map since T is a completely regular Hausdorff space. Hence $C(T,\underset{\sim}{F})/N_G$ is (algebraically) isomorphic to $C(G,\underset{\sim}{F})$ the isomorphism being the map $x+N_G \to x|_G$. Since $\dot{p}_G(x+N_G) = p_G(x) = p_G(x|_G)$, $C(T,\underset{\sim}{F})/N_G$ is seen to be isometrically isomorphic to $C(G,\underset{\sim}{F},c)$.

<u>Example 4.5-2. ANALYTIC FUNCTIONS ON THE UNIT DISC</u>. Let H denote the set of analytic functions on the open unit disc in the complex plane $\underset{\sim}{C}$. With respect to pointwise operations and the compact-open topology, H becomes an LMC algebra with topology generated by the saturated family of multiplicative seminorms $p_n(x) = \sup|x(C_n)|$ where $C_n = \{\mu \in C \,|\, |\mu| \leq 1 - \frac{1}{n}\}$. By the analytic identity theorem, the only function analytic on the open unit disc which vanishes on any of the C_n is the function which is identically 0 on the disc. Thus $p_n^{-1}(0) = \{0\}$ or, equivalently, each p_n is a norm. In this example too we wish to ascertain some facts about a set of factor algebras for H. Unlike the situation of the preceding example, however, the pre-factor algebras are not complete, so the completions must be determined. Noting that H is a first countable Hausdorff space, it follows that H is metrizable. Thus Cauchy <u>sequences</u> are adequate to describe completeness. Moreover, in view of

Weierstrass's theorem on uniformly convergent sequences of analytic functions* it follows that every Cauchy sequence converges and H is a Fre'chet algebra.

Similar to the approach used in the preceding example, we consider the map $A_n: H \to C(C_n, \underset{\sim}{C}, c)$, $x \to x|_{C_n}$ where $C(C_n, \underset{\sim}{C}, c)$ denotes the sup norm algebra** of continuous functions on C_n. A_n is a homomorphism whose kernel is $N_{p_n} = p_n^{-1}(0)$, which is $\{0\}$, so each A_n is an isomorphism. Since $p_n(x) = p_n(x + N_{p_n}) = p_n|_{C(C_n, \underset{\sim}{C}, c)}(x)$, H/N_{p_n} is seen to be isometrically isomorphic to $H|_{C_n} = \{h|_{C_n} \mid h \in H\}$ in $C(C_n, \underset{\sim}{C}, c)$.

---

* A.I.Markushevich, Theory of Functions of a Complex Variable, Vol. 1, Prentice-Hall, Englewood Cliffs, N.J., 1965,pp.327-330. (Weierstrass's theorem on uniformly convergent sequences of analytic functions). If the sequence (x_n) is uniformly convergent on each compact subset of a domain D (i.e. open and connected) and if each x_n is analytic on D, then the function x, determined at each $\lambda \in D$ according to $x(\lambda) = \lim x_n(\lambda)$ is also analytic on D. Moreover each sequence of derivatives $(x_n^{(k)})$ converges uniformly to $x^{(k)}$ on each compact subset of D.

** The compact-open topology on $C(T, \underset{\sim}{C})$ coincides with that generated by the sup norm when T is compact, so the notation "$C(C_n, \underset{\sim}{C}, c)$" is consistent with topologizing $C(C_n, \underset{\sim}{C})$ by the sup norm.

If $H|_{C_n}$ were a complete subset of $C(C_n,\underset{\sim}{C},c)$ it would certainly have to be closed, but this is not the case for in $C(C_n,\underset{\sim}{C})$ there is a function§ $x(\lambda) = \sum_{n=0}^{\infty} a_n\lambda^n$ which is analytic throughout C_n except for at least one point on the boundary. Hence x is certainly not the restriction of a function which is analytic on the entire open unit disc. The function x, however, is clearly the uniform limit on C_n of the polynomials $x_n(\lambda) = \sum_{j=0}^{n} a_j\lambda^j$ and each x_n is a restriction of a function which is analytic on the open unit disc. Thus $H|_{C_n}$ is not closed, hence not complete. Next we determine the completion.

If $x \in H$, then x is expressible as a power series about 0 which converges to x uniformly on any compact subset of the open unit disc. Thus, the completion of $H|_{C_n}$ in $C(C_n,\underset{\sim}{C},c)$ is seen to be the subalgebra $P(C_n)$ of those $x \in C(C_n,\underset{\sim}{C})$ which are uniform limits of polynomials on C_n.

4.6 Complete LMCH Algebras and Projective Limits.

One characteristic ((4.5-1)) of complete LMCH algebras X is that they may always be imbedded in a product of Banach algebras, namely a set of factor algebras for X. Another property of complete LMCH algebras is that they are projective limits of their factor algebras, and that is proved here. First some elementary discussion of projective limits is set forth. Somewhat dual to the notion of projective limit is the notion of

---

§ The function in mind is $x(\lambda) = \sum_{j=1}^{\infty}(1/j^2)(\lambda/(1-1/n))^j$. The result follows from Pringsheim's theorem and the Cauchy-Hadamard formula. cf. Markushevich, Theory of Functions of a Complex Variable, Vol. 1, Prentice Hall, Englewood Cliffs, N.J., 1965, p. 390.

inductive limit which is discussed in Chapter 6 in relation to the notion of LB-algebra.

Definition 4.6-1 PROJECTIVE SYSTEMS AND LIMITS. Let $(X_\alpha)_{\alpha \in \Lambda}$ be a family of sets indexed by the preordered set Λ. If for each pair $(\alpha,\beta) \in \Lambda\times\Lambda$ with $\alpha \leq \beta$ there is a mapping $h_{\alpha\beta}:X_\beta \to X_\alpha$ such that $h_{\alpha\alpha}$ is the identity function for each $\alpha \in \Lambda$ and $h_{\alpha\beta} \circ h_{\beta\gamma} = h_{\alpha\gamma}$ whenever $\alpha \leq \beta \leq \gamma$ then the family of sets together with the maps constitutes a projective system. The subset X of ΠX_α consisting of all (x_α) with the property that $h_{\alpha\beta}(x_\beta) = x_\alpha$ whenever $\alpha \leq \beta$ is called the projective limit of the system. In the event that each X_α is a topological space and each $h_{\alpha\beta}$ for $\alpha \leq \beta$ is continuous the projective system is called a topological projective system. The term topological projective limit is used to denote the associated projective limit X equipped with the product topology. Going one step further if each X_α is a topological algebra and the maps $h_{\alpha\beta}(\alpha\leq\beta)$ are continuous homomorphisms then we obtain a projective system of topological algebras and a projective limit of topological algebras, X. Certainly X is a topological algebra as each $h_{\alpha\beta}$ is an algebra homomorphism and so X is closed with respect to all the algebraic operations defined on ΠX_α.

Part (e) of the following result, first proved by Arens (1952) later reproved by Michael (1952) using projective limits, is of great importance in proving some of the results to follow. The force of it is that it enables us to convert certain questions allied to invertibility (such as spectral questions) in topological algebras into similar questions in Banach algebras. Since the latter theory has been heavily mined, this is quite desirable.

<u>Theorem 4.6-1 A COMPLETE LMCH ALGEBRA IS A PROJECTIVE LIMIT OF ITS FACTOR ALGEBRAS</u>. Let X be a complete LMCH algebra with topology generated by the saturated family of non-trivial semi-norms $P = (p_\alpha)_{\alpha\in\Lambda}$ with associated factor algebras $(X_\alpha)_{\alpha\in\Lambda}$. Then (a) The set Λ is directed by the relation: $\alpha \leq \beta$ iff $p_\alpha \leq p_\beta$.

(b) if $\alpha \leq \beta$ then the continuous extension $h_{\alpha\beta}: X_\beta \to X_\alpha$ of the mapping $x + N_\beta \to x + N_\alpha$, is a nontrivial homomorphism with $\| h_{\alpha\beta} \| \leq 1$.

(c) the collection $(X_\alpha)_{\alpha\in\Lambda}$ together with the mappings $(h_{\alpha\beta})_{\alpha\leq\beta}$ constitutes a projective system.

(d) if Y denotes the projective limit of the system of (c), equipped with pointwise multiplication and the induced product topology, then the mapping $T: X \to Y$, $x \to (x+N_\alpha)$ is an onto topological (algebra) isomorphism.

(e) <u>ARENS INVERTIBILITY CRITERION</u>, the element $x \in X$ is invertible iff $x+N_\alpha$ is invertible in X_α for each $\alpha \in \Lambda$.

<u>Proof</u>. (a) This follows directly from the fact that P is saturated.

(b) Clearly $h_{\alpha\beta}$ is well-defined. For $x \in X$ and $\alpha \leq \beta$, $\dot{p}(x+N_\alpha) = p_\alpha(x) \leq p_\beta(x) = \dot{p}_\beta(x+N_\beta)$. Hence $h_{\alpha\beta}$ restricted to X/N_β has norm less than or equal to 1; thus so does $h_{\alpha\beta}$.

(c) Clearly, if $\alpha \leq \beta \leq \gamma$ then $(h_{\alpha\beta} \circ h_{\beta\gamma})(x+N_\gamma) = h_{\alpha\beta}(x+N_\beta) = x+N_\alpha = h_{\alpha\gamma}(x+N_\gamma)$ and $h_{\alpha\alpha}(x+N_\alpha) = x+N_\alpha$. Since the maps $(h_{\alpha\beta})$ are continuous it follows that the collection is a topological projective system.

(d) First we note that by virtue of the fact that the mappings $h_{\alpha\beta}$ are homomorphisms, pointwise multiplication is a binary composition on Y. It is clear that T is a homomorphism. Suppose that $x \neq 0$; since X is Hausdorff, $p_\alpha(x)\neq 0$ for some $\alpha \in \Lambda$ and T is seen to be 1-1. Finally we show that T is onto. To this end let $y = (y_\alpha) \in Y$. Since each X_α is the completion of the corresponding normed linear space X/N_α there exists sequences $(x_{n\alpha}) \subset X$ such that for each α, $\dot{p}_\alpha(x_{n\alpha}+N_\alpha-y_\alpha) < \frac{1}{n}$ for $n \geq 1$. We prove that $(x_{n\alpha})$ is a Cauchy net in the complete space X and that its limit is the pre-image of y under T. The set of pairs (n,α) is clearly directed by the ordering: $(n,\alpha) \geq (m,\beta)$ iff $\alpha \geq \beta$ and $n \geq m$. To see that $(x_{n\alpha})$ is Cauchy it is enough to show that $p_\gamma(x_{n\alpha}-x_{m\beta})\to 0$ for each fixed γ. Indeed if $\alpha,\beta \geq \gamma$ and $\frac{1}{n}$, $\frac{1}{m} < \varepsilon/2$ then $p_\gamma(x_{n\alpha}-x_{m\beta}) = \dot{p}_\gamma((x_{n\alpha}+N_\gamma-y_\gamma) - (x_{m\beta}+N_\gamma-y_\gamma)) \leq \dot{p}_\gamma(x_{n\alpha}+N_\gamma-y_\gamma) + \dot{p}_\gamma(x_{m\beta}+N_\gamma-y_\gamma) \leq \dot{p}_\alpha(x_{n\alpha}+N_\alpha-y_\alpha) + \dot{p}_\beta(x_{m\beta}+N_\beta-y_\beta) < \frac{1}{n} + \frac{1}{m} < \varepsilon$. Let $x = \lim x_{n\alpha}$; it only remains to show that $x+N_\alpha = y_\alpha$ for each α. Fix α and suppose that $\beta \geq \alpha$. Then

$\dot{p}_\alpha(x+N_\alpha-y_\alpha) = \dot{p}_\alpha((x-x_{n\beta}+N_\alpha) + (x_{n\beta}+N_\alpha-y_\alpha)) \leq \dot{p}_\alpha(x-x_{n\beta}+N_\alpha) + \dot{p}_\alpha(x_{n\beta}+N_\alpha-y_\alpha)$. Since $z\to z+N_\alpha$ is a continuous mapping for each α, we have that $x_{n\beta}+N_\alpha \xrightarrow{(n,\beta)} x+N_\alpha$. Thus there exists $\delta \in$ and an integer N such that $\dot{p}_\alpha(x_{n\beta}-x+N_\alpha) < \varepsilon/2$ whenever $n \geq N$ and $\beta \geq \delta$. Furthermore, since $h_{\alpha\beta} \leq 1$ it follows that $\dot{p}_\alpha(x_{n\beta}+N_\alpha-y_\alpha) \leq \dot{p}_\beta(x_{n\beta}+N_\beta-y_\beta) < \frac{1}{n}$. Hence, if we take $\beta \geq \alpha,\delta$ and $n \geq N$, $2/\varepsilon$ we obtain $\dot{p}_\alpha(x+N_\alpha-y_\alpha) < \varepsilon$. Since ε is arbitrary, $\dot{p}_\alpha(x+N_\alpha-y_\alpha) = 0$ and $x+N_\alpha = y_\alpha$.

(e) Due to the fact that $x\to x+N$ is a nontrivial homomorphism, the invertibility of x implies that of $x+N_\alpha$ for each α. Conversely, if each $x+N_\alpha$ is invertible in X_α, then, as T is an isomorphism, we need only prove that $(x+N_\alpha)$ is invertible in Y. Moreover, since $((x+N_\alpha)^{-1})$ is the inverse of $(x+N_\alpha)$ in ΠX_α, we are only required to show that $((x+N_\alpha)^{-1}) \in Y$. To this end, recall that, for $\alpha \leq \beta$, $h_{\alpha\beta}$ maps the identity of X_β onto the identity of X_α. Thus $h_{\alpha\beta}[(x+N_\beta)^{-1}] = [h_{\alpha\beta}(x+N_\beta)]^{-1} = (x+N_\alpha)^{-1}$.∇

4.7 The Spectrum.

In Banach algebras, the notion of the spectrum of an element plays an important role. The fact that no element in a complex commutative Banach algebra with identity has an empty spectrum, in particular, is the cornerstone of the Gelfand theory of commutative complex Banach algebras. To define the spectrum one doesn't even need a topological algebra; our nomenclature on the subject is set forth in Definition 4.7-1 below.

Definition 4.7-1 SPECTRAL NOTIONS. Let X be an algebra. The element $x \in X$ is a unit (is regular) if x is invertible in X, i.e. there exists an element $y \in X$ such that $xy=e$. (When such a y exists it is unique and is denoted as x^{-1}.) Otherwise x is singular. The scalar μ is a regular point of $x \in X$ if $x-\mu e$ is regular. Otherwise μ is a singular point of x. The set $\sigma(x)$ of all singular points of x is the spectrum of x. Its complement-the set of all regular points of x-is denoted by $\rho(x)$. The extended real number $r_\sigma(x) = \sup_{\alpha\in\sigma(x)}|\alpha|$ is called the spectral radius of x and $U(\sigma)$ denotes the set of $x \in X$ with $r_\sigma(x) \leq 1$.

As has been remarked, when we say "Y is a subalgebra of X" this includes the requirement that the identity of X belongs to Y. In this context, it is necessary to distinguish between $\sigma_X(x)$, those μ such that $x-\mu e$ is singular in X, and $\sigma_Y(x)$, those μ for which $x-\mu e$ is singular in Y (for x's in Y, of course). Generally as the population of the algebra becomes larger, the opportunity for $x-\mu e$ to have an inverse increases so that:

(4.7-1) if $x \in Y$ where Y is a subalgebra of X, then $\sigma_X(x) \subset \sigma_Y(x)$.

We compute the spectrum of an element of $C(T,\underset{\sim}{F},c)$.

Example 4.7-1 SPECTRA IN $C(T,\underset{\sim}{F},c)$. Consider the LMCH algebra $C(T,\underset{\sim}{F},c)$ of continuous $\underset{\sim}{F}$-valued functions on the completely regular Hausdorff space T with compact-open topology (Examples 4.3-1 and 4.5-1). An element $y \varepsilon C(T,\underset{\sim}{F},c)$ is invertible iff $y(t) \neq 0$ for each $t \varepsilon T$. Thus $\mu \varepsilon \sigma(x)$ iff $y(t) = \mu$ for some $t \varepsilon T$ and $\sigma(y) = y(T)$.

Arens' invertibility criterion, Theorem 4.6-1(e), enables us to characterize spectra of elements in complete LMCH algebras in terms of spectra of elements in certain Banach algebras-a situation of some desirability because of the information already available about spectra in Banach algebras.

(4.7-2) A CHARACTERIZATION OF THE SPECTRUM. Let $P = (p_s)_{s \in S}$ be a saturated family of multiplicative seminorms for a complete LMCH algebra X. For each $s \in S$, let $N_s = p_s^{-1}(0)$. In the factor algebra X_s (Definition 4.5-1) let $\sigma_s(x)$ denote the spectrum of $x+N_s$. Then $\sigma(x) = \bigcup_{s \in S} \sigma_s(x)$.

Proof. The result follows immediately from Theorem 4.6-1(e).∇

We use Arens' invertibility criterion (i.e. Theorem 4.6-1 (e)) in this form to characterize spectra of elements in the algebra described in Section 4.5.

Example 4.7-2 SPECTRA IN H. In Example 4.5-2 we showed that a set of factor algebras for the Frechet algebra H of functions analytic on the open unit disc was given by the subalgebras $P(C_n)$--uniform limits of polynomials on C_n--of sup norm algebras of continuous functions $C(C_n, \underset{\sim}{C}, c)$, $n=1,2,\ldots,$ where C_n denotes the closed disc of radius $1-1/n$ about the origin.

Identify $x+N_n$ with its isometric isomorphic image $x|_{C_n}$ in $P(C_n)$. In the case at hand it is easy to compute the spectrum $\sigma(x|_{C_n})$ of $x|_C \in P(C_n)$ in $C(C_n, \underset{\sim}{C}, c)$: It is just $x(C_n)$. Thus $x(C_n) \subset \sigma(x|_{C_n})$. We contend that equality holds. To see this, suppose $\mu \notin x(C_n)$. Since $x-\mu$ is never 0 on C_n and C_n is compact, $|x-\mu|$ must have a positive lower bound, a say, on C_n. Thus $1/(x-\mu)$ is an analytic function on C_n, but the question is: Can it be expressed as a uniform limit of polynomials on C_n?

Since $x|_{C_n} \in P(C_n)$ there is a sequence (f_n) of Polynomials which converges to $x-\mu$ uniformly on C_n. Clearly we may assume that $\inf_m |f_m(C_n)| \geq b$ for some $b > 0$. The set of zeros of each f_m, therefore, has a positive distance from C_n, and as a result for each $m \in \underset{\sim}{N}$, $1/f_m$ is expressible as a power series whose radius of convergence is $> 1-1/n$. Moreover since $(1/f_m)$ converges to $1/(x-\mu)$ uniformly on C_n, it follows that $1/(x-\mu)$ is expressible as a uniform limit of polynomials on C_n. Hence if $\mu \notin x(C_n)$, then μ is a regular point of $x|_{C_n}$ in $P(C_n)$, i.e. $\mu \notin \sigma(x|_{C_n})$ which establishes the equality.

Now for any $x \in H$, $\sigma(x) = \cup x(C_n) = x(S_1(0))$. Thus, as was the case in the preceeding example, the spectrum of an element is just the set of values it assumes.

The results of the preceeding two examples make it clear that the spectrum of an element in a complete LMCH algebra need be neither closed nor bounded. In a Banach algebra, by contrast, $\sigma(x)$ may be expressed as the continuous image of a compact set and so is closed and bounded (cf. Section 4.8); it is always contained in the disc of radius $\|x\|$ about 0. Since the algebra H is a Frechet algebra, metrizability is seen not to be the critical ingredient.

4.8 Q-algebras and algebras with continuous inverse.

As is well known maximal ideals in Banach algebras are closed sets. This feature is unfortunately lost in the general case. It is recovered, however, in the special type of algebra we define below although we defer proving this until (4.10-1).

Definition 4.8-1 Q-ALGEBRAS. A topological algebra in which the set of units is open is called a Q-algebra.

Any Banach algebra is a Q-algebra (4.8-2); so is the space $\mathcal{D}$ of infinitely differentiable functions on $[a,b]$ as is proved after (4.8-1) below. An example of a topological algebra which is not a Q-algebra is given in Example 4.8-1 now.

Example 4.8-1 $C(T,\underset{\sim}{F},c)$ IS NOT GENERALLY A Q-ALGEBRA. Let $C(T,\underset{\sim}{F},c)$ be the topological algebra of continuous scalar-valued functions on the non-compact completely regular Hausdorff space T of Example 4.5-1, with compact-open topology. To show that $C(T,\underset{\sim}{F},c)$ is not a Q-algebra, we show that each neighborhood of the identity e contains a non unit. To this end consider a typical basic neighborhood of the identity $e+W(G,\varepsilon) = \{x \in C(T,\underset{\sim}{F},c) \mid \sup_{t \in G}|x(t)-1| < \varepsilon\}$, where G is a proper compact subset of T and $\varepsilon > 0$. By the Tietze extension theorem there is a continuous function y on T such that $y(G) = \{1\}$ and, for some fixed $t \notin G$, $y(t) = 0$. The non-unit y then is clearly a member of $e+W(G,\varepsilon)$.

<u>(4.8-1) Q-ALGEBRA IFF INT $Q \neq \emptyset$</u>. The topological algebra X is a Q-algebra iff its set Q of units has nonempty interior.

<u>Proof</u>. Clearly only the sufficiency of the condition need be proved. To this end, let $x \in \text{int } Q$ and let y be any unit. Consider the map $w \to yx^{-1}w$. As remarked in Section 4.1 such a map must be a homeomorphism which maps Q onto Q. Hence for $V \in \mathcal{V}(x)$ such that $V \subset Q$, $yx^{-1}V$ is a neighborhood of y which lies in Q.∇

As an application of (4.8-1) we now show that the space $\mathcal{D}$ of infinitely differentiable functions on $[a,b]$ with the topology generated by the seminorms $p_n(x) = \sup_{t \in [a,b]} |x^{(n)}(t)|$ $(n=0,1,\ldots)$ (first discussed in Example 4.3-2) is a Q-algebra. By (4.8-1) we need only show that the neighborhood $e+V_{p_o} \subset Q$ and to this end consider any $x \in e + V_{p_o}$. Since $e(t) = 1$ and $|x(t)-1| < 1$ for each $t \in [a,b]$, it follows that the function $1/x(t)$ is defined and infinitely differentiable on $[a,b]$ thereby implying that x is invertible.

<u>(4.8-2) BANACH ALGEBRAS ARE Q-ALGEBRAS</u>. If X is a Banach algebra then it is a Q-algebra. In particular if $\|e-x\| < 1$ then x is a unit.

Proof. Let X be a Banach algebra with set of units Q. To show that Q is open it suffices by (4.8-1) to show that $S_1(e) \subset Q$. To accomplish this we note that $\|e-x\| < 1$ implies that the sequence of partial sums of $\sum_{n=0}^{\infty} \|e-x\|^n$ is Cauchy. Hence the sequence of partial sums of $\sum_{n=0}^{\infty} (e-x)^n$ is Cauchy in X and the series therefore converges to y say. To see that $xy=e$, write x as $e-(e-x)$ and consider $(e-(e-x))(e+(e-x)+\ldots+(e-x)^n) = e-(e-x)^{n+1}$ from which the desired result follows.∇

As was mentioned in the previous section a pleasing feature of Banach algebras--the fact that all elements have compact spectra--is lost in the general case. Our next two results combine to recover this property in Q-algebras.

(4.8-3) IN Q-ALGEBRAS ALL ELEMENTS HAVE BOUNDED SPECTRA. For any topological algebra X, (a) X is a Q-algebra iff $U(\sigma)$ (the set of all x with $r_\sigma(x) \leq 1$) has non-empty interior;

(b) If X is a Q-algebra then the spectrum of each element is bounded.

Proof. (a) First we assume that x is an interior point of $U(\sigma)$, or equivalently, that there is a neighborhood $W \in V(0)$ such that $x+W \subset U(\sigma)$. Now $r_\sigma(y) \leq 1/2$ for each $y \in (1/2)x + (1/2)W$ so for each such y, $e+y$ is invertible. Thus $e + (1/2)x (1/2)W \subset Q$ and it follows, by (4.8-1), that Q is open.

Conversely if Q is open, choose a balanced neighborhood W of 0 such that $e + W \subset Q$. Then for each $x \in W$ and $\lambda \in \underset{\sim}{F}$ with $|\lambda| \geq 1$, it is clear that $\lambda e + x = \lambda(e + \lambda^{-1}x) \in \lambda(e+W) \subset Q$ so that $-\lambda \notin \sigma(x)$. Thus $r_\sigma(x) \leq 1$, $W \subset U(\sigma)$, and 0 is an interior point of $U(\sigma)$.

(b) Since there is a neighborhood $W \in V(0)$ such that $W \subset U(\sigma)$, it follows that for each $x \in X$ there is a $\lambda \in F$, with sufficiently small absolute value, such that $\lambda x \in W \subset U(\sigma)$. Thus since $r_\sigma(\lambda x) = |\lambda| r_\sigma(x)$ and $r_\sigma(\lambda x) \leq 1$ then $r_\sigma(x) < \infty$.∇

<u>(4.8-4) COMPACTNESS OF SPECTRA IN Q-ALGEBRAS</u>. If X is a Q-algebra then for each $x \in X$ (a) $\rho(x)$ is open, and (b) $\sigma(x)$ is compact.

<u>Proof</u>. By (4.8-3(b)) each $x \in X$ has bounded spectrum so we need only prove (a). To do this, let $\lambda \in \rho(x)$; then $\lambda e-x \in Q$ and, by the openness of Q, there is a neighborhood $W \in V(0)$ such that $\lambda e-x+W \subset Q$. Clearly for any $x \in X$ the map $\underset{\sim}{F} \to \underset{\sim}{F}e \to X$, $\alpha \to \alpha e \to \alpha e-x$ is continuous and it follows that there exists an $\varepsilon > 0$ such that $S_\varepsilon(\lambda)e-x \subset \lambda e-x + W \subset Q$. Thus $S_\varepsilon(\lambda) \subset \rho(x)$.∇

Since Banach algebras are Q-algebras [(4.8-2)], each element in a Banach algebra X has compact spectrum. Moreover, since $\| e-(e-x/\mu)\| = \|x\|/|\mu| < 1$ whenever $|\mu| > \|x\|$, it also follows by (4.8-2) that each such $\mu \in \rho(x)$. Thus $\sigma(x) \subset \{\lambda \in F \,|\, |\lambda| \leq \|x\|\}$.

Yet another property of Banach algebras--continuity of the map $x \to x^{-1}$ -- is lost in the general case. In fact it is not generally the case that inversion is continuous in Q-algebras (see (Exercise 4.7(c)) although LMC algebras do possess the property (we prove this in (4.8-6)). As mentioned in Example 4.8-1, $C(R,\underset{\sim}{F},c)$ is not a Q-algebra but $x \to x^{-1}$ is continuous anyway by (4.8-6).

<u>Definition 4.8-2 CONTINUOUS INVERSE</u>. A topological algebra X

in which the map $x \to x^{-1}$ is continuous at e is an <u>algebra with continuous inverse</u>. We also say that X <u>has continuous inverse</u>.

<u>(4.8-5) THE RESOLVENT MAP IS ANALYTIC</u>. In an algebra X with continuous inverse: (a) the map $x \to x^{-1}$ is continuous everywhere on Q, the set of units of X; (b) if X is a complex algebra then for any $x \in X$, if $\rho(x)$ is an open set (e.g. if X is a Q-algebra) the <u>resolvent map</u> $r_x : \rho(x) \to X$, $\lambda \to (x-\lambda e)^{-1}$ is analytic on $\rho(x)$*.

<u>Proof</u>. (a) Let $\mathcal{B}$ be a filterbase in Q convergent to $x \in Q$. We show that $\mathcal{B}^{-1} = \{B^{-1} | B \in \mathcal{B}\}$ converges to x^{-1}. As has been observed previously, if y is a unit the map $x \to xy$ is a homeomorphism of X onto X, for any topological algebra X. Thus, for any $x \in Q, y \to x^{-1}y$ is a homeomorphism. If the filterbase $\mathcal{B}$ in Q converges to x, then $x^{-1}\mathcal{B} \to e$. By the assumed continuity of the map $w \to w^{-1}$ at e, it follows that $x\mathcal{B}^{-1} \to e$, from which it readily follows that $\mathcal{B}^{-1} \to x^{-1}$.

(b) Let $r_x(\lambda) = (x-\lambda e)^{-1}$. The key observation in proving the analyticity of the resolvent map $\lambda \to r_x(\lambda)$ is that

(*) $r_x(\lambda) - r_x(\mu) = (\lambda-\mu) r_x(\lambda) r_x(\mu)$ $(\lambda, \mu \in \rho(x))$. To see this consider $r_x(\lambda)^{-1} r_x(\mu) = (x-\lambda e) r_x(\mu) = (x-\mu e+\mu e-\lambda e) r_x(\mu)$ from which (*) follows. By part (a), the resolvent map $\rho(x) \to Q \to Q$, $\lambda \to x-\lambda e \to (x-\lambda e)^{-1}$ is continuous. Thus the analyticity of $r_x(\lambda)$ on the open set $\rho(x)$ now follows from (*).∇

---

*The set $\rho(x)$ is open in any Q-algebra ((4.8-4(a))). We note, however, that it need not be connected. For example, consider the function $x(t)=t$ in $C(T,\underset{\sim}{C},c)$ where T is a closed annulus in $\underset{\sim}{C}$. Clearly $\sigma(x)=T$ and $\rho(x)$, being the complement of T, is not connected.

At the outset of the discussion on continuity of inversion we mentioned that not all topological algebras have continuous inverse. It is the case, however, that a very large class of topological algebras containing the normed algebras, namely the LMC algebras, possess the property.

(4.8-6) LMC ALGEBRAS ARE ALGEBRAS WITH CONTINUOUS INVERSE. If X is (a) a normed algebra, or, more generally, (b) an LMC algebra, then X has continuous inverse.

Proof. (a) To see that the map $x \to x^{-1}$ is continuous at e, suppose (x_n) is a sequence of units convergent to e. To show that $x_n^{-1} \to e$, we need only note that $\|x_n^{-1} - e\| \leq \|x_n^{-1}\| \, \|e - x_n\|$, the boundedness of $(\|x_n^{-1}\|)$ being apparent from the relation [in the completion of X] $x_n^{-1} = \sum_{k=0}^{\infty} (e - x_n)^k$ for $\|x_n - e\| \leq r < 1$ (see (4.8-2)) and the convergence of (x_n) to e.

(b) Let $\mathcal{B}$ be a filterbase of units convergent to e. We prove that $\mathcal{B}^{-1} = \{B^{-1} \mid B \in \mathcal{B}\}$ converges to e.

Let P be a family of multiplicative seminorms generating the topology on X and for each $p \in P$, let $N_p = p^{-1}(0)$. Since $\mathcal{B} \to e$ and the map $x \to x + N_p$ from X into the factor algebra X_p is continuous, it follows that $\mathcal{B} + N_p \to e + N_p$ in X_p. But X_p is a Banach algebra, so inversion is continuous and it follows that $\mathcal{B}^{-1} + N_p = (\mathcal{B} + N_p)^{-1} \to e + N_p$. In other words $\dot{p}(\mathcal{B}^{-1} - e + N_p) = p(\mathcal{B}^{-1} - e) \to 0$. Since p is arbitrary, it follows that $\mathcal{B}^{-1} \to e$. ∇

Since the algebra $C(T, \underset{\sim}{F}, c)$, T non-compact, completely regular, and Hausdorff, of Example 4.8-1 is an LMC algebra, it follows that inversion is continuous in $C(T, \underset{\sim}{F}, c)$; $C(T, \underset{\sim}{F}, c)$ is not a Q-algebra however, as is pointed out there.

4.9 Topological Division Algebras and the Gelfand-Mazur Theorem

A division algebra is a not necessarily commutative algebra in which each nonzero element has an inverse. The Mazur-Gelfand theorem,Theorem 4.9-1, for algebras with continuous inverse shows that there is essentially only one complex locally convex Hausdorff (LCH) division algebra with continuous inverse. Both parts of the theorem remain true when our general assumption of commutativity is lifted and one needs only to carefully preserve the order in which things are written down to see that this is so. Something which is essential to the validity of the theorem, however is the fact that the algebras be complex. The real LCH topological division algebra $\mathbb{R}$ in its usual topology is certainly not topologically isomorphic to $\mathbb{C}$, for example. The quaternions too constitute a real topological division algebra--in fact a Banach algebra, hence a topological algebra with continuous inverse by (4.8-6(a)--which is distinct from $\mathbb{C}$.

It was first established by Frobenius that any finite-dimentional division algebra over the real numbers is (topologically) isomorphic to $\mathbb{R}$,$\mathbb{C}$, or the quaternions. Subsequent generalizations of this result were made by S. Mazur (1938), G. Silov (1940), and I. Gelfand, for normed algebras. In addition to establishing the result for complex LCH algebras (Theorem 4.9-1(b)), Arens (1947a) also proved that essentially the only real LCH topological division algebras with continuous inverse are $\mathbb{R}$,$\mathbb{C}$, and the quaternions, part of which is established in Theorem 4.9-2.

The classical Liouville theorem states that a bounded entire function must be constant. With suitable analogs for "bounded" and "entire" we prove a version of Liouville's theorem for entire vector-valued functions $x:\underset{\sim}{C}\to X$. This version ((4.9-1)) is needed to prove our main result, Theorem 4.9-1.

If X is a topological vector space and G a subset of $\underset{\sim}{C}$ then $x:G\to X$ is bounded if x(G) is a bounded subset of X, 'bounded' subset of X in the sense that it is absorbed by any neighborhood of 0.

Definition 4.9-1 ANALYTICITY. Let G be an open subset of the complex plane and X be a topological vector space. The map $x:G\to X$ is analytic in G if the limit $\lim_{\mu\to\mu_o} \frac{x(\mu)-x(\mu_o)}{\mu-\mu_o} = x'(\mu_o)$ exists at each $\mu_o \varepsilon G$.

In the version of Liouville's theorem to follow, it is important that the vector space possess sufficiently many continuous linear functionals. In particular we want the topological vector space X to have enough continuous linear functionals so that the information that each continuous linear functional vanishes on a certain vector is enough to guarantee that the vector is 0. Whenever a subspace S of linear functionals on a vector space has the property that $f(x)=0$ for all $f\varepsilon S$ implies x=0, then the subspace is called total. In particular the set X' of all continuous linear functionals on a locally convex Hausdorff space is always total, as shown by the Hahn-Banach theorem.

(4.9-1) LIOUVILLE'S THEOREM. Let X be a topological vector space and suppose that its dual X' is total. If $x:\underset{\sim}{C}\to X$ is entire and bounded, then x must be constant.

Proof. Since the continuous linear image of a bounded set is bounded, if x is bounded, then so is fx for any $f \in X'$. Thus the standard Liouville theorem applies to each of the entire functions fx and we conclude that fx is constant for each $f \in X'$. Hence for any $\mu, \lambda \in \mathbb{C}$, $f(x(\mu)) = f(x(\lambda))$, so $f(x(\mu)-x(\lambda))=0$. Holding μ and λ fixed and noting that the last equality holds for every f in a total set of linear functionals, we conclude that $x(\mu) = x(\lambda)$. Since μ and λ are arbitrary, the constancy of x is proved.∇

<u>Theorem 4.9-1 COMPLEX LCH DIVISION ALGEBRAS WITH CONTINUOUS INVERSE</u>. (Gelfand-Mazur) Let X be a complex LCH algebra with continuous inverse. Then (a) for any $x \in X$, $\sigma(x) \neq \emptyset$ and (b) if each non-zero element in X has an inverse, X is topologically isomorphic to $\mathbb{C}$.

<u>Proof</u>. (a) For $x \in X$, suppose $\sigma(x) = \emptyset$. Then for each $\lambda \in \mathbb{C}$, letting $r_x(\lambda) = (x-\lambda e)^{-1}$, the map $\lambda \to r_x(\lambda)$ is seen to be entire by part (b) of (4.8-5). We show that $r_x(\lambda)$ is bounded. To this end, consider the filterbase formed by the sets $B_n = \{\lambda \in \mathbb{C} \mid |\lambda| \geq n\}$, n=1,2,... . For any seminorm $p \in P$, where P generates X's topology, $\lambda \in B_n$, and any $x \in X$, $p(\lambda^{-1}x) = |\lambda|^{-1}p(x) \leq n^{-1}p(x)$. Thus, letting $\mathcal{B}^{-1} = (B_n^{-1})$, it follows that (since p is arbitrary) $\mathcal{B}^{-1}x \to 0$. Hence $e-\mathcal{B}^{-1}x \to e$. Since X has continuous inverse and since $e-\lambda^{-1}x$ must be regular for any $\lambda \neq 0$, it follows that $(e-\mathcal{B}^{-1}x)^{-1} \to e^{-1}=e$ which implies $p((e-\mathcal{B}^{-1}x)^{-1}) \to p(e)$ for any $p \in P$.

Thus, for $\varepsilon > 0$, n sufficiently large and $|\lambda| \geq n$,

(*) $p((x-\lambda e)^{-1})=|\lambda|^{-1}p((e-x/\lambda)^{-1}) \leq |\lambda|^{-1}p(e)+\varepsilon \leq (1/n)p(e)+\varepsilon$,

the upshot of it all being that $p(r_x(\lambda))$ is bounded for $|\lambda|$ sufficiently large. Since r_x is a continuous function,

$p(r_x(\lambda))$ is certainly bounded for $|\lambda| \leq n$. Now, since p is arbitrary, it follows that r_x is bounded, i.e. $r_x(\underset{\sim}{C})$ is a bounded subset of the LCHS X. That is r_x is bounded entire function, so it must be constant by Liouville's theorem (4.9-1).

Certainly $p \cdot r_x$ is a bounded function too for any $p \in P$ and (*) implies that $p(r_x(\lambda))=0$ for each $\lambda \in \underset{\sim}{C}$. Since p is arbitrary and X is Hausdorff, it follows that $r_x(\lambda)=0$ for each λ which is absurd for how can 0 be the inverse of anything? Thus the assumption $\sigma(x)=\emptyset$ has led to a contradiction and the desired result, (a), follows.

(b) According to the result just proved in (a), given any $x \in X$, $x-\lambda e$ is singular for some λ. But if each non-zero element is regular, it must be that $x-\lambda e=0$ for some λ, that is, $x=\lambda e$ for some λ. Thus $X=\underset{\sim}{C}e$ and the map $x=\lambda e \to \lambda$ is clearly an algebra isomorphism from X onto $\underset{\sim}{C}$. As for its being a homeomorphism as well, suppose P is a saturated family of seminorms which generate X's topology. Since P is saturated, basic neighborhoods of 0 are of the form a V_p where $a > 0$ and $p \in P$. Now $x = \lambda e \in aV_p$ iff $p(\lambda e)=|\lambda|p(e) < a$. That is, for nontrivial seminorms p, $aV_p=\{\lambda e \mid |\lambda| < a/p(e)\}$ and the fact that the above map is a homeomorphism is now clear.∇

As a consequence of Theorem 4.9-1 and the fact that any LMC algebra has continuous inverse [(4.8-6)], the only complex LMCH division algebra is $\underset{\sim}{C}$. We can go slightly further and say that the only complex LMC division algebra is $\underset{\sim}{C}$. To see this, it suffices to show that $\sigma(x)\neq\emptyset$ for any x in a complex LMC division algebra X (with nontrivial topology). As the topology on X is not the trivial topology there exists a proper balanced m-convex neighborhood of 0 in X. Consequently there exists a nontrivial multiplicative seminorm p on X, and an associated factor algebra (See Sec. 4.5) X_p. Since X_p is a Banach algebra, $\sigma(x+N_p) \neq 0$ for any $x \in X$ by the previous theorem. Since the map $x \to x+N_p$ is a nontrivial homomorphism from X into X_p, then $\mu \in \sigma(x+N_p)$ implies $\mu \in \sigma(x)$.

For a time it seemed that there were probably no complex topological division algebras (cf. Kaplansky, 1948, p. 811) other than $\underset{\sim}{C}$. Williamson (1954) however showed that this was not the case by exhibiting a topology for the algebra $\underset{\sim}{C}(t)$, the quotient field of the polynomial algebra $\underset{\sim}{C}[t]$ of polynomials in t with complex coefficients, which is compatible with the algebraic operations. We present this construction in our next example.

Example 4.9-1 A COMPLEX TOPOLOGICAL DIVISION ALGEBRA DISTINCT FROM $\underset{\sim}{C}$. Let M(0,1) be the space of all Lebesgue measurable functions on (0,1) that assume finite complex values almost everywhere. The quotient field of the algebra of polynomials in t, $\underset{\sim}{C}(t)$, may be identified with the class of functions of M(0,1) consisting of ratios of polynomials with complex coefficients. Let $\mathcal{B}$ be the filterbase of sets of the form $B(k,\varepsilon) = \{f \in M(0,1) \mid m(I(|f| \geq k)) < \varepsilon\}$ where m is Lebesgue measure, the numbers ε and k are positive, and $I(|f| \geq k) = \{t \in (0,1) \mid |f(t)| \geq k\}$.

First we claim that $\mathcal{B}$ is a neighborhood base at 0 in M(0,1) for a compatible topology for the algebra M(0,1) called the topology of convergence in measure. A filterbase $\mathcal{F}$ on M(0,1) converges to 0 (in the topology of convergence in meas-

measure) if for any $B(k,\varepsilon) \in \mathcal{B}$ there exists $F \in \mathcal{F}$ such that $F \subset B(k,\varepsilon)$, i.e. for each $f \in F, m(I(|f| \geq k) < \varepsilon$.* To show compatibility, we show that the conditions of the Theorem 4.1-1 are satisfied by $\mathcal{B}$. Changing the order of things slightly we consider first the condition that to each $B(k,\varepsilon) \in \mathcal{B}$ there corresponds a neighborhood $B(k_1,\varepsilon_1)$ with the property that $B(k_1,\varepsilon_1)\, B(k_1,\varepsilon_1) \subset B(k,\varepsilon)$. If we choose $k_1=k^{1/2}$ and $\varepsilon_1=\varepsilon/2$ then $|f(t)g(t)| \geq k$ implies either $|f(t)| \geq k_1$ or $|g(t)| \geq k_1$, where f and $g \in M(0,1)$, and it follows that $I(|fg| \geq k) \subset I(|f| \geq k_1) \cup I(|g| \geq k_1)$. Thus if $f, g \in B(k_1,\varepsilon_1)$ both sets of the right side of the foregoing inclusion are of measure less than ε_1 and $fg \in B(k,\varepsilon)$. As for compatibility with the linear structure, we note first that $\mathcal{B}$ is closed with respect to multiplication by positive scalars. The conditions that the sets of $\mathcal{B}$ be balanced and that to each $B(k,\varepsilon)$ there corresponds a neighborhood $B(k_2,\varepsilon_2)$ such that $B(k_2,\varepsilon_2) + B(k_2,\varepsilon_2) \subset B(k,\varepsilon)$ are established with the aid of the relations $I(|\lambda f| \geq k) \subset I(|f| \geq k)$ for $f \in M(0,1)$ and $|\lambda| \leq 1$ and $I(|f+g| \geq k) \subset I(|f| \geq \frac{k}{2}) \cup I(|g| \geq \frac{k}{2})$ for f and $g \in M(0,1)$. To see that the elements of $\mathcal{B}$ are absorbent, let $B(k,\varepsilon) \in \mathcal{B}$ and $f \in M(0,1)$. Then the sets $I(|f| \geq n)$, $n \geq 1$, form a decreasing sequence of measurable sets, each with finite measure, and having an intersection of zero measure. Thus $m(I(|f| \geq n)) \to 0$ and, for λ with large enough absolute value, we have

---

* In the case of a sequence (f_n) from $M(0,1)$, $f_n \to 0$ in the topology of convergence in measure iff $f_n \to 0$, "in measure" in the classical measure-theoretic sense of the phrase.

$|\lambda|k \geq n$ where n is a fixed integer with the property that $m(I(|f| \geq n)) < \varepsilon$. Consequently for all such λ $m(I(|(1/\lambda)f| \geq k)) = m(I(|f| \geq |\lambda|k)) \leq m(I(|f| \geq n))$ and $(1/\lambda)f \in B(k,\varepsilon)$.

Now $\underset{\sim}{C}(t)$ with the subspace topology (also called the topology of convergence in measure) is a topological algebra and we contend that it is Hausdorff. To see this let $r(t) = p(t)/q(t) \in B(k,\varepsilon)$ for each $k > 0$ and $\varepsilon > 0$. Then for any $k > 0$, it follows that $m(I(|r| > k) = 0$. Thus $m(\{t \in (0,1) \mid r(t) \neq 0\}) \leq \sum_{i=1}^{\infty} m(I(|r| > 1/i)) = 0$. Hence $p(t) = 0$ almost everywhere on (0,1) so by the continuity of p(t) on (0,1), p(t) is identically zero.*

The topology of convergence in measure, however is not locally convex. To prove this it is sufficient to prove that inversion is continuous at e (the function identically equal to 1 on (0,1)) for if we assume that $\underset{\sim}{C}(t)$ is locally convex then the absurd conclusion follows, by Theorem 4.9-1(b) that $\underset{\sim}{C}(t)$ is isomorphic to $\underset{\smile}{C}$. To show that $\underset{\sim}{C}(t)$ has continuous inverse let

*On all of M(0,1), the topology of convergence in measure is not a Hausdorff topology, because there are non-zero measurable functions that are zero almost everywhere. To avoid this problem one usually defines M(0,1) to be the class of all equivalence classes of measurable functions on (0,1) generated by the equivalence relation of equality almost everywhere. The equivalence classes containing a rational function contain exactly one rational function and $\underset{\sim}{C}(t)$ may still be identified with a subalgebra of M(0,1). Furthermore, if $f \in B(k,\varepsilon)$ then so does every other function in the equivalence class generated by f.

$B(k,\varepsilon) \in \mathcal{B}$; we shall exhibit $k' > 0$ such that $1/f \in e + B(k,\varepsilon)$ whenever $f \in e + B(k',\varepsilon)$. This will be accomplished if k' satisfies (*) $\{t \in (0,1) \mid f(t) \neq 0\} \cap I(|1/f-e| \geq k) \subset I(|f-e| \geq k')$. Treating k' as an unknown and assuming that $f(t) \neq 0$ while $|f(t) - 1| < k'$ it follows that $|1/f(t) - 1| = |(1/f(t))(1-f(t))| = |1/f(t)||f(t) - 1| < |1/f(t)|k'$. To insure that the above inclusion holds we set $|1/f(t)|k' \leq k$. Since $|1/f(t)|k' \leq k$ iff $k' \leq |f(t)|k$ and $|f(t)| \geq 1-k'$ we have $k' \leq k/(1+k)$. Any such k' satisfies (*).

One might wonder: could there exist locally convex complex division algebras other than $\underset{\sim}{C}$? There can, and this is discussed in Exercise 4.6.

For multiplication to be continuous in $\underset{\sim}{C}(t)$ endowed with a linear topology $\mathcal{J}$, $\mathcal{J}$ can be neither too coarse nor too fine: As is discussed in Exercise 4.7, if $\mathcal{J}$ is locally convex, it cannot be a weak topology--i.e. there can be no linear space Y such that $(\underset{\sim}{C}(t),Y)$ is a dual pair for which $\mathcal{J} = \sigma(C(t),Y)$--and multiplication is discontinuous when $\underset{\sim}{C}(t)$ carries the finest locally convex topology.

In addition to Theorem 4.9-1 being interesting in its own right it plays an important role in the development of "Gelfand theory", roughly the consequences of topologizing the set of maximal ideals of an algebra in a certain way (see Sec. 4.10 and Sec. 4.12)).

---

*Con't: Thus the sets of $\mathcal{B}$ may be considered as being composed of equivalence classes rather than just functions, thereby making M(0,1) a Hausdorff topological algebra in the topology of convergence in measure.

The analog of Theorem 4.9-1 for real algebras is presented next. As usual, consideration is restricted to commutative algebras. If the (real) algebra is non-commutative however, it must be the quaternions, as discussed in Exercise 4.8.

Theorem 4.9-2 REAL LCH FIELDS WITH CONTINUOUS INVERSE. Every real LCH algebra X with continuous inverse in which each nonzero element has an inverse is topologically isomorphic to either $\mathbb{R}$ or $\mathbb{C}$. Furthermore X is topologically isomorphic to $\mathbb{R}$ iff for all $x,y \in X$, $x^2+y^2=0$ implies $x=y=0$; in this case X is called formally real.

Proof. We consider separately the cases when X is formally real and when it is not. When X is formally real, we introduce operations to $X\times X$ (identical in form to the operations one introduces to $\mathbb{R}\times\mathbb{R}$ to form $\mathbb{C}$) which make it a complex LCH algebra with continuous inverse and then apply Theorem 4.9-1 to conclude that $X\times X$ is topologically isomorphic to $\mathbb{C}$; the restriction of this topological isomorphism to $X\times\{0\}$ shows X to be topologically isomorphic to $\mathbb{R}$. If X is not formally real then X is shown to be topologically isomorphic to $\mathbb{C}$.

Suppose that X is formally real and let $Y = X\times X$. For (x,y), $(w,z)\in Y$ and $a+ib \in \mathbb{C}$ we define $(x,y)+(w,z)=(x+w,y+z)$ $(a+ib)(x,y)=(ax-by,ay+bx)$ and $(x,y)(w,z)=(xw-yz,yw+xz)$. With these operations and the product topology Y is easily seen to be a complex LCH algebra. (To verify that complex scalar multiplication is continuous, it is helpful to observe that the map $(x,y)\to(y,-x)$ is continuous.) To see that Y is a field, note first that $(e,0)$ is the multiplicative identity of Y and suppose that $(x,y)\neq(0,0)$. Then $x^2+y^2 \neq 0$, $(x^2+y^2)^{-1}$ exists in

X and $(x,y)[(x,-y)((x^2+y^2)^{-1},0)] = [(x,y)(x,-y)]((x^2+y^2)^{-1},0)$ $= (x^2+y^2,0)((x^2+y^2)^{-1},0) = (e,0)$ so that (x,y) is invertible and Y is seen to be a field.

In order to be able to call upon Theorem 4.9-1 to conclude that Y is topologically isomorphic to $\underset{\sim}{C}$, it only remains to show that Y has continuous inverse. To this end suppose that $((x_\mu,y_\mu))$ is a net convergent to $(e,0)$. Then $x_\mu \to e$ and $y_\mu \to 0$. As $(x_\mu,y_\mu)^{-1} = (x_\mu,-y_\mu)((x_\mu{}^2+y_\mu{}^2)^{-1},0)$ for each μ and X has continuous inverse, it follows that $(x_\mu,y_\mu)^{-1} \to (e,0)$ as well. Thus Y has continuous inverse and is topologically isomorphic to $\underset{\sim}{C}$ as established by the map $a+ib \to (a+ib)(e,0) = (ae,be)$ of Theorem 4.9-1. By restricting the map to X--more precisely to $X\times\{0\}$--it follows that X is topologically isomorphic to $\underset{\sim}{R}$.

What if X is not formally real? i.e. there are nonzero $x,y \in X$ such that $x^2+y^2=0$. If so, let $j=xy^{-1}$ and extend multiplication to multiplication by complex scalars as follows: $(a+ib)z = az + b(jz)$ $(z \in X)$. It is easily verified that X is now a complex LCH algebra. Theorem 4.9-1 may now be applied and it follows that X is topologically isomorphic to $\underset{\sim}{C}$. ∇

Following the proof of Theorem 4.9-1 we remarked that the only complex LMC division algebra with nontrivial topology is $\underset{\sim}{C}$. An analogous statement can be made for real algebras: the only real LMC fields with nontrivial topology are $\underset{\sim}{R}$ and $\underset{\sim}{C}$. Just as in the complex case, the fact that the topology on the real LMC field X is nontrivial guarantees the existence of a proper balanced m-convex neighborhood of 0 in X. Thus there is a nontrivial multiplicative seminorm p on X and an associated factor algebra X_p. Now $X/N_p = \underset{\sim}{F}(e+N_p)$ where $\underset{\sim}{F}=\underset{\sim}{R}$ or $\underset{\sim}{C}$ and the mapping $\mu(e+N_p)\to\mu$ is a topological isomorphism. It follows that the mapping $h:X\to X/N_p\to \underset{\sim}{F}$, $x\to x+N_p=\mu(e+N_p)\to\mu$ is a nontrivial homomorphism. Since all nonzero elements of X have inverses, $h^{-1}(0)=\{0\}$ and h is an isomorphism. That h is continuous is clear; that it is also open follows from the openness of the canonical homomorphism $x\to x+N_p$.

4.10 Maximal ideals and Homomorphisms.

A proper ideal which is not properly contained in any other proper ideal is a maximal ideal. If I is a proper ideal, a straightforward Zorn's lemma argument shows that there exists a maximal ideal containing I. In particular, if x is a singular element then $(x)=xX$ is an ideal containing x called the principal ideal generated by x. Thus any singular element is contained in a maximal ideal. In this section we prove two basic topological results about ideals in certain types of topological algebras and then investigate the connections between maximal ideals and homomorphisms. We also look at some examples.

(4.10-1) IDEALS IN Q-ALGEBRAS. If X is a Q-algebra then (a) the closure of a proper ideal is a proper ideal; (b) maximal ideals are closed.

Proof. Clearly we only need to prove (a). If I is an ideal in X, $x,y \in \text{cl } I$ and $z \in X$, it is easy to show by considering filterbases on I convergent to x and y that $x+y$ and $xz \in \text{cl } I$. The important part is to show that $\text{cl } I\neq X$. Since X is a Q-algebra, however, there exists a $V \in \mathcal{V}(e)$ such that V consists entirely of units. Thus $V\cap I=\emptyset$ and $e \notin \text{cl } I$.∇

In the algebra $C(\underset{\sim}{R},\underset{\sim}{F},c)$ as in Example 4.8-1, the set I of continuous functions x which vanish outside some compact set G_x constitutes a proper ideal which is clearly dense in $C(\underset{\sim}{R},\underset{\sim}{F},c)$. Thus no maximal ideal containing I can be closed.

In the result which follows, we consider (maximal) ideals I of an LMC algebra X such that the quotient topology on X/I is not the trivial topology. If I is a closed ideal (as is the case for maximal ideals in Q-algebras), then the quotient topology on X/I is Hausdorff, so the quotient topology is certainly not trivial in this case.

A weaker sufficient condition for nontriviality of the quotient topology when X is LMC is that I not be dense in X. More generally, if H is a linear subspace of the locally convex space X and $\text{cl } H \neq X$, then the quotient topology on X/H is not trivial and we now outline a proof of this fact. Indeed if H is not dense in X, then there is a convex neighborhood of the origin V and an element $x \in X$ such that $(x+V) \cap H = \emptyset$. If the quotient topology on X/H is trivial, then $(x+H) + V = X = -x-V+H$ (since $(x+V)+H$ is a neighborhood of $x+H$ in X/H). It follows that there exist elements $v,w \in V$ such that $x+v/2+w/2 \in H \cap (x+V)$, and this is a contradiction. As was mentioned after (4.10-1), $C(\underset{\sim}{R},\underset{\sim}{F},c)$ is an algebra containing dense ideals.

<u>(4.10-2) QUOTIENTS OF MAXIMAL IDEALS IN LMC ALGEBRAS</u>. In the event that X is a (commutative) real LMC algebra and M is a maximal ideal in X such that the quotient topology on X/M is not trivial then X/M is topologically isomorphic to either $\underset{\sim}{R}$ or $\underset{\sim}{C}$. In particular if X is a Banach algebra then X/M is topologically isomorphic to $\underset{\sim}{R}$ or $\underset{\sim}{C}$ if X is a real algebra or just $\underset{\sim}{C}$

if X is a complex algebra. Furthermore the quotient topology on X/M is induced by the inf norm: $\|x+M\| = \inf_{m \in M} \|x+m\|$.

Proof. As regards the main assertion all we need do is apply the remarks following Theorem 4.9-1 and Theorem 4.9-2 to the LMC division algebra X/M. If X is a Banach algebra then it is LMC and, being also a Q-algebra [(4.8-2)], the maximal ideal M is closed in X from which it follows that the quotient topology on X/M is Hausdorff, hence nontrivial. It is easy to verify that the quotient topology is induced by the inf norm. ∇

As mentioned before (4.10-2) it suffices for the maximal ideal M to be closed to guarantee that the quotient topology be nontrivial. In a Banach algebra, all maximal ideals are closed but this is not generally true for LMC algebras. There may be non-closed maximal ideals even in Frechet algebras, as shown by the discussion of $C(\underset{\sim}{R},\underset{\sim}{F},c)$ after (4.10-1). This property--existence of non-closed maximal ideals--constitutes, therefore, a major difference between LMC algebras and Banach algebras.

Notation. "Homomorphism" here means "complex or real homomorphism", depending on whether the algebra is real or complex, $\mathcal{M}$ (or $\mathcal{M}(X)$) denotes the maximal ideals of an algebra X, $\mathcal{M}_c$, (or $\mathcal{M}_c(X)$) denotes the closed maximal ideals of a topological algebra X and X^h denotes the nontrivial continuous homomorphisms of the topological algebra X. Note that $X^h \subset X'$, the continuous dual of the topological vector space X.

Our first result connecting the notions of maximal ideal and homomorphism is obvious.

(4.10-3) MAXIMAL IDEALS AND HOMOMORPHISMS. The kernel of any non-trivial homomorphism of any algebra X is a maximal ideal.

Clearly the kernel of any non-trivial continuous homomorphism is a closed maximal ideal in any topological algebra. It follows immediately from (4.10-2) and the remark following (4.10-2) that any closed maximal ideal in an LMC algebra is the kernel of a continuous homomorphism. We state these facts in our next result.

(4.10-4) $\mathcal{M}_c$ IN LMC ALGEBRAS. If X is a complex LMC algebra then a maximal ideal $M \subset X$ is closed iff M is the kernel of some continuous complex-valued homomorphism. If X is a real LMC algebra then M is closed iff M is the kernel of a real-valued homomorphism or a complex-valued homomorphism. Thus for complex LMC algebras there is a 1-1 correspondence between $\mathcal{M}_c$, the closed maximal ideals, and X^h, the continuous homomorphisms, namely that established by pairing M with the homomorphism $x \to x+M$.

Frequently it will be convenient to identify $\mathcal{M}_c$ and X^h when X is a complex LMC algebra. Thus, for example, we speak of $\mathcal{M}_c$ "with $\sigma(X',X)$-topology". An examination of the basic neighborhoods of 0 shows that this is the weakest topology on $\mathcal{M}_c$ with respect to which each of the <u>Gelfand maps</u> $\hat{x}: \mathcal{M}_c \to \underline{C}$, $M \to x+M$, (or $X^h \to \underline{C}$, $f \to f(x)$) continuous; the topology induced by $\sigma(X',X)$ on X^h (or $\mathcal{M}_c$) is called the <u>Gelfand topology</u>.

In Q-algebras all maximal ideals are closed [(4.10-1)], so in complex LMC Q-algebras each non-trivial homomorphism is continuous by (4.10-4). Furthermore, if X is a Banach algebra, real or complex, and f is a non-trivial homomorphism (real-valued if X is real, complex-valued if X is complex) on X, then for any $x \in X$ we have $f(x-f(x)e) = 0$ which implies that $f(x) \in \sigma(x)$. Thus, by the remark following, (4.8-4), $|f(x)| \leq \|x\|$ for every x and it follows that $\|f\| \leq 1$. We summarize these facts in our next result.

(4.10-5) HOMOMORPHISMS OF LMC Q-ALGEBRAS ARE CONTINUOUS. If X is an LMC Q-algebra, all homomorphisms are continuous. Furthermore, each homomorphism on a Banach algebra is not only continuous, it has norm less than or equal to one.

Questions concerning the continuity of homomorphisms of topological algebras into topological algebras are treated in Section 4.13. In particular a result similar to (4.10-5) is established in Theorem 4.13-1 where it is shown that any homomorphism of a complete barreled LMCH Q-algebra into a strongly semisimple fully complete LMCH algebra is continuous.

Unfortunately (4.10-5) does not remain true for an arbitrary LMC algebra. Our next result is of critical importance in Example 4.10-1 where we construct a whole class of LMCH algebras on which discontinuous homomorphisms exist.

(4.10-6) DISTINCT HOMOMORPHISMS ARE LINEARLY INDEPENDENT. Any collection of distinct non-trivial homomorphisms of an algebra X is linearly independent.

Proof. Suppose $\{f_1,\ldots,f_n\}$ is a linearly dependent set of distinct non-trivial homomorphisms of X into $\underset{\sim}{F}$ which is minimal in the sense that the removal of any one homomorphism from the set leaves a linearly independent set. It is clear that $n \neq 1$. Suppose that $\Sigma\alpha_i f_i=0$ where no $\alpha_i=0$. Since $f_n \neq f_1$ there exists $y \in X$ such that $f_n(y) \neq f_1(y)$. Holding y fixed and permitting x to be any vector in X we see that $\Sigma_{i=1}^{n}\alpha_i f_i(xy) = \Sigma_{i=1}^{n}\alpha_i f_i(y)f_i(x) = 0$, and $f_n(y)\Sigma_{i=1}^{n}\alpha_i f_i(x) = \Sigma_{i=1}^{n}\alpha_i f_n(y)f_i(x) = 0$. Subtracting the second equation from the first, we have $\Sigma_{i=1}^{n}\alpha_i(f_i(y)-f_n(y))f_i(x) = 0$ for all $x \in X$. Since $f_1(y)-f_n(y) \neq 0$ the last equation implies that $\{f_1,\ldots,f_{n-1}\}$ is linearly dependent which condradicts the minimality of $\{f_1,\ldots,f_n\}$.∇

Example 4.10-1 DISCONTINUOUS HOMOMORPHISMS. Let X be an algebra and H be a family of non-trivial homomorphisms on X with the property: (*) there exists $f_o \in H$ such that $\bigcap_{f\in H_o} \ker f = \{0\}$, where $H_o=H-\{f_o\}$. In this case f_o will be a discontinuous homomorphism when X carries $\sigma(X,H_o)$.

There are many algebras which satisfy (*). In particular if X is the complex algebra of continuous functions on $[a,b]$, $C([a,b],\underset{\sim}{F},c)$ with sup norm topology, then let H be the collection of homomorphisms determined by the evaluation maps $t^*:C[a,b]\to\underset{\sim}{C}$ $(t\in[a,b])$, $x\to x(t)$ and f_o to be $t_o{}^*$. To see that (*) is satisfied it is enough to note that a continuous function x which vanishes for all $t\in[a,b]-\{t_o\}$ must also vanish at t_o. Since the essential feature of t_o here is that it is not an isolated point of $[a,b]$ it is easy to see that, more generally, we can take x to be any B-* algebra, i.e. an algebra of continuous complex-valued functions $C(T,\underset{\sim}{C},c)$ on a compact space T with sup norm topology, and t_o to be any point of T which is not an isolated point.

Letting $[H_o]$ denote the linear span of H_o in X*, it follows by (*) that $(X,[H_o])$ is a dual pair. Furthermore, since the seminorms p_f defined at each $x\in X$ by $p_f(x)=|f(x)|$,

$f \in H_o$, which generate $\sigma(X,[H_o])$, are clearly multiplicative, $(X,\sigma(X,[H_o]))$ is an LMCH algebra. In view of that fact that $[H_o]$ is the continuous linear dual of $(X,\sigma(X,[H_o]))$, it follows that any non-trivial continuous homomorphism g of X must belong to $[H_o]$. If we assume that $g \notin H_o$ itself then $H_o \cup \{g\}$ is linearly independent by (4.10-6) and, therefore, $[H_o]$ is a proper subset of $[H_o \cup \{g\}]$. But this is ridiculous, so $g \in H_o$. Thus $X^h = H_o$ and, consequently, $f_o \notin X^h$.

As it happens the topological algebras of Example 4.10-1 are rarely, if ever, complete (see Exercise 4.5). An example of a complete LMCH algebra on which discontinuous homomorphisms exist is afforded by $C(T,\underset{\sim}{R},c)$ where $T=[0,\Omega)$ and Ω is the first uncountable ordinal. T is not replete (Example 1.5-1) so there are homomorphisms of $C(T,\underset{\sim}{R},c)$ which are not evaluation maps by (1.6-1). But all the continuous homomorphisms of $C(T,\underset{\sim}{R},c)$ are evaluation maps as shown in Example 4.10-2, so discontinuous homomorphisms exist on $C(T,\underset{\sim}{R},c)$. As T is an open subset of the compact space $[0,\Omega]$, it is locally compact and therefore also a k-space (Exercise 2.2(b)). Hence by Theorem 2.2-1, $C(T,\underset{\sim}{R},c)$ is complete. Other examples of complete LMCH algebras on which discontinuous homomorphisms exist are given in Exercise 2.2(g). Nevertheless there is still a close relationship between $\mathcal{M}$ and $\mathcal{M}_c$ in complex LMCH algebras; e.g. $\cup\mathcal{M} = \cup\mathcal{M}_c$ (see (4.10-9) and $\cap\mathcal{M} = \cap\mathcal{M}_c$ (see (4.11-1)).

Our next result is another example of how information about an LMC algebra can be gleaned from the factor algebras. It asserts that the continuous homomorphisms of an LMCH algebra can be "obtained" from the (continuous) homomorphisms of a set of factor algebras. First we state some conventions about certain things in topological vector spaces.

If X is a topological vector space then the <u>continuous dual</u> X' is the linear space of all continuous linear functionals on X. The <u>polar</u> B^o of a subset B of X is the set of $x'\varepsilon X'$ such that $\sup|\langle B,x'\rangle| \leq 1$. The <u>adjoint</u> A' of a linear map $A:X\to Y$ where X and Y are topological vector spaces is defined at $y'\varepsilon Y'$ according to $\langle Ax,y'\rangle=\langle x,A'y'\rangle$.

<u>(4.10-7) X^h FOR LMCH ALGEBRAS</u>. Let X be a LMCH algebra with topology generated by the saturated family (p_μ) of seminorms with associated factor algebras (X_μ). Then: (a) For each index μ, $\kappa_\mu'(X_\mu^h)=X^h\cap \overline{V}_\mu^{\,o}$ where κ_μ is the canonical homomorphism $x\to x+N_\mu$ (see Sec. 4.5) and $\overline{V}_\mu=\{x\in X|p_\mu(x) \leqq 1\}$; (b) $X^h= \cup\kappa_\mu'(X_\mu^h)$; (c) each $\kappa_\mu'(X_\mu^h)$ is compact in the Gelfand topology of X^h (i.e. the relative $\sigma(X',X)$- topology) and $\kappa_\mu'(X_\mu^h)$ and X_μ^h are homeomorphic via the correspondence $\kappa_\mu'(f_\mu)\leftrightarrow f_\mu$ when each space carries its Gelfand topology.

<u>Proof</u>. (a) Let $f\in \kappa_\mu'(X_\mu^h)$ so $f=\kappa_\mu'(f_\mu) = f_\mu\circ\kappa_\mu$ for some $f_\mu\in X_\mu^h$. Then for each $x\in X$ we have $f(x)=f_\mu(\kappa_\mu(x))=f_\mu(x+N_\mu)$. Thus $|f(x)|=|f_\mu(x+N_\mu)| \leqq \dot{p}_\mu(x+N_\mu) = p_\mu(x)$ since any homomorphism f_μ on a Banach algebra is continuous with norm one. Certainly then $|f(x)| \leqq 1$ whenever $x\in \overline{V}_\mu$ and so $f\in X^h\cap \overline{V}_\mu^{\,o}$. Then, as $N_\mu\subset\overline{V}_\mu$, f must vanish on N_μ and we may unambiguously define the nontrivial homomorphism $g_\mu:X/N_\mu\to \underset{\sim}{F}$ by the equation $g_\mu(x+N_\mu)=f(x)$. Since $\dot{p}_\mu(x+N_\mu)=p_\mu(x)$ for each $x\in X$, the boundedness of f on $\overline{V}_\mu$ implies the boundedness of g_μ on $\overline{V}_{\dot{p}_\mu}=\{\kappa_\mu(x)\in X/N_\mu|p_\mu(\kappa_\mu(x)) \leqq 1\}$. Thus g_μ is a continuous nontrivial homomorphism of X/N_μ; hence it can be extended by

continuity to a nontrivial homomorphism $f_\mu \in X_\mu{}^h$. For any $x \in X$ we have $f(x) = g_\mu(\kappa_\mu(x)) = f_\mu(\kappa_\mu(x)) = (\kappa_\mu'(f_\mu))(x)$ and it follows that $f = \kappa_\mu'(f_\mu) \in \kappa_\mu'(X_\mu{}^h)$.

(b) It sufficies to show that $X^h = \cup_\mu X^h \cap \bar{V}_\mu{}^o$. If $f \in X^h$ then f is bounded on $\bar{V}_\mu$ for some μ. Suppose that $|f(x)| > 1$ for some $x \in \bar{V}_\mu$. Then the contradictory facts follow that $\{|f(x^n)|\,|\,n \in \underset{\sim}{N}\}$ is unbounded while $x^n \in \bar{V}_\mu$ for all $n \in \underset{\sim}{N}$. Hence $|f(x)| \leq 1$ for each $x \in \bar{V}_\mu$ and $f \in X^h \cap \bar{V}_\mu{}^o$.

(c) It is easy to see that X^h is closed in the continuous dual X' of X so that the $\sigma(X',X)$-compactness of $\bar{V}_\mu{}^o$ [the polar of a neighborhood of 0 in any topological vector space X is always $\sigma(X',X)$-compact by Alaoglu's theorem] and part (a) combine to show that $\kappa_\mu'(X_\mu{}^h)$ is $\sigma(X',X)$-compact. As $X_\mu{}^h$ is Hausdorff in its Gelfand topology it suffices to show that $\kappa_\mu'(f_\mu) \to f_\mu$ is well-defined to conclude that it is a homeomorphism (it is clearly injective). Thus suppose that $\kappa_\mu'(f_\mu) = \kappa_\mu'(g_\mu)$ so that $f_\mu(x+N_\mu) = (\kappa_\mu'(f_\mu))(x) = (\kappa_\mu'(g_\mu))(x) = g_\mu(x+N_\mu)$ for each $x+N_\mu \in X/N_\mu$. As X/N_μ is dense in X_μ, $f_\mu = g_\mu$ and the proof is complete.∇

As an application of (4.10-7), in our next example we determine all the continuous homomorphisms of $C(T,\underset{\sim}{F},c)$ when T is a completely regular Hausdorff space. Note that the result is trivial if T is replete since in that case all the homomorphisms of $C(T,\underset{\sim}{F})$ into $\underset{\sim}{F}$ are evaluation maps by (1.6-1) and evaluation maps are always continuous when $C(T,\underset{\sim}{F})$ carries the compact-open topology.

<u>Example 4.10-2 $C(T,\underset{\sim}{F},c)^h = T^*$ FOR COMPLETELY REGULAR HAUSDORFF T.</u> Since T is a completely regular Hausdorff space, a set of fac-

tor algebras for $C(T,\underset{\sim}{F},c)$ is the set of Banach algebras $C(G,\underset{\sim}{F},c)$, where G is a compact subset of T [see Example 4.5-1]. We claim that $C(T,\underset{\sim}{F},c)^h=T^*$ follows from (4.10-7) provided we can show that $C(G,\underset{\sim}{F},c)^h=G^*$ for each compact set G. To see this let $f\in C(T,\underset{\sim}{F},c)^h$; then by (4.10-7) and the statement "$C(G,\underset{\sim}{F},c)^h=G^*$ for each compact G", there is a compact set $G\subset T$ and a point $t\in G$ such that for any $x\in C(T,\underset{\sim}{F},c)$ $f(x) = (\kappa_G'(t^*)(x)=t^*(\kappa_G(x))=t^*(x|_G)=x(t)$.

To show that $C(G,\underset{\sim}{F},c)^h=G^*$, when G is compact, it suffices to note that G* constitutes all the homomorphisms, for all homomorphisms on a Banach algebra are continuous, [(4.10-5)]. That G* constitutes all the homomorphisms follows immediately from (1.6-1).∇

<u>Example 4.10-3 $H^h=D^*$.</u> Consider the LMCH algebra H of analytic functions on the open unit disc D of the complex plane carrying the compact-open topology (See Examples 4.5-2 and 4.7-2).

We have already seen (Example 4.5-2) that H is a Frechet algebra with a set of factor algebras $(P(C_n))$ where $P(C_n)$ consists of all uniform limits of polynomials on $C_n=\{t\varepsilon \underset{\sim}{C} \mid |t| \leq 1-1/n\}$ endowed with sup norm topology. Thus, as in the previous example, once we show that the continuous homomorphisms of $P(C_n)$ are just the evaluation maps t^* for $t\varepsilon C_n$ it follows that $H^h=D^*$. Moreover, if $t\neq s$, then $t^*\neq s^*$ for they differ on the polynomial $x(u)=u-t$. Hence H^h is in 1-1 correspondence with D.

To see that the continuous homomorphisms of $P(C_n)$ are just evaluation maps let $h\varepsilon P(C_n)^h$ and set $t=h(y)$ where the polynomial $y(u)=u$. Since $y\varepsilon P(C_n)$, $\|y\|=1-1/n$, and $\|h\|=1$ since $P(C_n)$ is a Banach algebra, then $t=h(y)\varepsilon C_n$. For $x(u) = \Sigma_{k=0}^n a_k u^k$ we see that $x = \Sigma_{k=0}^n a_k y^k$ and $h(x) = \Sigma_{k=0}^n a_k (h(y))^k= \Sigma_{k=0}^n a_k t^k=x(t)=t^*(x)$. Thus $h=t^*$ on polynomial functions on C_n. Furthermore if $x\varepsilon P(C_n)$ there is a sequence (x_n) of polynomials which converges uniformly to x. Hence by the continuity of h, $h(x)=\lim_n h(x_n)=\lim_n x_n(t)=x(t)$ and $h=t^*$.∇

It is well known that in any complex Banach algebra the spectrum of an element x is just the set of values f(x) as f runs through the (continuous) homomorphisms or, put another way, the range of the Gelfand map $\hat{x}$ (see the remark immediately following 4.10-4). With the aid of the Arens invertibility criterion [Theorem (4.6-1)] we generalize this in our next result to complete complex LMCH algebras.

<u>(4.10-8) $\sigma(x) = X^h(x)$ IN COMPLETE COMPLEX LMCH ALGEBRAS</u>. If X is a complete complex LMCH algebra and x is any element in X then $\sigma(x) = \{f(x) \mid f\in X^h\}$.

Proof. For any $x \in X$ and $f \in X^h$, $f(x-f(x)e)=0$. Hence for any such f, $x-f(x)e$ is not invertible in X, and, therefore, $\{f(x) \mid f \in X^h\} \subset \sigma(x)$. To obtain the reverse inclusion let $\lambda \in \sigma(x)$; then for some factor algebra X_μ, the canonical image $\kappa_\mu(x-\lambda e)$ is not invertible in X_μ by Theorem 4.6-1. Consider a maximal ideal $M_\mu \subset X_\mu$ containing $\kappa_\mu(x-\lambda e)$. As X_μ is a Banach algebra, M_μ is closed (by(4.8-2) and (4.10-1)) and is in fact the kernel of a continuous nontrivial homomorphism f_μ of X_μ by (4.10-4). Thus $f_\mu(\kappa_\mu(x-\lambda e))=0$. Surely $f=f_\mu\kappa_\mu$ is a continuous nontrivial homomorphism of X and, as $f(x-\lambda e)=0$, $\lambda=f(x)$. Thus $\sigma(x) \subset \{f(x) \mid f \in X^h\}$.∇

We already know if x is a singular element then x lies in some maximal ideal. Our next result shows that in complete LMCH algebras, every singular element lies in some closed maximal ideal.

(4.10-9) $\cup\mathcal{M} = \cup\mathcal{M}_c$ IN A COMPLETE COMPLEX LMCH ALGEBRA. If X is a complete complex LMCH algebra, then $\cup\mathcal{M} = \cup\mathcal{M}_c$.

Proof. Clearly $\cup\mathcal{M}_c \subset \cup\mathcal{M}$. If $x \in \cup\mathcal{M}$ then x is not invertible and $0 \in \sigma(x)$. Thus by (4.10-8) there exists an $f \in X^h$ such that $f(x)=0$. Since the kernel of f is a closed maximal ideal, $x \in \cup\mathcal{M}_c$.∇

(4.10-9) shows that any proper principal ideal $(z) = \{xz \mid x \in X\}$ in a complete complex LMCH algebra X can be embedded in a closed maximal ideal. Some further information on when an ideal can be embedded in a closed maximal ideal is given in (4.10-10) below.

(4.10-10) EMBEDDING IDEALS IN CLOSED MAXIMAL IDEALS. Every proper nondense ideal I in a complex LMCH algebra X can be embedded in a closed maximal ideal.

Proof. Suppose I is not dense in X and P is a saturated family of multiplicative seminorms generating the topology on X. Since I is not dense in X, there is a $p \varepsilon P$ such that $I \cap (e+\varepsilon V_p) = \emptyset$ for some $\varepsilon > 0$ where $V_p = \{x \varepsilon X | p(x) < 1\}$. Consider the ideal $I+N_p$ in X. If $x \varepsilon I$ and $y \varepsilon N_p$ then $p(x+y-e) = p(x-e) \geq \varepsilon$. Hence $(I+N_p) \cap (e+\varepsilon V_p) = \emptyset$ and $J = I+N_p$ is not dense. Now denoting the factor algebra (Def. 4.5-1) associated with p by X_p, it is clear that $cl_{X_p}(\kappa(J))$ is an ideal in X_p (where κ is the canonical homomorphism of X into X_p). Moreover we claim that it is proper. Indeed if we make the assumption to the contrary then $e+N_p \varepsilon cl_{X_p}(\kappa(J))$. Thus, with ε as above, there must be elements $x_\varepsilon \varepsilon I$ and $y \varepsilon N_p$ such that $\dot{p}((x_\varepsilon+y+N_p)-(e+N_p)) < \varepsilon$. However $p(x_\varepsilon - e) = \dot{p}((x_\varepsilon+N_p)-(e+N_p)) = \dot{p}((x_\varepsilon+y+N_p)-(e+N_p)) < \varepsilon$ and this is a contradiction. Since $cl_{X_p}(\kappa(J))$ is a closed proper ideal in the Banach algebra X_p there is a (perforce closed) maximal ideal $M_p \supset cl_{X_p}(\kappa(J))$. We contend that $M = \kappa^{-1}(M_p)$ is the desired maximal ideal. Certainly M is a closed ideal containing I; it only remains to show that it is maximal. Letting h be the surjective complex homomorphism of X_p determined by M_p, it immediately follows that $M = \kappa^{-1}(h^{-1}(0))$ and M is the kernel of the nontrivial homomorphism $h \cdot \kappa$. Thus M is maximal and the proof is complete.∇

As mentioned earlier, an immediate consequence of (4.10-9) is that any proper principal ideal in a complete complex LMCH algebra X can be embedded in a closed maximal ideal. It is natural to inquire if this is true for any finitely generated ideal $(z_1, \ldots, z_n) = \{\Sigma x_i z_i | x_i \in X, i=1, \ldots, n\}$. In other words, when is $\overline{(z_1,\ldots,z_n)}$ proper? With the added condition of metrizability--i.e. if X is a complex Frechet algebra--then this is so, as is proved in (4.10-12). First, however, we must establish the following technical fact.

(4.10-11). Let h be a homomorphism between the topological algebras X and Y such that h(X) is dense in Y, and h(e)=e. Given elements $z_1,\ldots,z_n \in X$, suppose that elements $u_1,\ldots,u_n \in X$ exist such that $\Sigma u_i z_i = e$. It follows that there exist $y_1,\ldots,y_n \in Y$ ($y_i = h(u_i)$ for example) such that $\Sigma y_i h(z_i) = e$. Then for any such $y_1,\ldots,y_n$ and neighborhoods $U_1,\ldots,U_n$ of 0, there exist $v_1,\ldots,v_n \in X$ such that $\Sigma v_i z_i = e$ and $h(v_i) \in y_i + U_i$ for $i=1,\ldots,n$.

Proof. Suppose that $\Sigma u_i z_i = e$ for $u_1,\ldots,u_n \in X$. Then setting $v_i = w_i + u_i(e - \Sigma w_j z_j)$ for $i=1,\ldots,n$, it follows that $\Sigma v_i z_i = e$ regardless of the choice of the w's. By the hypothesis then, choosing $y_1,\ldots,y_n$ such that $\Sigma y_i h(z_i) = e$, (a): $h(v_i) - y_i = (h(w_i) - y_i) + h(u_i) h(e - \Sigma_j w_j z_j)$ for each i. Now let V and W be balanced neighborhoods of the origin such that $V+V \subset U_i$ for each i and $\Sigma_{i=1}^n W \subset V$. Note that $h(u_i) h(e - \Sigma_j w_j z_j) = h(u_i)(e - \Sigma_j h(w_j) h(z_j))$. Since $e = \Sigma y_i h(z_i)$, the right-hand side becomes $\Sigma_j h(u_i) h(z_j)(y_j - h(w_j))$. Thus, as multiplication is continuous and h(X) is dense in Y we can choose $w_1,\ldots,w_n$ such that $h(u_i) h(z_j)(y_j - h(w_j)) \in W$ and $h(w_i) - y_i \in V$, for $i,j=1,\ldots,n$. By (a) and the choice of V and W it follows that $h(v_i) - y_i \in U_i$.∇

(4.10-12). EMBEDDING FINITELY GENERATED IDEALS IN CLOSED MAXIMAL IDEALS. Let X be a complex Frechet algebra with topology generated by the saturated family (p_k) of seminorms (where it is assumed without loss of generality that $p_k \leq p_{k+1}$ for each k) and with associated factor algebras (X_k). Then (a) $(z_1,\ldots,z_n)$ is proper in X iff $(\kappa_k z_1,\ldots,\kappa_k z_n)$ is proper in X_k for some k; (b) any proper finitely generated ideal can be embedded in a closed maximal ideal.

Proof. (a) If $(z_1,\ldots,z_n)=X$ then $\Sigma x_i z_i = e$ for some $x_1,\ldots,x_n \in X$. Hence $\Sigma\kappa_k(x_i)\kappa_k(z_i)=e+N_k$ and it follows that $(\kappa_k z_1,\ldots,\kappa_k z_n)=X_k$ for each k. Conversely suppose that $(\kappa_k z_1,\ldots,\kappa_k z_n)=X_k$ for each k. By an induction process we construct a Cauchy sequence in X_k for each fixed i, $1 \leq i \leq n$, convergent to an element $u_i(k)$ such that $(u_i(k))_{k>0}$ is an element of the projective limit of the X_k's and $\Sigma_i u_i(k)\kappa_k(z_i)=e+N_k$ for each $k \geq 0$: We conclude the proof by invoking Theorem 4.6-1 to obtain elements $u_i \in X$ such that $\kappa_k(u_i)=u_i(k)$ for each i and k as $x_i=u_i$ turns out to be a solution of $\Sigma_i x_i z_i = e$.

We remind the reader that $h_{rs}(s \geq r)$ denotes the extension by continuity to X_s of the mapping $x+N_s \to x+N_r$ (see Theorem 4.6-1) and κ_r the canonical homomorphism of X into X_r. We choose $y_1^o,\ldots,y_n^o \in X_o$ such that $\Sigma_i y_i^o \kappa_o(z_i)=e+N_o$. Proceeding inductively we find that elements $y_1^m,\ldots,y_n^m \in X_m$ exist by (4.10-11) such that $\Sigma_i y_i^m \kappa_m(z_i)=e+N_m$ and $\dot{p}_{m-1}(h_{m-1,m}(y_i^m)-y_i^{m-1}) < 1/2^m$ $(i=1,\ldots,n)$. If $m>k$ then, since $h_{k,m-1}\cdot h_{m-1,m}=h_{km}$ and $\|h_{k,m-1}\| \leq 1$,

$$(*)\quad \dot{p}_k(h_{km}(y_i^m)-h_{k,m-1}(y_i^{m-1}))=\dot{p}_k(h_{k,m-1}(h_{m-1,m}(y_i^m)-y_i^{m-1}))$$
$$\leq \dot{p}_{m-1}(h_{m-1,m}(y_i^m)-y_i^{m-1}) < 1/2^m \quad (i=1,\ldots,n).$$

Set $u_i(k,m)=h_{km}(y_i^m)$ for $m>k$ and $1 \leq i \leq n$. Substituting in (*) we obtain

$\dot{p}_k(u_i(k,m)-u_i(k,m-1)) < 1/2^m$ for $m > k$. We now claim that $(u_i(k,m))_{m>k}$ is a Cauchy sequence in X_k. Indeed if $m > j > k$ we have $p_k(u_i(k,m)-u_i(k,j)) \leq \Sigma_{t=j+1}^m \dot{p}_k(u_i(k,t)-u_i(k,t-1)) \leq \Sigma_{t=j+1}^m 2^{-t}$. Since X_k is complete there exists an element $u_i(k) \in X_k$ such that $u_i(k,m) \overset{m}{\to} u_i(k)$. To see that the "tuple" $(u_i(k))_{k>0}$ is in the projective limit of the X_k's for each i we observe that for $m > k$ $h_{k,k+1}(u_i(k+1,m)) = h_{k,k+1}(h_{k+1,m}(y_i^m)) = h_{km}(y_i^m) = u_i(k,m)$. Thus as $m\to\infty$, using the continuity of $h_{k,k+1}$, we see that $h_{k,k+1}(u_i(k+1)) = u_i(k)$ for each i so $(u_i(k))_{k\geq 0}$ is indeed an element of the projective limit for each i. The fact that $\Sigma_i u_i(k)\kappa_k(z_i) = e+N_k$ for each k also follows by taking a limit with respect to m in the equation (recall that $h_{km}\cdot\kappa_m=\kappa_k$ and the way in which that y_i^m were chosen) $\Sigma_i u_i(k,m)\kappa_k(z_i)=h_{km}(\Sigma_i y_i^m \kappa_m(z_i)) = h_{km}(e+N_m)=e+N_k$. Finally elements $u_1,\ldots,u_n \in X$ exist by Theorem 4.6-1(d) such that $\kappa_k(u_i)=u_i(k)$ for each $k\geq 0$, so $p_k(\Sigma_i u_i z_i - e)=\dot{p}_k(\Sigma_i u_i(k)\kappa_k(z_i)-e+N_k))=0$ for each $k\geq 0$ and the concluding statement--$\Sigma_i u_i z_i = e$--follows since X is Hausdorff.

(b) If $(z_1,\ldots,z_n)$ is proper in X then, by (a), there is a $k\geq 0$ such that $(\kappa_k z_1,\ldots,\kappa_k z_n)$ is proper in X_k. Since X_k is a Banach algebra there exists a closed maximal ideal $M_k \supset (\kappa_k z_1,\ldots,\kappa_k z_n)$. Clearly $M=\kappa_k^{-1}(M_k)$ is a closed ideal in X containing $(z_1,\ldots,z_n)$. Since X_k is a complex Banach algebra there is a complex homomorphism h of X_k such that M_k is the kernel of h. Thus M is the kernel of $\kappa_k\cdot h$ and the maximality of M follows.∇

4.11 The Radical and Derivations

The radical of an algebra is an ideal of great utility in the study of the structure of topological algebras. Especially sharp statements can be made when the radical is the zero ideal. For example, if the complex algebra X has $\{0\}$ as its radical, there is at most one topology on X with respect to which X is a Banach algebra.

In this section we use the properties of the radical and derivations to produce an example of a Frechet LMC Q-algebra which is not a Banach algebra.

Definition 4.11-1 THE RADICAL AND SEMISIMPLICITY. The radical, Rad X, of an algebra X is the intersection of all maximal ideals in X. If Rad X = $\{0\}$, X is referred to as semisimple while a topological algebra is called strongly semisimple whenever $\cap \mathcal{M}_c = \{0\}$.

Thus in a strongly semisimple complex LMC algebra where there is a 1-1 correspondence between $\mathcal{M}_c$ and X^h, if every continuous homomorphism vanishes on an element x, then x must be 0. Moreover in such algebras, since the continuous homomorphisms seperate the points of X (i.e. if $x \neq 0$, there is some $f \in X^h$ such that $f(x) \neq 0$) the topology on X must be Hausdorff. Thus "strongly semisimple LMC" is the same as "strongly semisimple LMCH," for complex algebras.

Example 4.11-1 A STRONGLY SEMISIMPLE ALGEBRA. Consider the algebra $C(T,\underset{\sim}{F},c)$, where T is a completely regular Hausdorff space of Example 4.5-1. Since the continuous nontrivial homomorphisms of $C(T,\underset{\sim}{F},c)$ are just the evaluation maps on the points of T (See Example 4.10-2) if $x \in \cap \mathcal{M}_c$ then $x(t) = 0$ for all $t \in T$ and $x=0$.

Clearly X is semisimple whenever it is strongly semisimple. We now show that in complete complex LMCH algebras, semisimplicity implies strong semisimplicity.

(4.11-1) RAD X = $\cap \mathcal{M}_c$ IN COMPLETE COMPLEX LMCH ALGEBRAS. If X is a complete complex LMCH algebra then Rad X = $\cap \mathcal{M}_c$. Thus a complete complex LMCH algebra X is semisimple iff X is strongly semisimple.

Proof. Certainly Rad $X \subset \cap \mathcal{M}_c$. Suppose that $x \in \cap \mathcal{M}_c$. As $f(x)=0$ for all $f \in X^h$, $f(e-xy)=1$ for each $f \in X^h$ and $y \in X$, so by (4.10-8) $e-xy$ is invertible in X for each $y \in X$. If we suppose that $x \notin M$ for some $M \in \mathcal{M}$, then since, X is a commutative ring with identity, X/M is a field and there exists $y \in X$ such that $(x+M)(y+M) = e+M$. Thus $e-xy \in M$ contradicting the invertibility of $e-xy$. Thus $x \in$ Rad X and $\cap \mathcal{M}_c \subset$ Rad X. ∇

The notion of "derivation" defined below is purely algebraic. For the sake of the definition X needn't be commutative or possess an identity.

Definition 4.11-2 DERIVATIONS. Let X be an algebra. A linear map $D: X \to X$ is a derivation on X if for all $x, y \in X$, $D(xy) = xDy + (Dx)y$.

Clearly the trivial linear transformation is a derivation and so is the differentiation operator on spaces of infinitely differentiable functions. Let X be a linear space and $\mathcal{L}(X,X)$ the noncommutative algebra of all linear maps taking X into X [where $(AB)x = A(Bx)$]. Fix $B \in \mathcal{L}(X,X)$ and define $D_B(A)=AB-BA$ for any $A \in \mathcal{L}(X,X)$. The transformation $AB-BA$ is called the commutator of A and B and the derivation D_B is referred to as a commutator operator. Finally, let E be a field and X the algebra of all formal power series in a single variable t with coefficients from E. That is, $X = \{\sum_{n=0}^{\infty} \alpha_n t^n \mid \alpha_n \in E\}$ where addition and scalar multiplication are performed componentwise while multiplication is taken to be the Cauchy product.* Then $\sum_{n=0}^{\infty} \alpha_n t^n \to \sum_{n=0}^{\infty} n\alpha_n t^{n-1}$ is a derivation on X.

(4.11-3) CONTINUOUS DERIVATION MAPS A COMPLEX BANACH ALGEBRA INTO ITS RADICAL. If X is a (not necessarily commutative) complex Banach algebra and D a continuous derivation on X, then $D(X) \subset \text{Rad } X$.

Proof. As usual $L(X,X)$ denotes the complex Banach space of continuous (=bounded) linear transformations on X. Since D is bounded and $L(X,X)$ is a Banach space, the series $\sum_{n=0}^{\infty} (\alpha^n/n!)D^n$ is absolutely convergent in $L(X,X)$ for any $\alpha \in \underset{\sim}{C}$. We denote the sum of this series by $e^{\alpha D}$. Let f be any nontrivial homomorphism of X and define f_α to be $f \cdot e^{\alpha D}$. Since f must be continuous (see the discussion preceeding (4.10-5)), f_α is a continuous linear functional on X. To see that f_α is in fact a homomorphism we first note that since D is a derivation, a "Leibniz rule" holds: for any positive integer n,

$$\frac{D^n(xy)}{n!} = \sum_{i+j=n} \frac{D^i x}{i!} \frac{D^j y}{j!} .$$

Thus, since f is continuous,

$$f_\alpha(xy) = f\left(\sum_{n=0}^{\infty} \frac{\alpha^n}{n!} D^n(xy)\right) = \sum_{n=0}^{\infty} \frac{\alpha^n f(D^n(xy))}{n!}$$

$$(*) \qquad = \sum_{n=0}^{\infty} \alpha^n \sum_{i+j=n} \frac{f(D^i x)}{i!} \frac{f(D^j y)}{j!} .$$

* The Cauchy product of $\Sigma\alpha_n t^n$ and $\Sigma\beta_n t^n$ is $\Sigma\gamma_n t^n$ where $\gamma_n = \sum_{i+j=n} \alpha_i \beta_j$.

Since $f_\alpha(x)\ f_\alpha(y) = \sum_{i=0}^{\infty} \frac{\alpha^i f(D^i x)}{i!} \sum_{j=0}^{\infty} \frac{\alpha^j f(D^j y)}{j!}$ and the two series above converge absolutely, the product of these two series equals the Cauchy product(*) and $f_\alpha(xy) = f_\alpha(x) f_\alpha(y)$.

Since $\|f\| \leq 1$ (see the discussion preceeding (4.10-5)) and D is bounded the series $f_\alpha(x) = \Sigma \frac{\alpha^n f(D^n x)}{n!}$ converges absolutely for each fixed $x \in X$ and any α. Thus the mapping $\alpha \to f_\alpha(x)$ (with x held fixed) is an entire function of α. As was the case for f, $\|f_\alpha\| \leq 1$, and it follows that $\|f_\alpha(x)\| \leq \|x\|$ for each $\alpha \in \underset{\sim}{C}$. Hence $f_\alpha(x)$ is a bounded entire function of α and therefore is a constant by Liouville's theorem [(4.9-1)]. It follows from the identity theorem for power series that $f(D^n x)=0$ for all $n \geq 1$; thus, in particular, $f(Dx)=0$. Since f was any homomorphism we conclude that $Dx \in$ Rad X.∇

Next we state an immediate corollary of the preceeding proposition for ease of reference.

<u>(4.11-4) CONTINUOUS DERIVATIONS ON COMPLEX SEMISIMPLE BANACH ALGEBRAS ARE TRIVIAL</u>. If X is a semisimple complex Banach algebra then the only continuous derivation on X is the trivial one.

The fact that the commutator of a pair of bounded operators on a complex Banach space is never the identity transformation was first proved by H. Wielandt (1949-50). An independent proof can be given for matrices using the elementary properties of the trace function on matrices.

<u>(4.11-5) COMMUTATORS AND THE IDENTITY</u>.† If X is a complex

† The validity of the proof below depends on Theorem 4.6-1, (4.10-7), (4.10-8) and (4.11-3), all of which remain true without the assumption of commutativity as the reader may verify.

Banach space and $A, B \in L(X,X)$ then $AB-BA$ cannot be the identity map $x \to x$ on X.

Proof. Consider the derivation $D_B: L(X,X) \to L(X,X)$, $H \to HB-BH$. It is readily seen that D_B is bounded $(\|D_B(H)\| \leq 2\|B\| \, \|H\|)$. Since $L(X,X)$ is a complex Banach algebra, it follows by (4.11-3) that $D_B(L(X,X)) \subset \text{Rad } L(X,X)$. But $1 \notin \text{Rad } L(X,X)$, so $AB-BA = D_B(A) \neq 1$.∇

The results just obtained now enable us to exhibit a complete barreled semisimple Q-algebra which is not a Banach algebra.

Example 4.11-1 A BARRELED SEMISIMPLE FRECHET Q-ALGEBRA WHICH IS NOT A BANACH ALGEBRA. Consider the algebra $\mathcal{D}$ of all infinitely differentiable functions on $[a,b]$ of Example 4.3-2 with the LMCH topology generated by the seminorms $p_n(x) = \sup_{t \in [a,b]} |x^{(n)}(t)|$ $(n=0,1,\ldots)$. As was established in Example 4.3-2, $\mathcal{D}$ is a Frechet algebra. Since any complete metric space is a Baire space any LCS which is a Baire space is barreled (see Horvath, 1966, pp. 213-214), the algebra $\mathcal{D}$ is barreled. In a remark following (4.8-1) we proved that $\mathcal{D}$ was a Q-algebra by showing that $V_{p_o} = \{x \in \mathcal{D} \mid p_o(x) < 1\} \subset Q$. The fact that $\mathcal{D}$ is semisimple is apparent from the observation that any element $x \in \text{Rad } X$ must vanish on $[a,b]$ as the evaluation maps $t^*: \mathcal{D} \to \underset{\sim}{C}$, $x \to x(t)$ are continuous homomorphisms on $\mathcal{D}$.

To show that $\mathcal{D}$ is not a Banach algebra we have only to exhibit a nontrivial continuous derivation on $\mathcal{D}$ by (4.11-4). Clearly the differentiation operator $D: \mathcal{D} \to \mathcal{D}$, $x \to x'$ is a nontrivial derivation and it is continuous since $D^{-1}(V_{p_n}) = V_{p_{n+1}}$ for each $n \geq 0$.

We can also observe that (4.11-4) is no longer true if the condition that X be a Banach algebra is relaxed: It is not even true for barreled Frechet semisimple Q-algebras as this example illustrates.

Following the proof of (4.11-3) (Singer and Wermer (1955)) the suspicion grew that perhaps continuity of the derivation was not a necessary ingredient in the hypothesis of that result and this suspicion has been somewhat borne out. Curtis (1961) proved that <u>every</u> derivation on a <u>regular</u>* commutative semisimple Banach algebra with identity is continuous. This was subsequently generalized to semisimple Frechet algebras by Rosenfeld (1966). B.E. Johnson (1969) proved that every derivation on a semisimple commutative Banach algebra is continuous and hence trivial by (4.11-4).

Miller (1970) and Gulick (1970) have considered higher order derivations and some of Gulick's results subsume Rosenfeld's generalization of Curtis' theorem.

A number of the results just mentioned may be found in the Exercises for Chapter 5.

4.12 SOME ELEMENTS OF GELFAND THEORY - THE TOPOLOGIZING OF X^h AND THE MAPPING ψ.

In this section we consider a mapping useful in analyzing

* Let X be a commutative Banach algebra and $\mathcal{M}$ its space of maximal ideals. If $x \in X$, define $\hat{x}:\mathcal{M} \to \underset{\sim}{C}$ by the formula $\hat{x}(M)=h(x)$, where M=ker h, and provide $\mathcal{M}$ with the weakest topology with respect to which each $\hat{x}$ is continuous. If the functions $\hat{x}$, $x \in X$, separate points and closed sets in $\mathcal{M}$, i.e. for each closed $F \subset \mathcal{M}$ and $M \in \mathcal{M}$ there is an x such that $\hat{x}(M)=1$ and $\hat{x}(F)=\{0\}$, then X is called <u>regular</u>. For further discussion see Sec. 5.2.

the structure of LMCH algebras--the mapping Ψ sending $x \in X$ into the corresponding evaluation map on X^h taking a topological algebra X into $C(X^h, \underset{\sim}{F})$ the algebra of continuous functions on X^h equipped with the relative $\sigma(X', X)$-topology (see Definition 4.12-1). As we have seen it is frequently possible to reduce questions pertaining to the structure of LMCH algebras to related questions about the factor algebras with the aid of the canonical homomorphisms and some of their properties. In this section we shall see that in the event X^h is $\sigma(X', X)$-compact, Ψ can at times be expected to behave in a similar way in reducing questions concerning the LMCH algebra X to related questions about the Banach algebra $C(X^h, \underset{\sim}{F}, c)$. In order for Ψ to be useful in transforming questions from X into $C(X^h, \underset{\sim}{F}, c)$ it is sometimes desirable for Ψ to be continuous--our third result deals with this. We conclude the section by establishing certain conditions under which a topological algebra is a Banach algebra, by proving that Ψ is a topological isomorphism.

Michael (1952) has referred to LMCH algebras X for which Ψ is an isomorphism onto $C(X^h, \underset{\sim}{F})$ as <u>full algebras</u> (Definition 4.12-1). In Michael (1952) the question was raised: Is Ψ a <u>topological</u> isomorphism when X is a full Frechet algebra? Warner (1958) answered the question in the affirmative and his proof appears in (4.12-8). (In fact the analogous statement also holds for many topological algebras over nonarchimedean valued fields.) Thus Ψ can also be used to identify certain topological algebras as function algebras.

<u>Definition 4.12-1. THE MAPPING Ψ</u>. Let X be a complex topological algebra or a real topological algebra for which $X^h \neq \emptyset$

and X^h carry the Gelfand topology, i.e. topology induced on X^h by $\sigma(X',X)$ (see the discussion following (4.10-4)*). Then we define the mapping $\Psi: X \to C(X^h, \underset{\sim}{F})$, $x \to \hat{x}$ by the rule: $\hat{x}(f)=f(x)$ for each $f \in X^h$. As X^h carries the induced weak-* topology, the set $\{f \in X^h \mid |\Psi x(f) - \Psi x(f_o)| < \varepsilon\}$ is open in X^h for any $f_o \in X^h$ and $\varepsilon > 0$, and, therefore, Ψx is indeed an element of $C(X^h, \underset{\sim}{F})$. Although Ψ is seen to be a homomorphism, it is not generally 1-1 or onto. In the event that X is a LMCH algebra and Ψ is 1-1 and onto, X is referred to as a <u>full algebra</u>.

In order for Ψ not to be 1-1, there must exist a nonzero $x \varepsilon X$ such that $\Psi x=0$--in other words an x for which $f(x)=0$ for each $f \varepsilon X^h$. Thus Ψ is 1-1 iff X is strongly semisimple (Def. 4.11-1). In case X is a complete complex LMCH algebra so that Rad X = $\cap \mathcal{M}_c$ ((4.11-1)) and each $M \varepsilon \mathcal{M}_c$ is the kernel of some $f \varepsilon X^h$ ((4.10-4)), it follows that Ψ is 1-1 iff X is semisimple. We record these findings in our first result.

<u>(4.12-1) SEMISIMPLICITY $\Rightarrow \Psi$ 1-1</u>. If an LMCH algebra is (a) strongly semisimple or (b) semisimple, complete, and complex, then Ψ is 1-1.

As a consequence of (4.12-3) we see that Ψ takes any complete complex LMCH Q-algebra into a Banach algebra: We show that for such algebras X^h is weakly compact. The following technicality, (4.12-2), is a convenience.

(4.12-2) In any complete complex LMCH algebra $(X^h)^o = U(\sigma)$.

<u>Proof</u>. In a complete complex LMCH algebra X, $\sigma(x)=\{h(x) \mid h \in X^h\}$ for each $x \in X$ by (4.10-8). Thus

---

* If H is <u>any</u> collection of nontrivial homomorphisms on X then the topology induced on H by $\sigma(X^*,X)$ is also called the Gelfand topology (See Exercise 4.4).

$(X^h)^\circ = \{x \in X \mid \sup_{h \in X^h} |h(x)| \leq 1\} = \{x \mid r_\sigma(x) \leq 1\} = U(\sigma)$.∇

<u>(4.12-3) Q-ALGEBRAS AND COMPACTNESS OF X^h.</u> Let X be a complete complex LMCH algebra. If X is a Q-algebra, then $(X^h)^\circ$ is a neighborhood of 0 in X and X^h is $\sigma(X',X)$-compact. Conversely, if X is barreled and X^h is $\sigma(X',X)$-compact, then X is a Q-algebra.

<u>Proof</u>. Suppose that X is a Q-algebra. Then, by (4.8-3), the set $U(\sigma) = \{x \in X \mid r_\sigma(x) \leq 1\}$ has non-empty interior; in fact, $0 \in \text{int}(U(\sigma))$ (see the proof of (4.8-3) and $U(\sigma) \in V(0)$. By (4.12-2) $U(\sigma) = (X^h)^\circ$; hence $(X^h)^\circ$ is a neighborhood of 0 in X. Now $X^h \subset (X^h)^{\circ\circ}$, so we can utilize the fact that the polar of a neighborhood of 0 in X is $\sigma(X',X)$-compact together with the fact that X^h is $\sigma(X',X)$-closed in X' to complete this part of the argument.

Conversely, suppose that X is barreled and X^h is $\sigma(X',X)$-compact. Then X^h is $\sigma(X',X)$-bounded so $(X^h)^\circ$ is a barrel in X. Since X is complete and barreled, it follows that $U(\sigma) = (X^h)^\circ$ is a neighborhood of 0 in X. Thus X is a Q-algebra by (4.8-3).∇

(4.12-3) can be used to obtain still another realization of the Stone-Cech compactification βT of a completely regular Hausdorff space T, namely as the space of continuous nontrivial homomorphisms of the Banach algebra $C_b(T,\underset{\sim}{C},c)$ of bounded continuous $\underset{\sim}{C}$-valued functions on T with sup norm. The space $C_b(T,\underset{\sim}{C},c)^h$ of continuous homomorphisms is compact in its Gelfand topology by (4.12-3). So, identifying T with the space T* of evaluation maps on $C_b(T,\underset{\sim}{C},c)$, a possibility afforded by the complete regularity of the Hausdorff space T, we see that, in some sense, $C_b(T,\underset{\sim}{C},c)^h$ is a compact Hausdorff space containing T. If (a) T and T* can be identified as topological spaces, (b) T is dense in $C_b(T,\underset{\sim}{C},c)^h$, and (c) each $x\varepsilon C_b(T,\underset{\sim}{R},c)$ can be extended continuously to a function on $C_b(T,\underset{\sim}{C},c)^h$, then $C_b(T,\underset{\sim}{C},c)^h$ must be βT by Theorem 1.3-2. We now verify that these three conditions are indeed met.

By (4.10-4) we see that $C_b(T,\underset{\sim}{C},c)^h$ may be identified with the space $\mathcal{M}$ of maximal ideals of $C_b(T,\underset{\sim}{C},c)$. Moreover the Gelfand topology (first discussed after (4.10-4)) is the weakest topology for $C_b(T,\underset{\sim}{C},c)^h$ with respect to which the maps $\hat{x}:C_b(T,\underset{\sim}{C},c)^h \to \underset{\sim}{C}$, $f \to f(x)$ are continuous for each $x\varepsilon C_b(T,\underset{\sim}{C})$. Since T is a completely regular Hausdorff space, its topology is the initial topology determined by $C_b(T,\underset{\sim}{C})$ on T ((0.2-5)); a basic neighborhood of a point t_o in T is therefore a set of the form $V(t_o;x_1,\ldots,x_n,\varepsilon)=\{t\varepsilon T \mid |x_i(t)-x_i(t_o)| < \varepsilon, i=1,\ldots,n\}$ where $x_1,\ldots,x_n \varepsilon C_b(T,\underset{\sim}{C})$ and $\varepsilon > 0$.

A typical basic neighborhood of the evaluation map $t_o^* \varepsilon C_b(T,\underset{\sim}{C},c)^h$ in the relative Gelfand topology on T* would be $\{t^* \varepsilon\ T^* \mid |\hat{x}_i(t_o^*)-\hat{x}_i(t^*)| < \varepsilon, i=1,\ldots,n\} = \{t^* \varepsilon T \mid |x_i(t_o)-x_i(t)| < \varepsilon, i=1,\ldots,n\}$ where $x_1,\ldots,x_n \varepsilon C_b(T,\underset{\sim}{C},c)$ and $\varepsilon > 0$.

Hence the map $t \to t^*$ embeds T homeomorphically in $C_b(T,\underset{\sim}{C},c)^h$ and condition (a) is seen to be satisfied.

As for (c), given any $x\varepsilon C_b(T,\underset{\sim}{R})$, $\hat{x}$ is a continuous extension of x to $C_b(T,\underset{\sim}{C},c)^h$, i.e. $\hat{x}\varepsilon C(C_b(T,\underset{\sim}{C},c)^h,\underset{\sim}{C})$.

Last, it must be shown that T* is dense in $C_b(T,\underset{\sim}{C},c)^h$. If T* is not dense in $C_b(T,\underset{\sim}{C},c)^h$ there is some $f\varepsilon C_b(T,\underset{\sim}{C},c)^h$ and a basic neighborhood $V=V(f;x_1,\ldots,x_n,\varepsilon)$ of f such that $V \cap T^* = \emptyset$. Consider the functions $y_i=x_i-f(x_i)\varepsilon C_b(T,\underset{\sim}{C}), 1 \le i \le n$.

For each $t_o \varepsilon T$ there is some i such that $|y_i(t_o)| \geq \varepsilon$. Now, with $z = \Sigma_i y_i \bar{y}_i$, it is clear that $|z(t)| \geq \varepsilon^2$ for all $t \varepsilon T$: hence z is invertible in $C_b(T,\underset{\sim}{C},c)$. Yet, for each i, $f(y_i)=0$, so $f(z)=0$. In other words the maximal deal ker f contains the unit z which cannot be. We conclude that T* is dense in $C_b(T,\underset{\sim}{C},c)^h$ and therefore also that $\beta T=C_b(T,\underset{\sim}{C},c)$ or, equivalently, that $\beta T=\mathcal{M}$.

Next we establish a criterion for continuity of Ψ. An instance in which Ψ is discontinuous is given in Example 4.12-1

<u>(4.12-4) CONTINUITY OF Ψ</u>. If X is (a) a complete complex LMCH Q-algebra or (b) a barreled complex LMCH algebra then Ψ is continuous.

<u>Proof</u>. (a): By (4.12-3) we know that X^h is $\sigma(X',X)$-compact; consequently $C(X^h,\underset{\sim}{C},c)$ is a commutative complex Banach algebra with identity with the compact-open topology $\mathcal{T}_c$ generated by the supremum norm $\|\ \|$ over X^h. A neighborhood base at 0 for $\mathcal{T}_c$ is given by sets of the form $\{g\in C(X^h)\,|\,\sup_{f\in X^h}|g(f)| \leq \varepsilon\} = \{g\in C(X^h)\,|\,\|g\| \leq \varepsilon\}(\varepsilon> 0)$. Since $(X^h)^o$ is a neighborhood of the origin by (4.12-3), and $\Psi^{-1}(\{g\in C(X^h)\,|\,\|g\| \leq 1\}) = \{x\in X\,|\,\sup|<x,X^h>| \leq 1\} = (X^h)^o$, it follows that Ψ is continuous.

(b) If K is a $\sigma(X',X)$-compact subset of X^h, then, since X is barreled, K^o is a neighborhood of 0 in X. But $K^o = \{x\in X\,|\,|h(x)| \leq 1,\ h\in K\} = \{x\in X\,|\,|\Psi x(h)| \leq 1,\ h\in K\} = \Psi^{-1}(\{g\in C(X^h)\,|\,\sup_{f\in K}|g(f)| \leq 1)$. ∇

In the event that Ψ is 1-1--e.g., if the LMCH algebra X is strongly semisimple or semisimple, complete and complex ((4.12-1))--under what conditions will Ψ^{-1} be continuous?

<u>(4.12-5) CONTINUITY OF Ψ^{-1}.</u> If X is a LMCH algebra and its Gelfand map $\Psi: X \to C(X^h, \underset{\sim}{F}, c)$, $x \to \hat{x}$ is 1-1, then Ψ^{-1} is continuous whenever each equicontinuous subset of X' is contained in a multiple of the $\sigma(X',X)$-closed absolute convex hull of some compact subset of X^h.

<u>Proof</u>. Let $E \subset X'$ be equicontinuous and choose $n \in \underset{\sim}{N}$ and a compact set $K \subset X^h$ so that $E \subset n(\text{cl}_{\sigma(X',X)} K_{bc}) = nK^{oo}$. Then $E^o \supset (1/n)K^o$ and therefore $\Psi(E^o) \supset (1/n)\Psi(K^o)$. But $\Psi(K^o) = \{\Psi x \mid p_K(\Psi x) \leq 1\} = \bar{V}_{p_k} \cap \Psi(X)$ so that Ψ is a relatively open map.∇

Actually we are most interested in determining conditions under which Ψ is a topological isomorphism. The following well-known result moves us in that direction.

<u>(4.12-6) Ψ IS AN ISOMETRIC ISOMORPHISM $\Leftrightarrow \|(\ )^2\| = \|(\ )\|^2$.</u> If X is a complex Banach algebra, then Ψ is an isometric isomorphism into $C(X^h, \underset{\sim}{C}, c)$ iff $\|x^2\| = \|x\|^2$ for each $x \in X$.

<u>Proof</u>. We first establish the formula $r_\sigma(x) = \lim_n \|x^n\|^{1/n} (x \in X)$.

In any complex algebra with identity $\sigma(x^n) = \sigma(x)^n = \{\mu^n \mid \mu \in \sigma(x)\}$ for each $n \in \underset{\sim}{N}$ and each $x \in X$ since the polynomial $x^n - \mu e = \Pi_{i=1}^n (x - \mu_i e)$ where $\mu_1, \ldots, \mu_n$ are the n-th roots of μ. Thus $r_\sigma(x^n) = r_\sigma(x)^n$ for each $n \in \underset{\sim}{N}$ and $x \in X$. Now, by the discussion following (4.8-4), $r_\sigma(y) \leq \|y\|$ for each $y \in X$ and so $r_\sigma(x) = r_\sigma(x^n)^{1/n} \leq \|x^n\|^{1/n}$ $(n \in N, x \in X)$. To conclude the proof we show that $\limsup_n \|x^n\|^{1/n} \leq r_\sigma(x)$. Let a be an arbitrary real number larger than $r_\sigma(x)$. We shall see that $\|x^n\|^{1/n} \leq a$ for all but at most a finite number of indices n which will yield the desired conclusion.

Recall that the resolvent map $r_x, \mu \to (x - \mu e)^{-1}$, is analytic on $\rho(x)$ by (4.8-5). But whenever $|\mu| > r_\sigma(x), \mu \in \rho(x)$.

Thus, as was noted prior to (4.9-1), $f\cdot r_x$ is analytic on $\{\mu \mid |\mu| > r_\sigma(x)\}$ for each $f \varepsilon X'$. On the other hand, $r_x(\mu) = -\mu^{-1}\Sigma_{n\geq 0}\mu^{-n}x^n$ for each $|\mu| > \|x\|$. By the continuity of $f\varepsilon X'$, $f(r_x(\mu)) = (-\mu^{-1})\Sigma_{n>0}\mu^{-n}f(x^n)$ for each $|\mu| > \|x\|$. As Laurent expansions are unique, this expression is valid for all $|\mu| > r_\sigma(x)$. Hence if $r_\sigma(x) < b < a$ then $\Sigma_{n\geq 0} f(x^n/b^n)$ converges and $f(x^n/b^n) \to 0$ for each $f\varepsilon X'$. But a weakly convergent sequence in a Banach space is bounded so that a positive integer N exists such that $\|x^n/b^n\| \leq N$ for all $n\varepsilon \underset{\sim}{N}$. Therefore $\|x^n\|^{1/n} \leq (N^{1/n})b$ for each n and by choosing n sufficiently large, it follows that $\|x^n\|^{1/n} \leq a$.∇

Thus in Banach algebras where $\|(\ )^2\| = \|(\ )\|^2$, Ψ is a topological isomorphism. In complex LMCH algebras what happens if the topology is generated by seminorms which satisfy this condition? As we shall see, Ψ is a topological isomorphism for such algebras when $C(X^h, \underset{\sim}{C})$ carries a certain topology which is generally weaker than the compact-open topology. For ease of reference, we state some definitions, the first of which describes the aforementioned topology of $C(X^h, \underset{\sim}{C})$.

<u>Definition 4.12-2 THE WEAKENED COMPACT-OPEN TOPOLOGY</u>. Let X be a complex LMCH algebra and let X^h carry the $\sigma(X',X)$-subspace topology. The topology for $C(X^h, \underset{\sim}{C})$ generated by the seminorms $p_E, p_E(f) = \sup|f(E)|$, $f\varepsilon C(X^h, \underset{\sim}{C})$, where E runs over the closed equicontinuous subsets of $X^h (\subset X')$ is referred to as the <u>weakened compact-open</u> topology of $C(X^h, \underset{\sim}{C})$ and is denoted by $\mathcal{J}_{wc}$.

As X^h is $\sigma(X',X)$-closed, each closed equicontinuous subset E of X^h is $\sigma(X',X)$-compact, so the weakened compact-open topology is weaker than the compact-open topology of $C(X^h,\underset{\sim}{C})$. That this inclusion may be proper is demonstrated in Example 4.12-1.

Example 4.12-1 $\mathcal{T}_{wc} \neq$ COMPACT-OPEN TOPOLOGY. Let T be an uncountable compact Hausdorff space and $X=C(T,\underset{\sim}{C})$ carry the LMCH topology generated by the family of seminorms p_G where G is a countable compact subset of T. Just as in Example 4.5-1, it can be shown that the typical factor algebra $X_G=C(T,\underset{\sim}{C})/N_G$ $(N_G=\{x\varepsilon C(T,\underset{\sim}{C})\,|\,x(G)=\{0\}\})$ and is isometrically isomorphic to $C(G,\underset{\sim}{C},c)$ via the map $x+N_G \to x|_G$. Thus, as in Example 4.10-2, it follows by (4.10-7) that $X^h=T^*$. Furthermore we claim that the 1-1 map $t\to t^*$ from T onto T* is a homeomorphism. Since T is compact and T* is Hausdorff in the $\sigma(X',X)$-topology, it suffices to demonstrate continuity. Certainly if (t_μ) is a net from T converging to t then for each $x\varepsilon C(T,\underset{\sim}{C})$, $t_\mu^*(x)=x(t_\mu)\to x(t)=t^*(x)$ so that the map $t\to t^*$ is a homeomorphism. To determine the weakened compact-open topology of $C(X^h,\underset{\sim}{C})=C(T^*,\underset{\sim}{C})$ we must decide which of the compact subsets of T* are equicontinuous. Suppose K is compact and K* is equicontinuous. There is a neighborhood εV_{p_G} where $\varepsilon>0$ and G is a countable compact subset of T such that $\sup|K^*(\varepsilon V_{p_G})|\leq 1$. If we assume that K is uncountable then there is some $t\varepsilon K-G$. But T is completely regular and Hausdorff and G is compact so the contradictory conclusion follows that a function $x\varepsilon C(T,\underset{\sim}{C})$ exists such that $t^*(x)=x(t)=2$ and $x(G)=\{0\}$. Hence K must be countable in order for K* to be equicontinuous and the weakened compact-open topology is generated by the seminorms p_{K^*} where K is a countable compact subset of T. In other words the mapping

$\Psi: C(T,\underset{\sim}{C}) \to C(T^*,\underset{\sim}{C})$, $x \to \hat{x}$ is a topological isomorphism when $C(T^*,\underset{\sim}{C})$ carries the weakened compact-open topology. It follows by an argument similar to the one used to show that K* being equicontinuous implies K is countable that V_{p_T} is not a weakened compact-open neighborhood of 0 so that $\mathcal{T}_{wc}$ does not coincide with the compact-open topology. It is also worth noting that I is not continuous when $C(T^*,C)$ carries the compact-open topology.

The topology of the complex LMCH algebra of the preceding example is generated by a family of <u>square-preserving</u> multiplicative seminorms p_K, i.e., $p_K(x^2)=p_K(x)^2$, and, altho X and $C(X^h,\underset{\sim}{C},c)$ are not topologically isomorphic, X and $(C(X^h,\underset{\sim}{C}), \mathcal{T}_{wc})$ are. The theorem to follow, Theorem 4.12-1, generalizes the conclusion of the example to the extent that for algebras whose topology is determined by a family of square-preserving seminorms, Ψ is always a topological isomorphism into $C(X^h,\underset{\sim}{C})$ with the weakened compact-open topology.

<u>Definition 4.12-3. SQUARE ALGEBRAS</u>. A <u>square algebra</u> is a complex LMCH algebra whose topology is generated by a family P of square-preserving multiplicative seminorms.

It is simple to show that no loss of generality results from assuming the family P to be saturated, i.e., that P is closed with respect to the formation of finite maxima.

<u>Theorem 4.12-1 WHEN IS Ψ A TOPOLOGICAL ISOMORPHISM?</u> Let X be a complex LMCH algebra whose Gelfand map Ψ is 1-1 (e.g. if X is complete and semisimple or strongly semisimple). Then the following are equivalent.

(i) X is a square algebra;

(ii) there is a topological isomorphism of X into a product algebra $\Pi_{\mu\varepsilon M}C(G_\mu,\underset{\sim}{C},c)$ where each G_μ is a compact Hausdorff space;

(iii) there is a topological isomorphism of X into an algebra $C(T,\underset{\sim}{C},c)$ where T is a locally compact Hausdorff space;

(iv) there is a topological isomorphism of X into an algebra $C(T,\underset{\sim}{C},c)$ where T is a completely regular Hausdorff space;

(v) Ψ^{-1} is continuous when $C(X^h,\underset{\sim}{C})$ carries the weakened compact-open topology;

(vi) Ψ is a topological isomorphism when $C(X^h,\underset{\sim}{C})$ carries the weakened compact-open topology.

Furthermore if Ψ is continuous when $C(X^h,\underset{\sim}{C})$ carries the compact-open topology (e.g., if X is a complete Q-algebra or a barreled algebra) then we may replace (v) and (vi) by:

(v') $\Psi:X\rightarrow C(X^h,\underset{\sim}{C},c)$ is a topological isomorphism.

Proof. (i)$\Rightarrow$(ii) Let P be a saturated family of square-preserving multiplicative seminorms generating the topology on X. Recall that the factor algebra X_p is the completion of X/N_p $(N_p=p^{-1}(0))$ equipped with the norm $\dot{p}(x+N_p)=p(x)$ so that if we also denote the norm on X_p by $\dot{p}$, then $\dot{p}$ is square-preserving too. Thus, by (4.12-6), X_p is isometrically isomorphic to a subspace of $C(X_p{}^h,\underset{\sim}{C},c)$. Since (Theorem 4.6-1) (d)) X is topologically isomorphic to the projective limit of the X_p (a subalgebra of Π_pX_p), it follows that X is topologically isomorphic to a subalgebra of $\Pi_{p\varepsilon P}C(X_p{}^h,\underset{\sim}{C},c)$.

(ii)$\Rightarrow$(iii): Set $G_\mu'=\{\mu\}\times G_\mu$ and transfer the topology of G_μ to G_μ'; let $G=\cup_{\mu\varepsilon M}G_\mu'$. If G carries the topology $\mathcal{J}$ in which a subset U of G is open iff $U \cap G_\mu'$ is open in G_μ' for each $\mu \in M$, then G, referred to as the free union of the G_μ (Dugundji 1966, 131-133), is clearly locally compact and Hausdorff. Moreover since a function x defined on G is continuous iff each of its restrictions $x|_{G_\mu'}$ is continuous, the map $\Pi_{\mu\in M} C(G_\mu, \mathbb{C}, c) \to C(G, \mathbb{C}, c)$, $(x_\mu) \to x$ where $x(\mu,t) = x_\mu(t)$ for $(\mu,t) \in G_\mu'$, is a surjective topological isomorphism.

(iii)$\Rightarrow$(iv): Clear.

(iv)$\Rightarrow$(v): Let W be a topological isomorphism of X into $C(T, \mathbb{C}, c)$. Then for each $t \in T$ the functional $t^* \circ W \in X^h$. To see that the map $h, t \to t^* \circ W$ is continuous when X^h carries the $\sigma(X', X)$-topology, fix $x \in X$ and let (t_μ) be a net convergent to t. Then $h(t_\mu)(x) = t_\mu^*(W(x)) = W(x)(t_\mu) \to W(x)(t) = t^*(W(x)) = h(t)(x)$ for each $x \in X$, and h is seen to be continuous. Thus if K is a compact subset of T then h(K) is compact in X^h. Since W is a topological isomorphism, $U = \{x \in X \mid p_K(W(x)) \leq \varepsilon\}$ where $K \subset T$ is compact and $\varepsilon > 0$, is a typical basic neighborhood of 0 in X. But $\varepsilon(h(K))^o = \{x \in X \mid |h(t)(x)| \leq \varepsilon, t \in K\} = U$ so that a basis for the neighborhood system at 0 consists of multiples of polars of compact equicontinuous subsets of X^h. Now $\Psi(\varepsilon(h(K))^o) = \{\Psi x \mid p_{h(K)}(\Psi x) \leq \varepsilon\}$ i.e., Ψ is a relatively open map when $C(X^h, \mathbb{C})$ carries the weakened compact-open topology.

(v)$\Rightarrow$(vi): Since Ψ is an algebra isomorphism, it only remains to show that Ψ is continuous. Let K be a closed equicontinuous subset of X^h and consider the subspace neighborhood $V = \varepsilon \bar{V}_{p_K} \cap \Psi(X) = \{\Psi x \mid p_K(\Psi x) \leq \varepsilon\}$. Clearly $\Psi^{-1}(V) = \{x \in X \mid p_K(\Psi x) \leq \varepsilon\} = \{x \in X \mid |f(x)| \leq \varepsilon, f \in K\} = \varepsilon K^o$, a neighborhood of 0 in X, and Ψ is seen to be continuous.

(vi)⇒(i): Since Ψ is a topological isomorphism, the LMCH topology of X is generated by the family of multiplicative seminorms of the form $p_K \cdot \Psi$ where K is a closed equicontinuous subset of X^h. But $(p_K \cdot \Psi)(x^2) = p_K((\Psi x)^2) = p_K(\Psi x)^2 = ((p_K \Psi)(x))^2$ so that X is a square algebra.

(i)⇒(v'): Finally suppose that Ψ is continuous when $C(X^h, \underset{\sim}{C})$ carries the compact-open topology. If (i) holds, then so does (v). But $\mathcal{J}_{wc}$ is weaker than the compact-open topology, so $\Psi^{-1}: C(X^h, \underset{\sim}{C}, c) \cap \Psi(X) \to X$ is continuous and $\Psi: X \to C(X^h, \underset{\sim}{C}, c)$ is a topological isomorphism.

On the other hand, if (v') holds, then by essentially the same argument as in (vi)⇒(i), X is seen to be a square algebra.∇

Returning to the phenomenon observed in Example 4.12-1, we see that even tho the map $\Psi: X \to C(X^h, \underset{\sim}{C}, c)$ may fail to be a topological isomorphism, it is still possible for X to be algebraically and topologically embedded in an algebra $C(T, \underset{\sim}{C}, c)$. Indeed if X is the square algebra of Example 4.12-1, then, by (iii) of the preceding theorem, there is a topological isomorphism of X into an algebra $C(T, \underset{\sim}{C}, c)$ where T is a locally compact Hausdorff space. In Theorem 4.12-2 we focus attention on those complex topological algebras which are embeddable as complete subalgebras of algebras $C(T, \underset{\sim}{C}, c)$ which separate points in the compact Hausdorff space T. In this case $\Psi: X \to C(X^h, \underset{\sim}{C}, c)$ must be a topological isomorphism.

Let X be any topological algebra and suppose that $K \subset X^h$ is $\sigma(X', X)$-bounded so that K^o is absorbent. We denote the gauge of K^o by p_K. Note that X^h itself is $\sigma(X', X)$-bounded if

X is a complete complex LMCH Q-algebra by (4.12-3).

<u>(4.12-7) CONTINUITY OF THE SEMINORMS p_K AND r_σ.</u> If X is a barreled complete complex LMCH algebra then each of the seminorms p_K, where K is a $\sigma(X',X)$-bounded subset of X^h, is continuous. Thus if X is also a Q-algebra, then the gauge of $(X^h)^o$ is continuous: in this case it equals the spectral radius seminorm r_σ.

<u>Proof</u>. As K^o, the polar of the weak -* bounded subset K of X^h, is a barrel in the barreled space X, it is a neighborhood of 0 in X, and therefore, p_K is continuous. In the event that X is also a Q-algebra, X^h is $\sigma(X',X)$-compact by (4.12-3) so the gauge of $(X^h)^o$ is continuous. Thus it only remains to prove that r_σ is the gauge of $(X^h)^o$. Denoting the gauge of $(X^h)^o$ by p, we recall that $p(x) = \inf\{a>0 \mid x\in a(X^h)^o\}$ for each $x\in X$. If $r_\sigma(x)=0$, clearly then $p(x)=r_\sigma(x)=0$. Hence suppose that $r_\sigma(x)\neq 0$. Since $r_\sigma((r_\sigma(x))^{-1}x)=1$ and $(X^h)^o=U(\sigma)$ by (4.12-1) it follows that $(r_\sigma(x))^{-1}x\in (X^h)^o$ and $p(r_\sigma(x)^{-1}x)\leq 1$. Thus $p(x)\leq r_\sigma(x)$. However as $a(X^h)^o=\{x \mid r_\sigma(x)\leq a\}$, we have $r_\sigma(x)\leq a$ for each $x\in a(X^h)^o$ which, in turn, implies that $r_\sigma(x)\leq \inf\{a>0 \mid x\in a(X^h)^o\} = p(x)$ and the proof is complete.∇

In the next result we focus attention on complete subalgebras of function algebras on a compact set, with the aid of the mapping Ψ. Our next definition is a prerequisite for that result.

<u>Definition 4.12-4 UNIFORM ALGEBRAS</u>. If X is a complex topological algebra then X is a <u>uniform algebra</u>* provided that there

* Some authors, Browder 1969b for example, designate such algebras as <u>function algebras</u>.

exists a compact Hausdorff space T such that X is topologically isomorphic to a closed subalgebra of $C(T,\underset{\sim}{C},c)$ which separates the points of T, i.e. if t_1 and t_2 are distinct points of T there is a function χ in the closed subalgebra such that $\chi(t_1)\neq \chi(t_2)$.

In our next result, among other things, we obtain a necessary and sufficient condition for Ψ to be a topological isomorphism when X is a semisimple barreled complete complex LMCH Q-algebra. Warner proved that Ψ is a topological isomorphism when X is a full Frechet algebra ((4.12-8)).

<u>Theorem 4.12-2 CRITERION FOR WHEN X IS A UNIFORM ALGEBRA</u>. If X is a semisimple barreled complete complex LMCH Q-algebra then the following statements are equivalent:

(a) X is a uniform algebra,

(b) Ψ is a topological isomorphism,

(c) $\mathcal{T}_r$, the topology generated by the spectral radius norm r_σ, is a topology of the dual pair (X,X').

<u>Proof.</u> Since X is a semisimple complete complex LMCH algebra, X is strongly semisimple by (4.11-1). Thus $(\Psi x)(f)=f(x)=0$ for all $f\in X^h$ iff $x=0$, i.e. iff Ψ is an isomorphism. Furthermore as X is a barreled complex LMCH Q-algebra, it follows by (4.12-5) that r_σ is a continuous norm on X with respect to its original topology and so $\mathcal{T}_r\subset\mathcal{T}$. With these observations in mind we now proceed to establishing the necessary equivalences.

<u>(c)⇒(b)⇒(a):</u> Since the implication (b)⇒(a) is obvious we prove only that (c) implies (b). Suppose that $\mathcal{T}_r$ is a topology of the dual pair (X,X'). Since $\mathcal{T}_r$ is a locally convex metric topology it follows that $\mathcal{T}_r=\tau(X,X')$ (the Mackey topo-

logy). But X is barreled, so $\mathcal{J} = \tau(X,X')$ and $\mathcal{J} = \mathcal{J}_r$. Hence Ψ is a topological isomorphism.

(a)$\Rightarrow$(c): Suppose that X is a uniform algebra; then it may be viewed as a closed subalgebra of a function algebra $C(T,\underset{\sim}{C},c)$ where T is compact Hausdorff. As such it is a Frechet algebra and $\mathcal{J} = \tau(X,X')$. Having already observed that $\mathcal{J}_r \subset \mathcal{J}$, it will follow that $\mathcal{J}_r$ is a topology of the dual pair if it can be shown that $\sigma(X,X') \subset \mathcal{J}_r$. We accomplish this by proving that each $f \in X'$ is continuous in the $\mathcal{J}_r$ topology. If $f \in X'$ then there is a continuous linear functional $\hat{f}$ on $C(T,\underset{\sim}{C},c)$ extending f. By (2.2-1), $C(T,\underset{\sim}{C},c)'$ is the space spanned in the algebraic dual $C(T,\underset{\sim}{C})^*$ of $C(T,\underset{\sim}{C})$ by the $\sigma(C(T,\underset{\sim}{C})^*, C(T,\underset{\sim}{C}))$-closure of the balanced convex hull of the set $T^* = \{t^* | t \in T\}$ of evaluation maps. First consider an evaluation map $t_o^* \in T^*$. As $C(T,\underset{\sim}{C},c)$ is a complete complex LMCH algebra and the (continuous) homomorphisms of $C(T,\underset{\sim}{C},c)$ are just the elements of T^* (Example 4.10-2), we have $|t_o^*(x)| \leq \sup\{t^*(x) | t \in T\} = r_\sigma(x)$ for each $x \in X$ by (4.10-8). Clearly it follows that $|(\sum_{i=1}^{n} \alpha_i t_i^*)(x)| \leq r_\sigma(x)$ for each $x \in X$, whenever $\Sigma|\alpha_i| \leq 1$ and $t_i \in T$, $1 \leq i \leq n$. Furthermore, any continuous linear functional g in the $\sigma(C(T,\underset{\sim}{C})^*, C(T,\underset{\sim}{C}))$-closure of $(T^*)_{bc}$ is the limit of a filterbase $\mathcal{B} \subset (T^*)_{bc}$: $\mathcal{B}(x) \to g(x)$ for each $x \in X$. Thus $|g(x)| = \lim|\mathcal{B}(x)| \leq r_\sigma(x)$ for each $x \in X$. Hence the continuous linear functional $\hat{f}$, being a linear combination of elements g from the weak-* closure of $(T^*)_{bc}$, must satisfy an inequality of the form $|\hat{f}(x)| \leq K\, r_\sigma(x)$ for each $x \in X$, where K is some positive constant. Since $\hat{f}=f$ on X, the $\mathcal{J}_r$-continuity of f follows.∇

Michael (1952) proved that the Gelfand-topologized space X^h of nontrivial continuous scalar-valued homomorphisms of a full Frechet topological algebra X is hemicompact (Def. 2.1-1) and he raised the question as to whether it is a k-space (Sec. 2.3). Warner (1958) answered this question in the affirmative by showing that for such algebras the mapping Ψ is a topological isomorphism. We present these findings in our next result.

(4.12-8) FULL FRECHET ALGEBRAS. If X is a full Frechet algebra then the algebraic isomorphism Ψ, taking X onto $C(X^h,\underset{\sim}{F},c)$, is a topological isomorphism. Furthermore X^h is a hemicompact k-space.

Proof. Let Y denote $C(X^h,\underset{\sim}{F})$ carrying the complete metrizable locally convex Hausdorff topology $\mathcal{J}$ with neighborhood base at 0 of sets $\Psi(U)$ where U is a neighborhood of 0 in X. We shall show that X^h is a completely regular Hausdorff space in its Gelfand topology and $Y=C(X^h,\underset{\sim}{F},c)$. This suffices to prove the theorem as the contention that X^h is a hemicompact k-space follows by Theorem 2.1-1 and (2.3-4).

As the Gelfand topology of X^h is the relative $\sigma(X',X)$-topology, X^h is a completely regular Hausdorff space. The procedure for showing that $Y=C(T,\underset{\sim}{F},c)$ is as follows: First we show that $\mathcal{J}$ is finer than the compact-open topology by showing that the seminorms p_K which generate the compact-open topology are continuous on Y. Thus the identity mapping I taking Y onto $C(X^h,\underset{\sim}{F},c)$ will be a continuous isomorphism. Next we prove that $C(X^h,\underset{\sim}{F},c)$ is barreled so that I is almost open*.

* A linear map $A:X\to Y$, X and Y topological vector spaces, is almost open if for each neighborhood U of 0 in X, cl $A(U)$ is a neighborhood of 0 in $A(X)$.

Finally we shall conclude from the fact that Y is fully complete that I is open, i.e. a topological isomorphism.

To see that $\mathcal{T}$ is finer than the compact-open topology, consider the collection $(X^h)^*$ of evaluation maps on Y. If $h \in X^h$ then $h^*(\Psi x) = \Psi x(h) = h(x)$ so that h^* is continuous on Y. Hence $(X^h)^* \subset Y'$. Furthermore $(X^h)^*$ equipped with the relative $\sigma(Y', Y)$ topology is homeomorphic to X^h. Therefore if K is compact in X^h then K^* is equicontinuous on the barreled space Y as K^* is $\sigma(Y', Y)$-compact. Since $p_K(U) = \sup|U(K)| = \sup|K^*(U)|$ for each neighborhood U of 0 in Y, the continuity of p_K follows from the equicontinuity of K^* and $\mathcal{T}$ is stronger than the compact-open topology.

By our previous discussion it only remains to show that $C(X^h, \underset{\sim}{F}, c)$ is barreled. Since $C(X^h, \underset{\sim}{F}, c)$ is barreled whenever it is bornological ((2.6-1)) it suffices by Theorem 2.6-1 to show that X^h is replete. Thus, by Theorem 1.5-3(b) it suffices to show that X^h is σ-compact. Letting (U_n) be a countable base at 0 for the metrizable space X, it follows that $X' = \{0\}^o = (\bigcap_n U_n)^o = \bigcup U_n^o$. Thus X' as well as the closed subspace X^h are σ-compact and the proof is complete.∇

By scrutinizing the proof it is seen that a locally convex Hausdorff topology of $C(T, \underset{\sim}{F})$ which renders $C(T, \underset{\sim}{F})$ fully complete * and barreled and with respect to which all the evaluation maps are continuous, must be the compact-open topology. A somewhat more general result along these lines is (4.13-2).

---

* A locally convex Hausdorff space X is <u>fully complete</u> (or <u>Ptak</u>) if for each locally convex Hausdorff space Y, every

In Banach algebras with involution (Def. 4.12-5) the Gelfand map Ψ has more sharply defined properties. For example, for A*-algebras X, $\Psi(X)$ is dense in $C(X^h,\underset{\sim}{C},c)$ and if X is a B*-algebra, Ψ maps X isometrically isomorphically onto $C(X^h,\underset{\sim}{C},c)$. In the discussion to follow these notions are broadened and the results mentioned generalized.

Definition 4.12-5 ALGEBRAS WITH INVOLUTION. A complex topological algebra X is called an algebra with involution if there is a map $x \to x^*$ of X into itself such that for $x,y \in X$ and $\mu \in \underset{\sim}{C}$ (i) $(x^*)^*=x$, (ii) $(x+y)^*=x^*+y^*$, (iii) $(\mu x)^*=\bar{\mu}x^*$, and (iv) $(xy)^*=x^*y^*$. (It follows from (i) and (ii) that $x \to x^*$ is bijective.) X is a symmetric algebra with involution if $\Psi x^*= \overline{\Psi}x$ for each $x \in X$; sometimes "symmetric algebra with involution" will be shortened to "symmetric algebra." An algebra with involution X is a star algebra (or *-algebra) if there is a family P of multiplicative seminorms generating the topology on X such that for each $p \in P$ and $x \in X$, $p(xx^*)=p(x)^2$. As was the case for square algebras, no loss of generality is entailed by assuming P to be saturated.

If $X=C(T,\underset{\sim}{C},c)$ where T is any completely regular Hausdorff space, then X is an algebra with involution with respect to complex conjugation. Such an X is clearly also a *-algebra. By Example 4.10-2 and the definition of the Gelfand topology, X^h may be topologically identified with T. Moreover, identifying T and X^h, Ψx is just x for each $x \in X$, from which it follows

* Cont. continuous almost open map $A:X \to Y$ is relatively open in the sense that neighborhoods of 0 in X are mapped into neighborhoods of 0 in A(X). Every Frechet space is fully complete (Horvath 1966,p.299,Prop.3(a)). See also Exercise 2.3.

that X is symmetric.

In the class of complex Banach algebras, the symmetric algebras with involution are the A*-algebras and the star algebras are the B*-algebras.

The definition given above for a symmetric algebra with involution is motivated by our desire for such algebras to satisfy the condition that $\Psi(X)$ be dense in $C(X^h, \underset{\sim}{C}, c)$. Indeed to obtain this sort of density for complete X (Th. 4.12-3 (c)) we make use of the Stone-Weierstrass theorem which makes it necessary for $\Psi(X)$ to be closed under complex conjugation. Hence a natural condition to impose on X is that $\Psi x^* = \overline{\Psi}x$.

(4.12-9) SYMMETRIC ALGEBRAS. Let X be a complete LMCH algebra with involution and Ψ be the Gelfand map. Then the following are equivalent:

(a) For any $x \in X$, $xx^* + e$ is invertible.

(b) If x is self adjoint (i.e. $x=x^*$) then $\sigma(x) \subset \underset{\sim}{R}$;

(c) X is a symmetric algebra.

Proof. (a)⇒(b): Suppose that while (a) holds there is a self-adjoint element x whose spectrum $\sigma(x)$ is not wholly real. Choose $\mu=a+ib \in \sigma(x)$ where $b \neq 0$ and consider

$$(x-(a+ib)e)(x-(a-ib)e) = (x-ae)^2 + b^2e = b^2((x-ae/b)^2 + e).$$

Now $w = (x-ae)/b = (x^*-ae^*)/b = ((x-ae)/b)^* = w^*$. Since (a) holds, $ww^* + e = ((x-ae)/b)^2 + e$ has an inverse, say y. But then $(x-(a+ib)e)((x-(a-ib)e)(y/b)^2) = e$ which contradicts the assumption that $\mu=a+ib \in \sigma(x)$.

(b)⇒(c): Clearly if $x \in X$ then, by taking u and v to be the self-adjoint elements $(x+x^*)/2$ and $(x-x^*)/2i$ respectively, $x=u+iv$ and $x^*=u-iv$. Thus, for $h \in X^h$, $h(x)=h(u)+ih(v)$

while $h(x^*)=h(u)-ih(v)$. Since $h(u)\varepsilon\ \sigma(u)$ and $h(v)\varepsilon\ \sigma(v)$ and $h(u)$ and $h(v)$ are real, it follows that $h(x^*)=\overline{h(x)}$ and therefore $(\Psi x^*)(h) = h(x^*) = \overline{h(x)} = \overline{\Psi x(h)}$ for each $x\varepsilon X$.

(c)$\Rightarrow$(a): If $h\varepsilon X^h$ and $x\varepsilon X$ then $h(xx^*+e) = h(x)h(x^*) + h(e) = (\Psi x)(h)(\Psi x^*)(h) + 1 = (\Psi x)(h)\overline{(\Psi x)}(h) + 1 = |h(x)|^2 + 1 > 0$. As X is complete, $\sigma(xx^*+e) = \{h(xx^*+e)|h\varepsilon X^h\}$ by (4.10-8) and therefore $0 \notin \sigma(xx^*+e)$. Hence xx^*+e is invertible.∇

Prior to discussing the connection between star algebras and symmetric algebras we record some elementary properties of star algebras.

(4.12-10) PROPERTIES OF STAR ALGEBRAS. If X is a star algebra with P a family of seminorms such that $p(xx^*) = p(x)^2$ for each $p\varepsilon P$ and generating the topology on X (cf. Def. 4.12-5) then

(a) for each $x\varepsilon X$ and $p\varepsilon P$, $p(x) = p(x^*)$ and

(b) X is a square algebra.

Proof. (a) Certainly for any $x\varepsilon X$ $(p(x))^2 = p(xx^*) \leq p(x)p(x^*)$ and $p(x^*)^2 = p(x^*x^{**}) = p(xx^*) \leq p(x)p(x^*)$. Thus $p(x)=0$ iff $p(x^*)=0$. If $p(x)$ and $p(x^*)$ are not zero then it follows from the equations above that $p(x) = p(x^*)$.

(b) For $x\varepsilon X$ and $p\varepsilon P$ consider $p(x^2)^2 = p(x^2(x^2)^*) = p(x^2(x^*)^2) = p((xx^*)^2) = p((xx^*)(xx^*)^*) = (p(xx^*))^2 = p(x)^4$. Thus $p(x^2) = p(x)^2$ and X is a square algebra.∇

(4.12-11) COMPLETE STAR ALGEBRAS ARE SYMMETRIC. A complete star algebra is symmetric.

Proof. Suppose that X is a Banach algebra, i.e., $P = \{\|\ \|\}$ and $(X, \|\ \|)$ is complete. By (4.12-9)(b) it suffices to show that each self-adjoint element has a real spectrum. Let x be self-adjoint; we claim that $i \notin \sigma(x)$. As the spectrum of an element $y \varepsilon X$ coincides with $(\Psi y)(X^h)$ by (4.10-8), it follows that for any polynomial p with complex coefficients, $\sigma(p(x)) = p(\sigma(x)) = \{p(\mu) | \mu \varepsilon \sigma(x)\}$. Hence if we assume that $i \varepsilon \sigma(x)$, then, for any $a \varepsilon \underset{\sim}{R}$, $(1+a)i \varepsilon \sigma(x+aie)$. Since e is self-adjoint an element $y \varepsilon X$ is invertible iff y^* is invertible. Therefore, since $(x+aie)^* - (-(1+a)ie) = (x+aie - (1+a)ie)^*$ and $(1+a)i \varepsilon \sigma(x+aie)$, $-(1+a)i \varepsilon \sigma((x+aie)^*)$. In the proof of (4.12-6) it was shown that $r_\sigma(y) = \lim_n \|y^n\|^{1/n}$ for any $y \varepsilon X$. Since X is a square algebra by (4.12-10), i.e., $\|y^2\| = \|y\|^2$, it follows that $r_\sigma(y) = \|y\|$ for each $y \varepsilon X$. Thus $(1+a)^2 \leq r_\sigma(x+aie) r_\sigma((x+aie)^*) = \|x+aie\| \|(x+aie)^*\| = \|x+aie\|^2 = \|(x+aie)(x+aie)^*\| = \|x^2+a^2e\| \leq \|x^2\| + a^2$ since $\|e\| = r_\sigma(e) = \sup\{|h(e)| | h \varepsilon X^h\} = 1$ by (4.10-8). But then $1+2a \leq \|x\|^2$ for each $a \varepsilon \underset{\sim}{R}$--an obvious contradiction. We conclude therefore that $i \notin \sigma(x)$. Next suppose that $\mu = a+ib \varepsilon \sigma(x)$ where $b \neq 0$. Then, for $p(c) = (c-a)/b$, $i = p(\mu) \varepsilon \sigma(p(x)) = \sigma((x-ae)/b)$ and $(x-ae)/b$ is self-adjoint. Thus each self-adjoint element of X has real spectrum and X is a symmetric algebra.

Suppose now that X is a star algebra with P as in Def. 4.12-5 and $p \varepsilon P$. Then the norm $\dot{p}(x+N_p) = p(x)$ $(N_p = p^{-1}(0))$ satisfies the condition $\dot{p}((x+N_p)(x^*+N_p)) = \dot{p}(x^2+N_p)$ for each $x \varepsilon X$. If we define $(x+N_p)^* = x^*+N_p$ then X/N_p is an algebra with involution. Moreover it is also clear from the fact that $\dot{p}((x+N_p)^*) = \dot{p}(x+N_p)$ that this involution has a unique extension to X_p, the completion of X/N_p, so that $\dot{p}(zz^*) = (p(z))^2$ for each $z \varepsilon X_p$. Hence $(X_p, \dot{p})$ is a symmetric algebra for each

$p\varepsilon P$. Suppose that $x\varepsilon X$ is self-adjoint. Then for each $p\varepsilon P$, $x+N_p$ is self-adjoint in X_p and $\sigma(x+N_p) \subset \underset{\sim}{R}$. Since X is complete, we know by (4.7-2) that $\sigma(x) = \cup_{p\varepsilon P}\sigma(x+N_p) \subset \underset{\sim}{R}$. Thus X is symmetric.∇

In our next result some well-known results about A*- and B*- algebras are generalized.

A homomorphism H between algebras with involution X and Y is a <u>star</u> <u>homomorphism</u> (or <u>*-homomorphism</u>) if $Hx^* = (Hx)^*$ for each $x\varepsilon X$.

<u>Theorem 4.12-3</u>. <u>X"=" $C(X^h,\underset{\sim}{C},\mathcal{J}_{wc})$</u>. Let X be an LMCH algebra with involution. Then

(a) if X is a symmetric algebra, $\Psi(X)$ is dense in $C(X^h,\underset{\sim}{C},c)$;

(b) If X is a star algebra, Ψ is a topological star isomorphism of X into $C(X^h,\underset{\sim}{C},\mathcal{J}_{wc})$;

(c) if X is a complete star algebra then X is a symmetric algebra and Ψ is a topological star isomorphism of X onto $C(X^h,\underset{\sim}{C},\mathcal{J}_{wc})$.

<u>Proof</u>. (a) $\Psi(X)$ is a subalgebra of $C(X^h,\underset{\sim}{C},c)$ containing the constant functions, separating the points of X^h, and closed under complex conjugation. The desired result follows from the Stone-Weierstrass theorem (Dugundji 1966, pp. 282-293): A subalgebra of $C(T,\underset{\sim}{C},c)$, T a Hausdorff space, closed under conjugation, separating points of T, and containing constants is dense in $C(T,\underset{\sim}{C},c)$.

(b) The result of (4.12-10)(b) already shows that a star algebra is a square algebra. Hence Ψ is a topological isomorphism into $C(X^h,\underset{\sim}{C},\mathcal{J}_{wc})$ by Theorem 4.12-1. It remains to show that Ψ is a star isomorphism. In the proof of (4.12-11) it was argued that each of the factor algebras X_p is a star algebra (B*-algebra) with respect to the involution obtained by (uniquely) extending the involution defined by $(x+N_p)^* = x^* + N_p$ on X/N_p. Let Ψ_p be the Gelfand map taking $X_p{}^h$ into $C(X_p{}^h, \underset{\sim}{C}, c)$. It was also established in the proof of (4.12-11) that the norm in a B*-algebra coincides with the spectral radius r_σ. Thus Ψ_p is an isometry and $\Psi_p(X_p)$ is complete, hence symmetric by (4.12-11). Hence we may conclude by (a) that $\Psi_p(X_p) = C(X_p{}^h, \underset{\sim}{C}, c)$. We also claim that Ψ_p is a star isomorphism. Indeed $\Psi_p(X_p)$ is a symmetric algebra so that $\Psi_p z^* = \overline{\Psi_p z}$ for each $z \in X_p$. Finally, recalling that $X^h = \cup_{p\in P} \kappa_p'(X_p{}^h)$, where κ_p is the canonical map $x \to (x+N_p)$ by (4.10-7), it follows that for any $h = \kappa_p'(h_p) \in \kappa_p'(X_p{}^h)$, $(\Psi x^*)(h) = \Psi x^*(\kappa_p' h_p) = \kappa_p'(h_p(x^*) = h_p(x^*+N_p) = h_p((x+N_p)^*) = \overline{h_p(x+N_p)} = \overline{\kappa_p'(h_p)(x)} = \overline{(\Psi x)}(h)$. Hence Ψ is a star isomorphism and the proof of (b) is complete.

(c) We know from (4.12-11) that any complete star algebra is a symmetric algebra. Ψ is a topological star isomorphism by (b) and the surjectivity of Ψ follows from (a) and the fact that the compact-open topology is stronger than the weakened compact-open topology.∇

It is well known that for T_1 spaces, paracompactness is equivalent to the existence of partitions of unity (defined in a footnote to a part of the proof of (2.4-4)) subordinate to any given open cover of the space. When $\Psi(X) = C(X^h, \underset{\sim}{C}, c)$, for example, figuring that Ψ will be reasonably kind to partitions of unity on X, it may be that existence of partitions of unity on X play a role in characterizing the algebras X for which X^h

is paracompact. But first: What is a partition of unity for an LMC algebra?

Definition 4.12-6 PARTITIONS OF UNITY. Let X be an LMC algebra with topology generated by a saturated family P of multiplicative seminorms. A family $(x_p)_{p\varepsilon P}$ of elements x_p from X is called a P-partition of unity if (1) for each $q\varepsilon P$ there is a finite subset F_q of P such that $q(x_p)=0$ iff $p \notin F_q$ (i.e., each q vanishes on all but finitely many x_p's) and $q(x_p) \leq 1$ for all $p\varepsilon P$; (2) for each $p\varepsilon P$, $q(x_p)=0$ for each $q \notin F_p$; (3) for each $q\varepsilon P$ and $x\varepsilon X$, $q(x-\Sigma_{p\varepsilon F_q} xx_p) = 0$.

The force of (3) is that the net of elements $\Sigma_{p\varepsilon F} xx_p$ indexed by the directed (by set inclusion) family of finite subsets F of P converges to x so that we may sensibly write $\Sigma_{p\varepsilon P} xx_p = x$ for each $x\varepsilon X$. In particular $\Sigma_p x_p = e$.

(4.12-12) WHEN IS X^h A LOCALLY COMPACT PARACOMPACT SPACE? Let X be a complete LMCH algebra with involution and Ψ be the Gelfand map taking X into $C(X^h,\underset{\sim}{C},c)$. Then Ψ is a surjective topological star isomorphism and X^h is locally compact and paracompact iff X is a star algebra whose topology is generated by a family P of multiplicative seminorms (as in Def. 4.12-5) for which a P-partition of unity exists.

Proof. Suppose that Ψ is a surjective topological star isomorphism and X^h is a locally compact paracompact space. Then X is a star algebra generated by the family of seminorms $p_K \cdot \Psi$ where K is a compact subset of X^h. As X^h is locally compact, each $h\varepsilon X^h$ has a relatively compact open neighborhood U_h. By the paracompactness of X^h, a neighborhood finite refinement of $(U_h)_{h\varepsilon X^h}$ exists consisting of relatively compact open sets.

Let $\mathcal{G}$ be the family of closures of the sets in the refinement. $\mathcal{G}$ is clearly a neighborhood finite cover of X^h. Furthermore we may assume that $\mathcal{G}$ is closed with respect to the formation of finite unions without altering the fact that $\mathcal{G}$ is a neighborhood finite cover of X^h. Now paracompactness of X^h also guarantees the existence of a partition of unity $(f_G)_{G\varepsilon\mathcal{G}}$ subordinate to $\mathcal{G}$, i.e., for each $f_G \geq 0$, $f_G(h)=0$ whenever $h \notin G$ and $\Sigma_{G\varepsilon\mathcal{G}} f_G$ is the identity element of $C(X^h,\underset{\sim}{C})$.

Let K be an arbitrary compact subset of X^h. Since $\mathcal{G}$ is neighborhood finite, only finitely many G's meet K. But $\mathcal{G}$ covers K and is closed with respect to the formations of finite unions. Thus K is contained in some $G\varepsilon\mathcal{G}$ and therefore the saturated families $\{p_G | G\varepsilon\mathcal{G}\}$ and $P = \{p_G \cdot \Psi | G\varepsilon\mathcal{G}\}$ generate the topologies of $C(X^h,\underset{\sim}{C},c)$ and X respectively. Moreover if for each $G\varepsilon\mathcal{G}$ we set $F_G = \{G'\varepsilon\mathcal{G} | G' \cap G \neq \emptyset\}$ then $p_G(f_{G'}) = 0$ for all $G \notin F_{G'}$ so $p_G(f - \Sigma_{G'\varepsilon F_G} ff_{G'}) = 0$ for each $G\varepsilon\mathcal{G}$ and $f \;\varepsilon C(X^h,\underset{\sim}{C})$. Hence it clearly follows that the family $X = \{\Psi^{-1}(f_G) | G\varepsilon\mathcal{G}\}$ serves as a P-partition of unity in X.

Conversely suppose that X is a star algebra with topology generated by P and $(x_p)_{p\varepsilon P}$ is a P-partition of unity for X. Then Ψ is a star isomorphism of X onto $C(X^h,\underset{\sim}{C},\mathcal{T}_{wc})$ by Theorem 4.12-3(c). First we show that X^h is locally compact. Let $G_p = \{h\varepsilon X^h | h(x_p) \neq 0\} = C((\Psi x_p)^{-1}(0))$ for each $p\varepsilon P$. Clearly each G_p is open. Since $\cup_{p\varepsilon P} G_p = X^h$ (if $h\varepsilon X^h$ vanished at each x_p, then h would vanish on e and be trivial) it suffices to show that each G_p has compact closure to conclude that X^h is locally compact. Indeed, $G_p \subset X^h = \cup_{q\varepsilon P} \kappa'_q(X_q{}^h)$ by (4.10-7) (b). Now, if $q \notin F_p$, then $q(x_p)=0$ and therefore $x_p \;\varepsilon\; N_q$.

Thus for each $h_q \in X_q^h$, $(\Psi x_p)(h_q \cdot \kappa_q) = (h_q \cdot \kappa_q)(x_p) = h_q(x_p + N_q) = 0$ and $\kappa_q'(X_q^h) \subset (\Psi x_p)^{-1}(0)$. Therefore $G_p \cap \kappa_q'(X_q^h) = \emptyset$ whenever $q \notin F_p$. This means that $G_p \subset \cup_{q\varepsilon F_p} \kappa_q'(X_q^h)$-a compact set by (4.10-7)(c).

To see that X^h is a paracompact, we begin by establishing the fact that each G_p meets only finitely many G_q's. For fixed p, suppose $G_p \cap G_q \neq \emptyset$, i.e., there is an $h\varepsilon G_p$ which is also in G_q. Now $h\varepsilon\kappa_s'(X_s^h)$ for some $s\varepsilon P$, whence $h(N_s) = \{0\}$. Thus neither x_p nor x_q belong to N_s so $s\varepsilon F_p \cap F_q$. As F_p is finite, if we assume that $F_p \cap F_q \neq \emptyset$ for infinitely many q's, then there must be an element $s\varepsilon F_p$ which also belongs to infinitely many F_q's. This means that s fails to vanish on infinitely many members of $(x_p)_{p\varepsilon P}$ which is contradictory. Hence there are only finitely many G_q's that meet G_p.

Now suppose that $\mathcal{U}$ is an open cover of X^h. For each $p\varepsilon P$ choose $C_{1,p},\ldots,C_{n_p,p} \varepsilon \mathcal{U}$ which meet and cover G_p (recall that cl G_p is compact). Next for each $1 \leq i \leq n_p$, let $U_{i,p} = C_{i,p} \cap G_p$. To see that $X^h = \cup_{p\varepsilon P} \cup_{i=1}^{n_p} U_{i,p}$, let $h\varepsilon X^h$. Then $h\varepsilon G_p$ for some $p\varepsilon P$ and therefore to $C_{i,p}$ for some $1 \leq i \leq n_p$. But $U_{i,p} = G_p \cap C_{i,p}$ so $h\varepsilon U_{i,p}$. Finally we claim that the $U_{i,p}$'s are a neighborhood refinement of $\mathcal{U}$. G_p meets each $U_{i,p}$, $i=1,\ldots,n_p$ and if G_p meets G_q, it is possible for G_p to meet $U_{i,q}$, $1 \leq i \leq n_q$. On the other hand suppose that $G_p \cap G_q = \emptyset$. Then, for each $1 \leq i \leq n_q$, $U_{i,q} \subset G_q$ so that $U_{i,q} \cap G_p = \emptyset$. Thus G_p can meet only finitely many $U_{i,q}$'s. Next, consider an arbitrary $U_{i,p}$. By definition $U_{i,p} \subset C_{i,p} \varepsilon \mathcal{U}$. Hence $\mathcal{U}$ has a neighborhood finite refinement and X^h is paracompact.

By Theorem 4.12-3(c), the final thing to establish is

that $\mathcal{T}_{wc}$ and the compact-open topology coincide. Suppose that K is a compact subset of X^h. We know that the G_p's are neighborhood finite, they form an open cover of X^h, and each G_p meets only finitely many G_q's. Hence K can meet only finitely many G_p's and is therefore contained in some finite union of the G_p's. On the other hand we saw that each G_p meets only finitely many of the covering sets $\kappa_q'(X_q^h)$, $q \in P$. Thus K is contained in some finite union of $\kappa_q'(X_p^h)$, each one of which is equicontinuous by (4.10-7)(a). In conclusion K is equicontinuous and the weakened compact-open topology coincides with the compact-open topology.∇

4.13 Continuity of Homomorphisms.

In Section 4.10 we proved that any (complex) homomorphism in a complex LMC Q-algebra is continuous (4.10-5) and gave examples of complex LMCH algebras on which discontinuous homomorphisms exist (Ex. 4.10-1). In this section we discuss continuity of homomorphisms between complex LMCH algebras. We prove that any homomorphism taking a complex barreled Q-algebra into a strongly semisimple fully complete (See Sec. 4.12) complex LMCH algebra is continuous. From this we obtain the fact that there is at most one topology on a complex algebra with respect to which it is a fully complete barreled LMCH Q-algebra--a result which subsumes the well known fact that there is at most one Banach algebra topology for a complex semisimple algebra.

In proving the theorem on the continuity of homomorphisms mentioned above it is first established that the given homomorphism is continuous when the range space carries a special topology, called the homomorphism topology. We define this topology now.

Definition 4.13-1 HOMOMORPHISM TOPOLOGY. Let X be a topological algebra and X^h denote the nontrivial continuous homomorphisms of X. The initial topology on X generated by the linear span $[X^h]$ of X^h in X' is denoted by $\sigma(X,X^h)$ and referred to as the homomorphism topology. A base of neighborhoods of 0 for $\sigma(X,X^h)$ consists of sets of the form $V(0,f_1,\ldots,f_n,\varepsilon) = \{x \in X \mid |f_i(x)| < \varepsilon,\ f_i \in X^h, 1 \leq i \leq n\}$ where $\varepsilon > 0$. If the linear transformation A maps X into the topological algebra Y and A is continuous when X and Y carry their homomorphism topologies then A is called homomorphically continuous.

Suppose that X is a complex LMC algebra. Clearly if it is strongly semisimple then X^h separates the points of X and $\sigma(X,X^h)$ is Hausdorff. Conversely, if $\sigma(X,X^h)$ is Hausdorff then for any non-zero $x \in X$ there is some $g \in [X^h]$ where $g = \Sigma\alpha_i f_i$ and $f_i \in X^h$, such that $g(x) \neq 0$. Thus $f_i(x) \neq 0$ for some i and the kernel of f_i, a closed maximal ideal, does not contain x. Hence X is strongly semisimple. We summarize these observations below.

(4.13-1) HAUSDORFF HOMOMORPHISM TOPOLOGIES. If X is a complex LMC algebra then $\sigma(X,X^h)$ is Hausdorff iff X is strongly semisimple.

The main result of this section is in three parts. The first two parts are concerned with continuity of a homomorphism when one or both spaces carry their homomorphism topologies. The third, which is established with the aid of the closed graph theorem is the result mentioned in the introductory remarks of this section on continuity of homomorphisms.

Theorem 4.13-1. CONTINUITY OF HOMOMORPHISM. Let A be a homomorphism taking the complex topological algebra X into the complex topological algebra Y. Then:

(a) A is homomorphically continuous if it is continuous when X and Y carry their original topologies;

(b) If X is a complete LMCH Q-algebra and Y is a topological algebra then A is continuous when Y carries its homomorphism topology and X its original topology;

(c) If the complete LMCH Q-algebra X is barreled and the LMCH algebra Y is strongly semisimple and fully complete, then A is continuous.

Proof. (a) For any $f_1,\ldots,f_n \in Y^h$, note that $f_1A,\ldots,f_nA \in X^h$. Clearly for each $\varepsilon > 0$, $A(V(0,f_1A,\ldots,f_nA,\varepsilon) \subset V(0,f_1,\ldots,f_n\varepsilon)$, so A is homomorphically continuous.

(b) Since X is a complete LMCH Q-algebra, it follows from (4.12-3) that $(X^h)^o$ is a neighborhood of 0 in X. Consider any finite collection $f_1,\ldots,f_n \in Y^h$; then by (4.10-5) we have that $f_1A,\ldots,f_nA \in X^h \cup \{0\}$. Thus for any $\delta > 0$ and $x \in \delta(X^h)^o, |f_iA(x)| \leq \delta$ for $i=1,\ldots,n$ and we see that $A(\delta(X^h)^o) \subset V(0,f_1,\ldots,f_n,\delta)$. Hence A is continuous when Y carries $\sigma(Y,Y^h)$.

(c) By (b) A is continuous when Y carries $\sigma(Y,Y^h)$. Since Y is strongly semisimple, $\sigma(Y,Y^h)$ is a Hausdorff topology, so the graph of the continuous homomorphism A is closed in the product topology on $X \times Y$ when X carries its original topology and Y carries $\sigma(Y,Y^h)$. [Any continuous map $f:S \to T$ where S is a topological space and T a Hausdorff space has a closed graph in $S \times T$]. Now this product topology is clearly coarser than the one induced by the original topologies of X and Y, so the graph of A remains closed in the product of the original topologies. Finally, since X is barreled and Y is fully complete, the closed graph theorem implies that A is continuous when X and Y carry their original topologies.∇

As an application of (c) above, suppose that X is a semisimple fully complete, barreled, complex LMCH Q-algebra when X carries either of the topologies $\mathcal{J}$ or $\mathcal{J}'$. Then the identity homomorphism $(X,\mathcal{J})\to(X,\mathcal{J}')$, $x\to x$ is bi-continuous by part (c) of the previous theorem and $\mathcal{J}=\mathcal{J}'$. Hence we may state:

<u>(4.13-2) UNIQUENESS OF THE TOPOLOGY OF A SEMISIMPLE, FULLY COMPLETE BARRELED COMPLEX LMCH Q-ALGEBRA</u>. If X is a complex semisimple algebra then there is at most one topology with respect to which X is a fully complete, barreled, LMCH Q-algebra. Thus there is at most one topology making such an X a Banach algebra.

Exercises 4.

4.1 WEAKLY TOPOLOGIZED ALGEBRAS. Let X be a real or complex algebra and X' a total subspace of the algebraic dual X*.

(a) COMPATIBILITY OF $\sigma(X,X')$. The following are equivalent:

(1) $(X, \sigma(X,X'))$ is an LMC algebra.

(2) $(X, \sigma(X,X'))$ is a topological algebra.

(3) For each $x' \in X'$, $x'^{-1}(0)$ contains a $\sigma(X,X')$-closed ideal of finite codimension.

(4) The map $(x,y) \to xy$ is continuous at 0 when X carries $\sigma(X,X')$ and $X \times X$ its product topology.

Hint. We sketch proofs of two preliminary statements, A and B.

(A) Let $x' \in X'$, W be a $\sigma(X,X')$-neighborhood of 0 and $W \supset Y$ a subspace of X. If $W \cup W^2 \subset \{x'\}^o$ then $Y, xY, Yx \subset x'^{-1}(0)$ for each $x \in X$. Moreover if W^3 is also contained in $\{x'\}^o$ then $xYx \subset x'^{-1}(0)$ for all $x \in X$. To see that $Yx \subset x'^{-1}(0)$ let n be a positive integer and $y \in Y$, $x \in X$ and consider $nyx = \mu^{-1}ny(\mu x)$ where $\mu x \in W$ for small enough μ. Thus $nyx \in \{x'\}^o$ for each such n and it follows that $yx \in x'^{-1}(0)$.

(B) If V is a $\sigma(X,X')$-neighborhood of 0 then $L = \cap\{x'^{-1}(0) \mid x' \in V^o\}$ is a $\sigma(X,X')$-closed subspace of finite codimension. (To prove the finite codimensionality of L choose $\{x'_1,\ldots,x'_n\} \subset X'$ such that $\{x'_1,\ldots,x'_n\}^o \subset V$. It follows that $L \supset \cap_{i=1} x_1'^{-1}(0)$ is of finite codimension.)

We now provide a hint to the proof that (1) implies (3). For $x' \in X'$ there exists an absolutely m-convex closed $\sigma(X,X')$-neighborhood of the origin V such that $V \subset \{x'\}^o$. Letting $L = \cap \{y'^{-1}(0) \mid y' \in V^o\}$, a $\sigma(X,X')$-closed subspace of finite codimension by (B), it follows that $L \subset V^{oo} = V$. To see that L is an ideal, consider nxy for a positive integer n, $x \in X$, and $y \in L$. Then $nxy = \mu x(n\mu^{-1}y) \in VL \subset V^2 \subset V$ for sufficiently large μ; since n is arbitrary, $xy \in L$. Consider the $\sigma(X,X')$-closed subspace of finite codimension $Y = L \cap x'^{-1}(0)$. Since $Y \subset L \subset V = V \cup V^2 \subset x'^{-1}(0)$, it follows by (A) that $YX, XY \subset x'^{-1}(0)$--thus $YX, XY \subset Y$. We now provide a hint to proving that (3) implies (1). Given $x' \in X'$ it suffices to obtain a multiplicative neighborhood V of 0 such that $V \subset \{x'\}^o$. If I is a $\sigma(X,X')$-closed ideal of finite codimension, then it follows that X/I is a finite-dimensional normed algebra. Let $\dot{x}'(x+I) = x'(x)$, since $\dot{x}'$ is continuous, choose a multiplicative neighborhood U of 0+I in X/I such that $U \subset \{\dot{x}'\}^o$. Then $V = \kappa^{-1}(U)$ (κ the canonical homomorphism of X onto X/I) is the desired neighborhood.

Finally we provide a hint to the proof that (4) implies (3). By (4), the composite map $(x,y,z) \to (xy,z) \to xyz$ is continuous when X is endowed with the $\sigma(X,X')$ topology and the products with their product topologies. It follows that there is an absolutely convex $\sigma(X,X')$-closed neighborhood of 0,W, such that $W \subset W \cup W^2 \cup W^3 \subset \{x'\}^o$. If $L = \cap\{y'^{-1}(0) \mid y' \in W^o\}$ then $L \subset W^{oo} = W$ and L is a $\sigma(X,X')$-closed subspace of finite codimension. Since the ideal I generated by L coincides with the subspace spanned by $L \cup XL \cup LX \cup XLX$, it follows by (A) that $I \subset x'^{-1}(0)$. Once again if κ is the canonical homomorphism of X onto X/I, the proof is completed by observing that $\kappa(I)$ is a closed subspace of X/I and $I = I+L = \kappa^{-1}(\kappa(I))$.

In the discussion to follow it is shown that multiplication in an LMCH algebra X with continuous dual X' is separately continuous in the $\sigma(X,X')$ topology but may not be jointly continuous with respect to that topology.

(b) SEPARATE VERSUS JOINT WEAK CONTINUITY OF MULTIPLICATION.

(i) If X is an LMCH algebra with continuous dual X' then multiplication is separately $\sigma(X,X')$-continuous, i.e., the maps $x \to yx$ and $x \to xy$ are continuous. (ii) If X is a complex semi-simple A*-algebra (see after Def. 4.12-5) with continuous dual X', then $(X,\sigma(X,X'))$ is a topological algebra iff X is finite-dimensional.

Hint to (ii): Suppose that $\sigma(X,\sigma(X,X'))$ is a topological algebra and X is infinite-dimensional. Since X is semi-simple, $\{0\} = \{f^{-1}(0) \mid f \in X^h\}$, and X is infinite-dimensional, then X^h is infinite. Now X^h is a subset of the unit sphere of the Banach space X' [(4.10-5)]; so $f = \Sigma_{n\in N} 2^{-n} h_n \in X'$ where (h_n) is any distinct sequence of elements of X^h. By (3) of (a), $f^{-1}(0)$ contains a $\sigma(X,X')$-closed ideal I of finite co-dimension. For each $x \in I$, $x^* x \in I$ and it follows that $h_n(x) = 0$ for each n. Since each h_n can be written as $\hat{h}_n \cdot \kappa$ where κ is the canonical homomorphism of X onto X/I with $\hat{h}_n(x+I) = h_n(x)$ for each x, it follows that $(X/I)'$ is infinite-dimensional which is a contradiction.

4.2 $\mathcal{M} = \mathcal{M}_c$ IN COMPLETE COMPLEX ALGEBRAS WHENEVER $\mathcal{M}_c$ IS FINITE. If $\mathcal{M}_c$ is finite in a complete complex LMCH algebra, then $\mathcal{M} = \mathcal{M}_c$ and all homomorphisms of X are continuous.

Hint: Let $\mathcal{M}_c = \{M_1, \ldots, M_n\}$ and suppose that $M \notin \mathcal{M}_c$ for some $M \in \mathcal{M}$. By (4.10-9), $M = \bigcup_{i=1}^{n} M \cap M_i$. Assuming that $M \cap M_j \not\subset M \cap M_i$ for any $j \neq i$ we may find $x_i \in M \cap M_i$ such that $x_i \notin M \cap M_j$. Let $y = x_i + \prod_{j \neq i} x_j$. As $y \in M \cap M_k$ for some k, consideration of the possibilities $i \neq k$ and $i = k$ leads to a contradiction.

<u>4.3 THE CLOSED IDEALS OF $C(T,\underset{\sim}{F},c)$</u>. If A is a closed subset of the Hausdorff space T, then $I_A = \{x \in C(T,\underset{\sim}{F},c) \mid x(A) = 0\}$ is a closed ideal in $C(T,\underset{\sim}{F},c)$. Thus the mapping $A \to I_A$ establishes a correspondence from the class of all closed subsets of T into the class of all closed ideals of $C(T,\underset{\sim}{F},c)$.

(a) If T is Hausdorff then the closed subsets of T and the closed ideals of $C(T,\underset{\sim}{F},c)$ are in 1-1 correspondence via the mapping $A \to I_A$ iff T is completely regular.

Hint: Consider $A = \bigcap_{x \in I} x^{-1}(0)$ and note that $I \subset I_A$. If G is any compact subset of T then $I' = I|_G = \{x|_G \mid x \in I\}$ and $I'_A = I_A|_G$ are ideals in $C(G,\underset{\sim}{F},c)$. I' is dense in I'_A and thus I is dense in I_A.

<u>4.4 FUNCTIONALLY CONTINUOUS COMPLEX ALGEBRAS</u>. A complex topological algebra X is <u>functionally continuous</u> if X^h constitutes all of the nontrivial homomorphisms of X. In the exercises to follow, X^h carries the Gelfand topology $\sigma(X',X)$ restricted to X^h).

(a) If X^h is not compact and there exists a closed (in X^h) non-compact $\sigma(X^*,X)$-bounded subset of X^h, then X is not functionally continuous.

(b) <u>A CLASS OF ALGEBRAS WHICH ARE NOT FUNCTIONALLY CONTINUOUS</u>. Let T be a completely regular Hausdorff topological space which is locally compact and pseudocompact but not compact (e.g. the ordinal space $[0,\Omega)$ where Ω is the first uncountable ordinal). Then $C(T,\underset{\sim}{C},c)$ is a complete LMCH algebra which is not functionally continuous.

(c) A COMPLEX ALGEBRA X WHERE X^h IS COMPACT AND DISCONTINUOUS HOMOMORPHISMS EXIST. Let X be the algebra of polynomial functions on [0,1] with sup norm topology.

Hint: In this case $X^h = [0,1]^*$; consider also homomorphisms defined by evaluation at points not belonging to [0,1].

(d) A CLASS OF FUNCTIONALLY CONTINUOUS COMPLEX ALGEBRAS X FOR WHICH X^h IS COMPACT. Let X be a complex topological algebra for which X^h is compact in the Gelfand topology. If X is a symmetric algebra (Def. 4.12-5) and $\cup\mathcal{M} = \cup\mathcal{M}_c$ then X is functionally continuous.

Hint: As any homomorphism f on X induces a homomorphism $\hat{f}$ on $\Psi(X)$, (Sec. 4.12) via the formula $\hat{f}(\Psi(x)) = f(x)$, it suffices to show that the nontrivial homomorphisms of $\Psi(X)$ are just the evaluation maps determined by the points of X^h. To do this first note that the hypothesis implies (1) if $F \in \Psi(X)$ then $\bar{F} \in \Psi(X)$ where $\bar{F}(h) = \overline{F(h)}$ for $h \in X^h$; and (2) if $\Psi(x)(h) \neq 0$ for each $h \in X^h$ then $1/\Psi(x) \in \Psi(X)$. Now use (1) and (2) and imitate the proof that $C(T,\underset{\sim}{C},c)^h = T^*$ when T is compact (see Example 4.10-2).

4.5 COMPLETENESS OF INITIAL TOPOLOGIES GENERATED BY HOMOMORPHISMS. Let X be a complex algebra and H be a collection of nontrivial complex homomorphisms on X with Gelfand topology (see footnote to Def. 4.12-1). Letting F(H) denote the algebra of all complex-valued functions on X, the map $\Psi: X \to F(H)$, $x \to \Psi x$ has its range in $C(H,\underset{\sim}{C})$, the continuous complex-valued functions on H. [H] denotes the linear span of H in X*.

(a) The algebra X is $\sigma(X,[H])$-complete iff $\Psi(X) = C(H,\underset{\sim}{C}) =$

$F(H)$.

Hint: First note that X is $\sigma(X,[H])$-complete iff: $\Psi(X)$ is a complete subspace of $F(H,p)$ where $F(H,p)$ denotes $F(H)$ endowed with the point-open topology, as discussed in Example 4.3-1. Note that $F(H,p)$ is complete. Show that $\Psi(X)$ is dense in $F(H,p)$.

(b) X is $\sigma(X,[H])$-complete iff the Gelfand topology on H is discrete and $\Psi(X) = C(H,\underset{\sim}{C})$.

(c) If X is a Banach algebra and $H=X^h$, then X is $\sigma(X,[H])$ complete iff H is finite. Furthermore if X is also semisimple then $(X,\sigma(X,[H]))$ is complete iff X is finite-dimensional.

(d) Let H be a set of nontrivial homomorphisms of X separating the points of X and containing at least two non-isolated points f and f_1 (when H carries the Gelfand topology). Then X is not $\sigma(X,[H_o])$ complete with $H_o = H - \{f_o\}$.

Hint: Note that $x=0$ if $f(x)=0$ for all $f \in H_o=H-\{f_o\}$ and recall from Example 4.10-1 that f_o is a $\sigma(X,[H_o])$-discontinuous homomorphism. Use the presence of f_1 in H_o to show that $C(H_o,\underset{\sim}{C}) \neq F(H_o)$.

4.6 <u>A LOCALLY CONVEX TOPOLOGY FOR THE ALGEBRA $C(t)$</u>. Let $C(t)$ be the quotient field of the polynomial algebra $C[t]$ (Example 4.9-1) of polynomials in t with complex coefficients. Let $r \in C(t)$ and consider r as a complex-valued function of a complex variable. As such it has a Laurent series expansion $r(t) = \sum_{s=-\infty}^{\infty} a_s t^s$. Let S be the collection of all double sequences w_n, where $n \geq 1$ and $w_n(s) = (-s+1)^{n(-s+1)}$ if $s \leq -1$; 1 if $s=0$; and $(s+1)^{-(s+1)/n}$ if $s \geq 1$. We define $p_n(r) = \sum_{s=-\infty}^{\infty} |a_s| w_n(s)$.

(a) The metric locally convex topology generated by the family of norms $P = (p_n)$ is compatible with respect to the algebraic structure of C(t).

Hint: Prove that $w_{4n}(r)w_{4n}(s) \geqq w_n(r+s)$ for all integers r ans s and then show that $(\varepsilon^{1/2}V_{p_{4n}})^2 \subset \in V_{p_n}$ for each $n \geqq 1$.

(b) C(t) is not complete in the above topology.

Hint: Show that the sequence $(r_n(t)) = (\sum_{s=1}^{n}(s+1)^{-s(s+1)}t^{-s})$ is Cauchy but does not converge.

4.7 BOUNDS ON COMPATIBLE TOPOLOGIES FOR C(t). Let C(t) be as in the previous exercise. In order for multiplication to be continuous with respect to a locally convex topology it can neither be too weak (part (a)) or too strong (part (b)).

(a) If $\mathcal{T}$ is a compatible topology for C(t) then for each neighborhood of the origin there exists a neighborhood U such that $U \subset V$ and $\sup\{|\alpha| \mid \alpha r \in U\} < \infty$ for each $r \in C(t)$. Thus if $\mathcal{T}$ is locally convex there exists a base of continuous norms generating $\mathcal{T}$ and any weak topology $\sigma(C(t),N)$, where $(C(t),N)$ is a dual pair, fails to be compatible.

(b) Multiplication in C(t) is separately continuous with the finest locally convex topology applied, but this topology is not compatible.

Hint: First observe that if ϕ is a real-valued function on a complete metric space which never vanishes, then the set of λ_o such that $\lim_{\lambda\to\lambda_o} \phi(\lambda) = 0$ is nowhere dense. Next realize that each $r \in C(t)$ has a unique representation (by partial fractions) in a finite sum of the form $\sum_{s\geq 0} b_{\infty,s}t^s + \sum_\lambda \sum_{r\geq 1} b_{\lambda,r}(t-\lambda)^{-r}$. Let B be the set of all linear functionals f such that $f((t-\lambda_o)^{-r}) = 1$ for a particular λ_o and all $r \geqq 1$, $f(t^s)=0$ for each $s \geqq 0$, and $f((t-\lambda)^{-r}) = 0$ for each $\lambda \neq \lambda_o$ and $r \geqq 1$. Then $V=B^o$ is a neighborhood in the finest locally convex topology on C(t) and $|b_{\lambda,1}| \leqq 1$ for all $r(t) \in V$ and all λ. Assume multiplication is continuous so that there exists a neighborhood of the origin U such that $U^2 \subset V$. Define $\phi(\lambda) = \sup \{|\alpha| \mid \alpha(t-\lambda)^{-1} \in U\}$. Since $\phi(\lambda)$ is never zero it follows that there exists λ_1 and $\delta > 0$ such that for each $\epsilon > 0$ there exists $\lambda \neq \lambda_1$ with $|\lambda-\lambda_1| < \epsilon$ and $|\phi(\lambda)| \geqq \delta$. Using this, show that $2^{-1}\phi(\lambda_1)\phi(\lambda)(t-\lambda_1)^{-1} \in U^2 \subset V$ but the coefficients $b_{\lambda,1}$ of these rational functions are not bounded.

(c) C(t) endowed with the topology of Exercise 4.6 is a locally convex Q-algebra which is not an LMC algebra.

Hint: Show that inversion is not continuous.

<u>4.8 NONCOMMUTATIVE DIVISION ALGEBRAS WITH CONTINUOUS INVERSE</u>. (Arens 1947a). In Sec. 4.9 it was seen that the only commutative real or complex locally convex Hausdorff division algebras with continuous inverse are $\underset{\sim}{R}$ and $\underset{\sim}{C}$. Here it is established that <u>if X is a real noncommutative LCH division algebra with continuous inverse, then X is topologically isomorphic to the quaternions</u>.

To show this, proceed as follows. The center Z of X is a commutative division algebra over $\underset{\sim}{R}$ with continuous inverse and so by Theorem 4.9-2, Z is isomorphic to $\underset{\sim}{R}$ or $\underset{\sim}{C}$. Since X is not commutative, X possesses an element x which does not belong to Z. The subalgebra Y generated by x and Z is a commutative real division algebra with continuous inverse. As such, Y must be isomorphic to $\underset{\sim}{R}$ or $\underset{\sim}{C}$. Z, moreover, cannot be isomorphic to

$\underset{\sim}{C}$ since x is outside Z. Furthermore each $x \varepsilon X$ must satisfy a real polynomial equation of degree not greater than two. Clearly X is at least a two dimensional space over $\underset{\sim}{R}$. If $\{e, y_1\}$ is linearly independent in X, then y_1 satisfies an irreducible polynomial of the form t^2+at+b. Since $4b-a^2 > 0$ we may set $x_1 = (2/(4b-a^2)^{1/2})(y_1+ae/2)$. It follows that $x_1^2=-e$. Furthermore this relation makes it impossible for $\{e, x_1\}$ to be a basis for X for in this event X is seen to be isomorphic to $\underset{\sim}{C}$. Let $\{e, x_1, y_2\}$ be linearly independent in X such that $y_2^2=-e$. Then x_1+y_2 and x_1-y_2, neither of which is a multiple of e, must have minimal polynomials of degree 2. Hence there must be real numbers r, s, u and v such that $(x_1+y_2)^2 = r(x_1+y_2) + se = x_1y_2 + y_2x_1 - 2e$ and $(x_1-y_2)^2 = u(x_1-y_2) + ve = -(x_1y_2+y_2x_1) - 2e$. By adding these equations and using the linear independence of $\{e, x_1, y_2\}$, we obtain $r=u=0$ and $s+v=4$. Thus (1) $x_1y_2 + y_2x_1 = (s+2)e = -(v+2)e$. Since the minimal polynomials of x_1+y_2 and x_1-y_2 are irreducible, s and v must be negative. It follows that $4< s< 0$, so that $-s^2-4s> 0$. Let $x_2 = ((s+2)/(-s^2-4s)^{1/2})x_1+(2/(s^2-4s)^{1/2})y_2$ so that $\{e, x_1, x_2\}$ is linearly independent, $x_2^2=-e$, and $x_1x_2 = -x_2x_1$. Finally, select $x_3=x_1x_2$. To see that $\{e, x_1, x_2, x_3\}$ is linearly independent, suppose that $x_3=fe+gx_1+hx_2$. Then $x_1x_3 = x_1^2x_2 = -x_2 = fx_1+gx_1^2 + hx_1x_2 = h(fe+gx_1+hx_2)$ so that $f=g=h=0$. Thus $\{e, x_1, x_2, x_3\}$ is linearly independent. Furthermore $x_1^2=x_2^2=x_3^2=x_1x_2x_3 = -e$. It only remains to show that e, x_1, x_2, and x_3 span X. Suppose that $x \varepsilon X$ and $x \notin \underset{\sim}{R}e$. Then there is no loss in generality in assuming that $x^2=-e$ so that equations analogous to (1) may be obtained for the pairs (x_1,x), (x_2,x), and (x_3,x), i.e. there are real numbers k,m, and n such that $x_1x=xx_1=ke, x_2x=xx_2=me$, and $x_3x=xx_3=ne$. Since $x_3=x_1x_2=-x_2x_1$, it follows that $xx_3 = (xx_1)x_2 = kx_2 - x_1(xx_2) = kx_2-mx_1 + (x_1x_2)x = kx_2-mx_1 + x_3x = kx_2-mx_1 + x_3e - xx_3$ which implies that $2y = -(kx_1+mx_2+nx_3)$. Hence $\{e,x_1,x_2,x_3\}$ spans X and the proof is complete.

<u>4.9 A-NORMED ALGEBRAS</u>. Let X be a complex algebra which is a LCHS. A seminorm p defined on X will be called an <u>A-seminorm</u> if for each $x \in X$ and all $y \in X$ there exists a real number $m_{p,x} > 0$ such that $p(xy) \leq m_{p,x}p(y)$. If the topology on X is generated by a single A-seminorm, then we refer to X as <u>A-normed</u>. If the topology on X is generated by a family of A-seminorms, then we refer to X as <u>locally A-convex</u>.

(a) If X is a complete A-normed algebra, then X is a Banach algebra.

Hint: Consider the linear space isomorphism $H:X \to L(X,X)$, $x \to A_x$ where $L(X,X)$ is the Banach algebra of bounded linear transformations on X and $A_xy = xy$ for all $y \in X$. Show that the mapping H has a continuous inverse and that H(X) is closed in $L(X,X)$. Then, assuming that X is complete, apply the closed graph theorem.

(b) If X is an A-normed division algebra, then X is topologically isomorphic to the complex numbers.

(c) If X is an A-normed algebra and I a closed ideal in X, then the algebra X/I with quotient norm is an A-normed algebra. A maximal ideal in an A-normed algebra X is the kernel of a continuous homomorphism of X onto $\underset{\sim}{C}$ iff it is closed in X.

(d) Let C[0,1] denote the algebra of continuous complex-valued functions on the closed interval $[0,1]$. Consider the function $z(t) = t,\ 0\leq t\leq 1/2;\ 1-t,\ 1/2\leq t\leq 1$. For any $x\in C[0,1]$ let $\|x\| = \sup_{t\in[0,1]} z(t)x(t)$. Show that $C[0,1]$ is A-normed but not normed.

Hint: Consider the functions $x_n(t) = 1-nt,\ 0\leq t\leq 1/n;$ $0,\ 1/n< t$. Show that the functions $x_n\to 0$ but the linear transformations $A_{x_n} \not\to 0$. Consider $y_n(t) = n^2t,\ 0\leq t\leq 1/n^2;$ $2-n^2t,\ 1/n^2\leq t\leq 2/n^2;\ 0,\ 2/n^2\leq t$ and show that $\|A_{x_n}(y_n)\| / \|y_n\|\geq 1-1/n$. (e) A subset U of X is said to be an <u>A-set</u> if for each $x\in X$ and some real number $a_x> 0$, $xU\subset a_xU$. Prove that all of the following are A-sets:

(a) the closure of an A-set;

(b) the balanced hull of an A-set;

(c) the convex hull of an A-set;

(d) the balanced convex hull of an A-set;

(e) the intersection of two A-sets.

(f) Prove that the gauge of an absolutely convex and absorbing A-set is an A-seminorm. If p is an A-seminorm show that $V_p = \{x\in X|p(x)\leq 1\}$ is an A-set.

(g) Prove that X is locally A-convex iff there exists a base of neighborhoods of 0 consisting of A-sets.

(h) (Michael, 1952) Prove that a barreled locally A-convex space is an LMC algebra.

Hint: Let U be an absolutely convex closed A-set which is a neighborhood of 0 in X and a be a real number such that $a> 0$ and $e\in aU$. Consider $V = \{x\in X|xU\subset U\}$. Show that $V\subset aU$, that V is a barrel in X, and that $V^2\subset V$.

(i) Let $C_o^+(\underset{\sim}{R})$ be the strictly positive functions in $C_b(\underset{\sim}{R},\underset{\sim}{C})$

which vanish at infinity. For each $x \in C_o^+(R)$ let p_x denote the A-seminorm $p_x(y) = \sup_{t \in \mathbb{R}}\{|x(t)y(t)|\}$. Show that the family of seminorms $\{p_x\}$ cannot be replaced by a family of multiplicative seminorms generating the same topology.

Hint: Let x_1 and x_2 be two functions in $C_o^+(R)$ and suppose that for some multiplicative seminorm p the set inclusions $\overline{V}_{p_{x_2}} \subset \overline{V}_p \subset \overline{V}_{p_{x_1}}$ hold. Let $a = \min\{1, \max_{t \in \mathbb{R}} x_2(t)\}$ and let b be a real number such that $0 < b < a$. Then for some $t_o \in \mathbb{R}$, $x_2(t_o) = b \geq x_1(t_o)$ and for some positive integer n, $b^n < x_1(t_o)$. Consider the function y defined by $y(t) = (t - t_o + 1/x_2(t),\ t_o - 1 \leq t \leq t_o;\ (-t + t_o + 1)/x_2(t), t_o \leq t \leq t_o + 1;\ 0,\ 1 < |t - t_o|$. Show that $p_{x_2}(y) = 1$ and $p_{x_1}(y^n) \geq y(t_o)^n x_1(t_o) = b^{-n} x_1(t_o) > 1$. But as $p_{x_2}(y) = 1$, $p(y) \leq 1$ and therefore $p(y^n) \leq p(y)^n \leq 1$. But then $p_{x_1}(y^n) \leq 1$ which is a contradiction.

(j) Let p be an A-seminorm on X. Then the null space N_p of p is an ideal in X.

(k) Let p be an A-seminorm on X and N_p the null space of p. On X/N_p let a norm be defined by $\|x + N_p\| = p(x)$. Show that X/N_p is an A-normed algebra.

(l) An algebra X is a locally A-convex algebra iff it is topologically isomorphic to a subspace of a product of A-normed algebras.

(m) Let X be an A-normed algebra whose topology is generated by a family $(p_\alpha)_{\alpha \in \Lambda}$ of A-seminorms. Suppose that the algebras X/N_{p_α} are complete for all α. Then show that X is an LMCH algebra.

Suppose that T is a completely regular Hausdorff space

such that to each point t of T there corresponds a Banach algebra B(t). Let F(T) denote the subset of the product of the algebras B(t) such that for each $x \in F(T)$, the function $|x|(t) = \|x(t)\|$ ($\| \|$ denoting the norm of x(t) in B(t)) is a continuous real-valued function on T. If H is a subset of F(T) and H is an algebra (operations between functions in H being the usual pointwise operations), then we refer to H as an <u>algebra of vector-valued functions</u>. On H the topology of convergence on compact sets has as a base at 0 sets of the form (K being a compact subset of T, $\epsilon > 0$) $V(0;K,\epsilon) = \{x \in H \,|\, |x|(t) \leqq \epsilon$ for all $t \in K\}$. On all Banach algebras we assume that the norm of the identity is equal to 1.

(n) Let H be defined as above. Suppose that for each $t \in T$, $H_t = \{x(t) \,|\, x \in H\} = B(t)$, and for each continuous real-valued function f on T, fx belongs to the closed ideal generated by x for each $x \in H$ where $fx(t) = f(t)x(t)$ for all $t \in T$. Then the following statements hold:

(1) Every closed ideal in H is of the form $\{x \in H \,|\, x(t) \in I_t\}$ where I_t is a closed ideal in B(t) for every $t \in T$. Conversely, any such collection of closed ideals I_t in B(t) yields a closed ideal in H.

(2) Every closed maximal ideal in H is of the form $\{x \in H \,|\, x(t) \in I_t\}$ where I_t is maximal in B(t) for each $t \in T$.

(3) If B(t) contains a unique maximal ideal for each $t \in T$, then every closed ideal in H is the set of functions that vanish on a closed subset of T. Conversely every closed subset of T yields a closed ideal in H.

(4) If B(t) contains a unique maximal ideal for each $t \in T$,

then there is a 1-1 correspondence between the points of T and the closed maximal ideals of H where $t \to M_t = \{x \in H \mid x(t)=0\}$ establishes the correspondence.

In the previous four statements generalizations of theorems concerning the nature of closed ideals in $C(T,\underset{\sim}{C})$ have been presented. Indeed if $B(t)=\underset{\sim}{C}$ for all t, $F(T)=C(T,\underset{\sim}{C})$.

(o) Let P denote a saturated set of continuous A-seminorms defining the topology on the locally A-convex algebra X. Let $V(p;x_1,\ldots,x_n,\epsilon) = \{q \in P \mid |p(x_i) - q(x_i)| \leq \epsilon\}$. These sets form a base for a topology on P with respect to which P is a completely regular Hausdorff space.

(p) Assume that for each $p \in P$, the factor algebra X/N_p can be completed to a Banach algebra X_p. Then show that X is an LMC algebra.

(q) Assume the condition of (p) applies. Consider the algebra H of continuous functions on P taking values in the family of Banach algebras X_p where $H = \{f_x \mid f_x(p) = x+N_p, x \in X\}$. Let G be a map taking X into H where $G(x) = f_x$ for each $x \in X$; then show that G is an algebra isomorphism between X and H, and that G^{-1} is continuous.

In the final part of this problem, a representation of X as an algebra of vector-valued functions is obtained.

(r) Show that if X is barreled, then G is continuous.

4.10 P-NORMED SPACES (ZELAZKO 1965). In this exercise we consider a generalization of the concept of a Banach algebra. A p-normed space X is a topological vector space X whose topology is generated by a p-norm $\| \ \|_p$, $0 < p \leq 1$, where a p-norm satisfies:

1) $\|x\| \geqq 0$, $\|x\| = 0$ if and only if $x=0$.

2) $\|x+y\| \leqq \|x\| + \|y\|$.

3) $\|\mu x\| = |\mu|^p \|x\|$ for all scalars and any $x \in X$.

A locally p-convex topological vector space is a space whose topology at 0 is generated by a family of balanced neighborhoods V each of which satisfies the relationship $\sum_{i=1}^{n} \mu_i x_i \in V$ for any μ_i such that $\sum_{i=2}^{n} |\mu_i|^p \leqq 1$ and $x_i \in V$. Such a neighborhood V is called absolutely p-convex.

(a) A topological vector space X is p-normed iff there exists a bounded absolutely p-convex neighborhood of 0.

Hint: $\|x\|_p = \inf\{r^p | x \in rV, r \geqq 0\}$.

(b) A locally bounded (there exists a bounded neighborhood of 0) topological vector space is p-normed.

Hint: For every bounded balanced neighborhood U of 0 there exists a real number $k \geqq 2$ such that $U+U \subset kU$. The greatest lower bound of all such numbers k is called the module of concavity of U and the greatest lower bound k(X) of all modules of concavity is called the module of concavity of X. Of course $k(X) \geqq 2$. Let $k(X) = 2^{1/p_o}$ and consider any $p < p_o$. Then there exists a bounded balanced neighborhood U of 0 such that $U+U \subset 2^{1/p}U$. This relationship can be extended (Köthe 1966) to $\sum_{i=1}^{n} 2^{-k_i/p}U \subset U$ for $\sum_{i=1}^{n} 2^{-k_i} \leqq 1$. Now the absolutely p-convex hull $\Gamma_p(U) = \{\sum_{i=1}^{n} \mu_i x_i | x_i \in U$ and $\sum_{i=1}^{n} |\mu_i|^p \leqq 1\}$ of U can be shown to satisfy the relationship $\Gamma_p(U) \subset 2^{1/p}U$ and the gauge $\| \ \|_p$ of $\Gamma_p(U)$ generates the topology.

(c) Let X be a complete metrizable algebra in which multiplication is separately continuous. Then the following are equivalent:

(1) There exists a metric ρ generating the topology such that $\rho(xy,0) \leqq \rho(x,0)\ \rho(y,0)$.

(2) X is locally bounded.

(3) X is a p-normed algebra ($\|xy\|_p \leqq \|x\|_p \|y\|_p$) for all $x,y \in X$.

Hint: Assuming that X is p-normed (or equivalently, by (b) and (a), locally bounded) then the topology on X is generated by a p-norm $\|\ \|_p$ which is not necessarily submultiplicative. However, as multiplication is separately continuous, letting A_x denote the continuous linear map $y \to xy$, $y \in X$, $\|A_x\| = \sup_{y \neq 0} \frac{\|yx\|_p}{\|y\|_p} < \infty$. Letting $\|x\|_p' = \|A_x\|$, $\|\ \|_p'$ is the desired norm.

(d) Show that the algebra ℓ_p $(0 < p \leqq 1)$ of all two-sided sequences $x = (\mu_n)$ of scalars satisfying $\|x\|_p = \sum_{n=-\infty}^{n=\infty} |\mu_n|^p$, is a p-normed algebra when multiplication is defined as Cauchy product.

(e) Show that the algebra X of all holomorphic functions in the closed unit disc $x(\lambda) = \sum_{n=0}^{\infty} \mu_n \lambda^n$ with $\|x\|_p = \sum_{n=0}^{\infty} |\mu_n|^p$ is a p-normed algebra.

In the next series of exercises we develop a concept which leads to an analog of Gelfand theory for p-normed algebras. Let X be a complex complete p-normed algebra and $K_s = \{x \in X \mid \lim \|x^n\|_p = 0\}$. The <u>spectral norm</u> $\|x\|_s = (\sup\{|\lambda| \mid \lambda x \in K_s\})^{-p}$.

(f) The spectral norm has the following properties:

(1) $\|x\|_s < 1$ iff $x \in K_s$

(2) $\|\lambda x\|_s = |\lambda|^p \|x\|_s$

(3) $\|x+y\|_s \leqq \|x\|_s + \|y\|_s$

(4) $\|xy\|_s \leqq \|x\|_s \|y\|_s$

(5) $\|x^n\|_s = \|x\|_s^n$

(6) $\|x\|_s \leqq \|x\|_p$

(7) If $x^{-1} \in X$, $\|x\|_s > 0$ (8) $\|x\|_s = \lim_n (\|x^n\|_p)^{1/n}$

Hint: To prove (3) it is sufficient to show that if $\|x\|_s + \|y\|_s < 1$, then $\|x+y\|_s < 1$. Let $\alpha > \|x\|_s$, $\beta > \|y\|_s$, $\alpha+\beta < 1$; as $\|(x+y)^n\|_p = \|(\alpha^{1/p}\alpha^{-1/p}x+\beta^{1/p}\beta^{-1/p}y)^n\|_p \leq \sum_{k=0}^{n} \binom{n}{k}^p \alpha^{n-k}\beta^k \|(\alpha^{-1/p}x)^{n-k}(\beta^{-1/p}y)^k\|_p$; by (2) $\|(\alpha^{-1/p}x)^n\|_p \to 0$ and $\|(\beta^{-1/p}y)^n\|_p \to 0$. Letting n_o be such that if $N > n_o$ then $\|(\alpha^{\bar{1}/p}x)^{N-k}(\beta^{1/p}y)^k\|_p < 1$ for $k=0,\dots,N$, then $\|(x+y)^N\|_p \leqq \sum_{k=0}^{N} \binom{N}{k}^p \alpha^{N-k}\beta^k \leqq \sum_{k=0}^{N} \binom{N}{k} \alpha^{N-k}\beta^k = (\alpha+\beta)^N < 1$. Thus $\|x+y\|_s < 1$.

As in the Banach algebra case, the invertible elements of a complete commutative p-normed algebra with identity are an open set, and inversion is continuous on the invertible elements.

(g) If X is a complex complete commutative p-normed field and Y is a closed subalgebra of X, then Y is a field.

Hint: Let x be a non-zero element of Y. Assuming that $x \neq \lambda e$ for any scalar λ we observe that $\Lambda = \{\lambda \mid (x-\lambda e)^{-1} \in Y\}$ is a closed and open subset of the complex plane, hence $\Lambda = \mathbb{C}$. Thus $(x-(1/n)e)^{-1} \in Y$ for all n and therefore $x^{-1} \in Y$.

(h) A p-normed field over the complex numbers is topologically isomorphic to $\mathbb{C}$.

Hint: First prove the following statement: If X is a p-normed field, then for every $\lambda \neq 0$, $\epsilon > 0$, $x \in X$ and $x \neq \lambda e$, there exists a sequence of polynomials $W_n(x)$ such that $\| e-(e+xW_n(x))^{-1} \|_p < \epsilon$ for large n. To do this consider the closed algebra X(x) generated by x and e. By (g) X(x) is a field. Now use continuity of inversion.

Now let $x \in X$ and $f(\lambda) = \|(x^{-1}+\lambda e)^{-1}\|_s$. Show that as

$|\lambda|\to\infty$, $f(\lambda)\to 0$. Letting $f(\lambda_o)$ be the maximum of f, if $y = (f(\lambda_o))(x^{-1}+\lambda_o e)^{-1}$, then $\|y\|_s = 1$ and $\|(y^{-1}+\lambda e)^{-1}\|_s \leqq 1$ for each $\lambda \in \underset{\sim}{C}$. Let $V_n(\lambda) = \lambda^n+\alpha_1\lambda^{n-1}+\ldots+\alpha_n$ be a sequence of polynomials. Then $V_n(y^{-1}) = (y^{-1}-\beta_1 e)\ldots(y^{-1}-\beta_n e)$ for suitable $\beta_j \in \underset{\sim}{C}$ and $\|V_n(y^{-1})\|_s \leqq 1$. However $[V_n(y^{-1})]^{-1} = y^n[e+yW_n(y)]^{-1}$ where $W_n(y) = \alpha_1+\alpha_2 y+\ldots+\alpha_n y^{n-1}$. Choose V_n such that $\|\lambda e-(e+yW_n(y))^{-1}\|_p < \epsilon$. Let $z = \lambda e-[e+yW_n(y)]^{-1}$ so that $\|z\|_s \leqq \|z\|_p < \epsilon$. Then $\|y^n(\lambda e-z)\|_s = \|y^n(e+yW_n(y))^{-1}\|_s \leqq 1$ and therefore $1 \geqq \|\lambda y^n-zy^n\|_s \geqq |\lambda|^p\|y^n\|_s - \|z\|_s\|y^n\|_s \geq (|\lambda|^p-\epsilon)\|y^n\|_s$ so that $\|y^n\|_s \leqq 1/(|\lambda|-\epsilon) < 1$ for suitable ϵ and $|\lambda|$, and $\|y\|_s < 1$. But this contradicts $\|y\|_s = 1$.

From the previous result, the basic theorems of Gelfand theory concerning maximal ideals and continuous complex-valued homomorphisms of a Banach algebra X hold for p-normed algebras (see Sec. 4.12).

<u>4.11 RADICALS OF P-NORMED ALGEBRAS</u>. In this exercise we consider the relationships between the spectral norm $\| \ \|_s$ of a complete commutative p-normed algebra X with identity and the radical of X. The result of principal importance (part (c)) is that $\text{rad } X = \{x \in X \mid \|x\|_s = 0\}$.

(a) $K = \{x \in X \mid \|x\|_s \leqq 1\}$ is convex and therefore $\| \ \|^* = \| \ \|_s^{1/p}$ is a seminorm on X.

Hint: As K is a closed subset of X, it is sufficient to show that if $x, y \in K$ then $x+y/2 \in K$. This is equivalent to showing that if $\|x\|_s < 1$ and $\|y\|_s < 1$, then $\|x+y/2\|_s \leqq 1$. But $\|x+y/2\|_s = (1/2)^p \lim_n(\|(x+y)^n\|_p)^{1/n} \leqq (1/2)^p \lim_n (\sum_{k=0}^{n} \binom{n}{k}^p \|x^k y^{n-k}\|_p)^{1/n}$. As $\|x\|_s < 1$ and $\|y\|_s < 1$, then if n is made sufficiently large, for all $0 \leqq k \leqq n$ $\|x^k y^{n-k}\|_p < 1$

and $\|x+y/2\|_s \leqq (1/2)^p \lim_n (\sum_{k=0}^{n} \binom{2n}{k}^p)^{1/2n} = (\frac{1}{2})^p \lim_n ((2n+1) \binom{2n}{n}^p)^{1/2n} = 2^{1/p} \lim_n (\binom{2n}{n}^p)^{1/2n} = 1.$

(b) Every (necessarily continuous) multiplicative linear functional f on X is continuous with respect to $\| \ \|_s$ and $|f(x)|^p \leqq \|x\|_s \leqq \|x\|_p$.

Hint: Show that if $\|x\|_s < |f(x)|^p$ for some $x \in X$ then letting $y = x/|f(x)|$, $y \in K_s$ (see preceding exercise) and $y^n \to 0$, violating continuity of f.

(c) Let X^h be the set of all (continuous) complex-valued multiplicative linear functionals on X. Then $\|x\|_s = \sup_{f \in X^h} |f(x)|^p$. Hence $x \in \text{Rad } X$ iff $\|x\|_s = 0$.

Hint: From (a) $\| \ \|^* = \| \ \|_s^{1/p}$ is a seminorm on X and as $\| \ \|_s$ is continuous, $I = \{x \in X \mid \|x\| = 0\}$ is a closed ideal in X. Let $\hat{X} = X/I$ with a norm defined on $\hat{X}$ by $\|x+I\| = \inf_{y \in I} \|x+y\| = \|x\|$. Then $\hat{X}$ is a normed algebra with the property that all multiplicative linear functionals $\hat{f}$ on $\hat{X}$ are of the form $\hat{f}\,(x+I) = f(x)$ for some $f \in X^h$. Hence all linear functionals on $\hat{X}$ are continuous. From standard Gelfand theory $\sup_{\hat{f} \in \hat{X}^h} |\hat{f}(x+I)| = \sup_{f \in X^h} |f(x)| = \lim_n (\|(x+I)^n\|)^{1/n} = \lim_n (\|x^n\|^*)^{1/n} = \lim_n (\|x\|_s^{n/p})^{1/n} = \|x\|_s^{1/p} = \|x\|^*$.

<u>4.12 F-ALGEBRAS</u>. An F-space X is a complete metrizable topological vector space. In Dunford and Schwartz 1958 (p. 51) it is shown that the topology of X can be generated by an <u>F-norm</u>, i.e. a mapping $\| \ \|$ of X into R satisfying

(1) $\|x\| \geqq 0$, $\|x\| = 0$ iff $x=0$.

(2) $\|x+y\| \leqq \|x\| + \|y\|$.

(3) $\|\mu x\| = \|x\|$ for $|\mu| = 1$.

(4) if $\mu_n \to 0$, then $\lim \|\mu_n x\| = 0$ for all $x \in X$.

(5) If $x_n \to 0$, then $\lim \|\mu x_n\| = 0$ for all μ.

(6) $d(x,y) = \|x-y\|$ is a metric generating the topology of X and X is d-complete.

If X is an F-space and an algebra in which multiplication is separately continuous, then multiplication is jointly continuous.

Hint: If $C_r(x_o) = \{x \in X \mid \|x_o - x\| \leq r\}$, then $C_{1/n}(0) = U_n, n \in N$, V is a base of neighborhoods of 0. If U is a neighborhood of 0 in X and $A_n = \{x \in X \mid xU_n \subset U\}$, then by separate continuity of multiplication in X, $\cup A_n = X$ and as X is a Baire space and each A_n is closed, there exists A_{n_o} such that int $A_{n_o} \neq \emptyset$. Thus for some $x \in X$ and $r > 0$, $C_r(x) \subset A_{n_o}$. If $x \in C(0, r_o)$ then $xU_{n_o} \subset U+U$. Let V be an arbitrary neighborhood of 0 and U be chosen such that $U+U \subset V$. Then $C_{r_o}(0) U_{n_o} \subset V$ and also $[C_{\min(r_o, 1/n_o)}(0)]^2 \subset V$.

<u>4.13 FRECHET ALGEBRAS</u>. We consider in this exercise an algebra X carrying a topology with respect to which it is a Frechet space. Having shown in the previous exercise that if multiplication in this algebra is separately continuous, then it is jointly continuous, we show in (a) that a family of seminorms generating the topology of X can be produced which are "almost multiplicative". Interestingly enough, the property they do have is sufficiently strong so that it can be said that these algebras resemble LMC algebras quite strongly.

(a) Let X be a Frechet space and an algebra in which multiplication is separately (therefore by Exercise 4.1 jointly) continuous. Then the seminorms generating the topology of X can be chosen so that they are increasing and $p_i(xy) \leqq p_{i+1}(x) p_{i+1}(y)$ for all i and $x, y \in X$.

Hint: Let (p_n) be an increasing family of seminorms generating the topology and p be any continuous seminorm. Then there exists $C > 0$ such that $p(xy) \leqq Cp_n(x)p_n(y)$ for some $n > 0$; otherwise there exist sequences (x_n) and (y_n) such that $p(x_n y_n) > n^2 p_n(x_n)p_n(y_n) \geqq 0$. First we deal with the case where $p_n(x_n)p_n(y_n) = 0$ for infinitely many n. If infinitely many $p_n(x_n) = 0$, then $x_n \to 0$ and in fact for any m, $p_m(x_n)$ ultimately is equal to 0. Thus $x_n/p(x_n y_n) = z_n \to 0$. With no loss of generality the sequence (y_n) could have been chosen so that $y_n \to 0$. Now $p(z_n y_n) = 1$ while $z_n \to 0$ and $y_n \to 0$. This is a contradiction. For the case where we can assume that $p_n(x_n)$ and $p_n(y_n)$ are never zero, we simply let $z_n = x_n/np_n(x_n)$ and $w_n = y_n/np_n(y_n)$. Once again $w_n \to 0$ and $z_n \to 0$ but $p(z_n w_n) \geqq 1$. Now let $q_1 = p_1$ and for the first i such that $q_1(xy) \leqq Cp_i(x)p_i(y)$, set $q_2 = \sqrt{C}p_i$. Continuing in this way we generate the desired seminorms.

A Frechet space which is an algebra in which multiplication is separately continuous (or in which there is a family of seminorms generating the topology which satisfy the conditions of (a)) is called a Frechet algebra.

(b) Show that a complex Frechet division algebra is topologically isomorphic to $\mathbb{C}$.

Hint: Using Theorem 4.9-1 we need only prove that inversion is continuous. Since the units are open they are a G_δ-set. Then Theorem 7.4 of Zelazko 1965 may be applied.

(c) Let X be a complex Frechet algebra and suppose that there exists a closed maximal ideal in X. Show that a system of seminorms (p_i) can be found satisfying the conditions of (a) such that $p_i(e) = 1$ for all i.

Hint: Let (q_i) be a family of seminorms generating the topology of X and satisfying the condition of (a). If M is a closed ideal in X then $X = M + \{\lambda e | \lambda \in \mathbb{C}\}$. If $x \in X$ then $x = m+\lambda e$ with $\lambda \in \mathbb{C}$. Define $w_i(x) = q_i(m) + |\lambda|$ for each i. If $y = m' + \mu e$, then $w_i(xy) = q_i(mm'+\lambda m'+\mu m) + |\lambda|\ |\mu| \leqq q_i(mm') + |\lambda| q_i(m') + |\mu| q_i(m) + |\lambda|\ |\mu| \leqq q_{i+1}(m)q_{i+1}(m') + |\lambda| q_{i+1}(m') + |\mu| q_{i+1}(m) = w_{i+1}(x)w_{i+1}(y)$. Clearly $w_i(e) = 1$ for all i and the family (w_i) generates the topology of X.

4.16 ES-ALGEBRAS. In this exercise we consider complete topological algebras X with the property that every continuous homomorphism on a closed subalgebra Y of X can be extended to a continuous linear functional on X. Such algebras are referred to as ES-algebras. We list first a number of statements (see Narici, Beckenstein, and Bachman 1971, Rickart 1960, and Naimark 1964.)

(1) If T is a locally compact Hausdorff space and T is totally disconnected, then T is ultraregular.

(2) If S is a compact totally disconnected subset of $\mathbb{C}$, then int $S=\emptyset$. S is totally disconnected iff it contains no continuum.

(3) If X is a Banach algebra and Y is a closed subalgebra, then $\sigma_X(x) \subset \sigma_Y(x)$ for all $x \in Y$ and $\mathrm{bd}\sigma_Y(x) \subset \mathrm{bd}\sigma_X(x)$.

(4) If X and Y are as in (3) then $f \in Y^h$ is extendible to $f \in X^h$ iff f is in the Shilov boundary of Y ($f \in X^h$ is in the Shilov boundary of X iff for every neighborhood U of f in

the Gelfand topology, there exists $x \in X$ such that the Gelfand function Ψx fails to achieve its maximum absolute value on CU.)

(5) If X is as in (3) and $\mathcal{M}(X) = U_1 \cup U_2$ where U_1 is open and $U_1 \cap U_2 = \emptyset$, then there exists $x \in X$ such that $x^2=x$ and the Gelfand function Ψx is the characteristic function of U_1.

(a) Let X be a Banach algebra and Y a closed subalgebra. If $\sigma_X(x)$ is totally disconnected for $x \in Y$, then $\sigma_X(x) = \sigma_Y(x)$.

(b) Let X be a Banach algebra and suppose that for some $x \in X$, $\sigma_X(x)$ contains a continuum. Then there exists a closed subalgebra $Y \subset X$ and $f \in Y^h$ such that f is not extendible to $f \in X^h$.

Hint: If there is a continuum joining λ_1 and λ_2 in $\sigma_X(x)$ let $y = x-\lambda_1 e/(\lambda_2-\lambda_1)$ and $z = e^{\pi/2iy}$. Then z is invertible in X and there is a continuum in $\sigma_X(z)$ joining 1 and i. Let Y be the closed subalgebra of X generated by z^4 and e. Let $f(p(z^4)) = p(0)$ for any polynomial p in z^4. By the maximum modulus principle of complex variable theory, $|f(p(z^4))| \leq \|z^4\|$ hence f can be extended continuously to a homomorphism on Y. However as z^4 is invertible in X and $f(z^4) = 0$, f cannot be extended to X.

(c) A Banach algebra X is an ES-algebra iff for any subalgebra $Y \subset X$, Y^h is totally disconnected in the Gelfand topology.

(d) A complex Banach algebra X is an ES-algebra iff $\sigma_X(x)$ is totally disconnected for all $x \in X$.

Hint: If for some $x \in X$, $\sigma_X(x)$ is not totally disconnected, taking Y to be the closed subalgebra generated by x and e, since $\mathcal{M}(Y)$ is homeomorphic to $\sigma_Y(x)$, we see that X is not an ES-algebra.

(e) If T is a compact Hausdorff space and $C(T,\underset{\smile}{C},c)$ is an ES-algebra, then for any complex Banach algebra X such that $\mathcal{M}(X)=T$, X is an ES-algebra.

(f) If G is a compact abelian topological group, then $L_1(G)$ with convolution as the multiplication in $L_1(G)$ is an ES-algebra and any multiplicative linear functional on $L_1(G)$ is of the form $f(x) = \int x(t)\chi(t)dt$ where χ is a continuous character and integration is with respect to Haar measure on G.

Hint: Although $L_1(G)$ does not have an identity, it can be complexified and an identity adjoined. The multiplicative linear functionals on $L_1(G)$ and the algebra with identity adjoined can be placed in 1-1 correspondence and the spectrum of each element in $L_1(G)$ is the same in both algebras.

(f) A Frechet algebra X is an ES-algebra iff for each $x \in X$, $\sigma_X(x)$ is totally disconnected.

Hint: If (X_n) is a denumerable family of factor algebras associated with a collection of seminorms generating the topology and K is a continuum in $\sigma_X(x)$, then as $\sigma_X(x) = \bigcup_{n \in N} \sigma_{X_n}(\kappa_n(x))$, it follows that for some n, $\sigma_{X_n}(\kappa_n(x))$ contains a continuum. Thus X_n is not an ES-algebra and there exists a closed subalgebra Y_n of X_n and a homomorphism f on Y_n which cannot be extended to X_n. Let Y be the projective limit of the algebras X_k $(k \neq n)$ and Y_n. Then Y is a subalgebra of X and f induces a homomorphism on Y which cannot be extended to X.

<u>4.17 PERMANENTLY SINGULAR ELEMENTS IN TOPOLOGICAL ALGEBRAS</u>. In this exercise X is a complete topological algebra and we ex-

plore the nature of permanently singular elements of X, i.e., elements $x \in X$ such that x has no inverse in any superalgebra of X (Arens 1958a, Zelazko 1971, Kuczma 1958, Suffel, Beckenstein and Narici 1974).

A topological divisior of 0 c in a normed algebra X over a valued field is an element $c \in X$ such that for some sequence $x_n \in X, cx_n \to 0$ while $x_n \not\to 0$. This is equivalent to requiring that the multiplicative operator $y \mapsto xy$ not have a continuous inverse. Another form of this is that $a = \inf_{x \neq 0} \|cx\| > 0$.

(a) An element of a normed algebra X has no inverse in any superalgebra of X iff it is a topological divisor of 0.

Hint: Let $t > 0$ and $X(z;t)$ (z transcendental over X) be the algebra of formal power series in z with coefficients from X such that for any such series $f(z) = \sum_{n=0}^{\infty} x_n z^n$, $\sum_{n=0}^{\infty} \|x_n\| t^n < \infty$. Let $\|f(z)\| = \sum_{n=0}^{\infty} \|x_n\| t^n$ and J be the smallest closed ideal in $X(z;t)$ generated by $e-cz$. Let $Y = X(z;t)/J$ be the quotient algebra with $\|f(z)+J\| = \inf_{j \in J} \|f(z)+j(z)\|$. We show that for suitable t X is isometrically embedded in Y. Let $t_o \|cx\| \geq \|x\|$ for all $x \in X$ $(a \geq t_o^{-1})$ with $t_o > 0$. Let $x \in X$ and $g(z) = x-(e-cz)\sum_{n=0}^{\infty} x_n z^n = (x-x_o) + (cx_o-x_1)z + (cx_1-x_2)z^2 + \ldots$. As $\|g(z)\| = \|x-x_o\| + \|cx_0-x_1\|t + \|cx_1-x_2\|t^2 + \ldots \geq \|x\| - \|x_o\| + (t_o^{-1}\|x_o\| - \|x_1\|)t + \ldots = \|x\| + (tt_o^{-1})\|f(z)\|$, if we choose $t \geq t_o$ the proof is done.

A nonarchimedean normed algebra over a (necessarily) nonarchimedean valued field is a normed algebra such that $\|x+y\| \leq \max(\|x\|, \|y\|)$ for all $x, y \in X$. A nonarchimedean LMCH algebra is a topological algebra over a (necessarily) nonarchimedean valued field $\underset{\sim}{K}$ whose Hausdorff topology is generated by a a family P of submultiplicative seminorms such that $p(x+y) \leqq \max(p(x), p(y))$ for all $x,y \in X$, and each $p \in P$. The factor algebras associated with these seminorms are nonarchimedean Banach algebras.

(b) Let X be a commutative nonarchimedean normed algebra with identity over a complete nontrivially valued field. Then an element $c \in X$ is <u>permanently singular</u> in X (has no inverse in any nonarchimedean superalgebra) iff $a = \inf_{x \neq 0} \|cx\| > 0$ (c is a topological divisor of 0 in X).

Hint: Consider t_o such that $a \geqq t_o^{-1}$ and t such that $t\, t_o^{-1} \geqq 1$. Apply to $X(z;t) = \{\sum_{n=0}^{\infty} x_n t^n \mid \|x_n\| t^n \to 0\}$ a nonarchimedean norm $\|f(z)\| = \sup_n \|x_n t^n\|$. As in (a) we show that X is isometrically embedded in $Y = X(z;t)/J$ as follows. Let $x \in X$. Then $x-(e-cz)f(z) = x-x_o + (cx_o-x_1)z+\ldots$. If $\|x_o\| \neq \|x\|$, then $\|x-(e-cz)f(z)\| \geqq \|x\|$. If $\|x_o\| \neq \|x\|$ then show that $\|x_{n-1}c\| = \|x_n\|$ cannot occur for all n without violating the requirement that $\|cx_{n-1} - x_n\| t^n \to 0$.

FIVE
Hull-Kernel Topologies

THIS CHAPTER IS directed mainly toward investigating the question: When may the space $\mathcal{M}$ of maximal ideals of a complex LMC algebra X be viewed as a Wallman compactification of the space $\mathcal{M}_c$ of closed maximal ideals of X? As to why such a question would be considered, one could look back to algebras $C(T,\underset{\sim}{F},c)$ of continuous functions with compact-open topology. The closed maximal ideals $\mathcal{M}_c$ of $C(T,\underset{\sim}{F},c)$ could be identified with the points of T by Example 4.10-2, while all the maximal ideals $\mathcal{M}$ were in 1-1 correspondence with βT. Thus, for $C(T,\underset{\sim}{F},c)$, it could be said that $\mathcal{M}=\beta\mathcal{M}_c$, and we have the first suggestion that the maximal ideals of a topological algebra might be a compactification of the closed maximal ideals in more general situations. Of course to begin to consider the question $\mathcal{M}$ and $\mathcal{M}_c$ must be endowed with topologies and this is the subject to which Sec. 5.1 is devoted. The ramifications of imposing normality type conditions on these topologies are considered in Sec. 5.2.

In Sec. 5.3 a sufficient condition - called condition hH - on a topological algebra for $\mathcal{M}$ to equal $\beta\mathcal{M}_c$ is discussed. The way in which this - realizing $\mathcal{M}$ as $\beta\mathcal{M}_c$ - is done in Sec. 5.4 is to show that $\mathcal{M}$ can be realized as a Wallman compactification of $\mathcal{M}_c$ in the presence of condition hH.

Experience with algebras $C(T,\underset{\sim}{R})$ of continuous functions suggests a way to determine which maximal ideals of a topological algebra X are kernels of homomorphisms. For $C(T,\underset{\sim}{R})$ those maximal ideals which corresponded (under the correspondence of the Gelfand-Kolmogorov theorem, Theorem 1.4-1) to points of the repletion υT of T were kernels of homomorphisms. The developments of Sec. 5.4 show an analogy between T and $\mathcal{M}_c$, so perhaps those maximal ideals of X which are kernels of homomorphisms are determined by some sort of repletion of $\mathcal{M}_c$. Such a repletion is defined in Sec. 5.5 and a result of the type just mentioned is obtained there in Theorem 5.5-1.

In Sec. 5.6 the well-known result that Banach algebras are regular iff they are normal is generalized to Frechet algebras.

<u>5.1 Hull-kernel topologies</u> Our goal is to relate the topological properties of $\mathcal{M}$ endowed with hull-kernel topology to the algebraic and analytic properties of X. To define the topology we consider two operations - computing hulls and kernels - on ideals and then show that the one followed by the other constitutes a closure operator.

Definition 5.1-1 KERNELS AND HULLS The $\mathcal{M}$-kernel K(S) of a subset S of $\mathcal{M}$ is $\bigcap_{M\in S}M$. The $\mathcal{M}_c$-kernel k(S) of a subset S of $\mathcal{M}_c$ is $\bigcap_{M\in S}M$. If $S=\emptyset$, then $k(S)=K(S)=X$. The $\mathcal{M}$-hull H(I) of an ideal I of X is $\{M\in\mathcal{M} \mid I\subset M\}$; the $\mathcal{M}_c$-hull h(I) of I is $\{M\in\mathcal{M}_c \mid I\subset M\}$.

Below are some of the basic properties of hulls and kernels.

(5.1-1) PROPERTIES OF HK AND hk The operators hk and HK defined on the subsets S of $\mathcal{M}_c$ and $\mathcal{M}$ respectively satisfy the following relations.

(1) $hk(\emptyset)=HK(\emptyset)=\emptyset$.

(2) $S\subset HK(S)$ and $S\subset hk(S)$.

(3) $HKH(I)=H(I)$ and $hkh(I)=h(I)$ for each ideal I of X.

(4) $HKHK(S)=HK(S)$ and $hkhk(S)=hk(S)$.

(5) $HK(S\cup T)=HK(S)\cup HK(T)$ and $hk(S\cup T)=hk(S)\cup hk(T)$.

(6) $HK(\{M\})=\{M\}$ and $hk(\{M\})=\{M\}$ for any $M\in\mathcal{M}$ in the first case, any $M\in\mathcal{M}_c$ in the second.

(7) $H([\bigcup_\mu I_\mu])=\bigcap_\mu H(I_\mu)$ and $h([\bigcup_\mu I_\mu])=\bigcap_\mu h(I_\mu)$ where $[\bigcup_\mu I_\mu]$ denotes the ideal generated by $\bigcup_\mu I_\mu$.

Proof Due to the similarity of the arguments we only give proofs for HK.

(1) $HK(\emptyset)=H(X)=\emptyset$.

(2) If $M\in S$ then, since $K(S)\subset M$, $M\in HK(S)$.

(3) Since $I\subset M$ for each $M\in H(I)$, $I\subset KH(I)$. Thus if $M\in HKH(I)$, $I\subset KH(I)\subset M$ and $M\in H(I)$, so that $HKH(I)\subset H(I)$. Conversely if $M_0\in H(I)$, then $KH(I)=\bigcap_{M\in H(I)}M\subset M_0$ and $M_0\in HKH(I)$. Thus $HKH(I)=H(I)$.

(4) Follows from (3) with $I=K(S)$.

(5) Clearly $K(S\cup T)=K(S)\cap K(T)$. If $M\in HK(S\cup T)$, then $M\supset K(S)\cap K(T)\supset K(S)K(T)$. As a maximal ideal is a prime ideal, $M\supset K(S)$ or $M\supset K(T)$. It then follows that $M\in HK(S)\cup HK(T)$. Hence $HK(S\cup T)\subset HK(S)\cup HK(T)$. The reverse inclusion follows from the facts that $S\cup T\supset S$ and $S\cup T\supset T$.

(6) Clearly $K(\{M\})=\{M\}$. Since M is a maximal ideal, $H(M)=\{M\}$.

(7) Suppose $M\in\bigcap_\mu H(I_\mu)$. Then $M\supset I_\mu$ for all μ and thus $M\supset[\bigcup_\mu I_\mu]$. Conversely if $M\in H([\bigcup_\mu I_\mu])$ then $M\supset I_\mu$ for each μ and therefore $M\in\bigcap_\mu H(I_\mu)$. ∇

Definition 5.1-2 HULL-KERNEL TOPOLOGIES As a consequence of (5.1-1) the operator HK defined on the collection of all subsets of $\mathcal{M}$ is a closure operator and generates a topology $\mathcal{T}_{HK}$ on called the $\mathcal{M}$-hull-kernel topology or just the hull-kernel topology. The restriction of $\mathcal{J}_{HK}$ to $\mathcal{M}_c$ is called the $\mathcal{M}_c$-hull-kernel topology (or just hull-kernel topology, depending upon the context) and is denoted by $\mathcal{J}_{hk}$.

It is easy to verify that $\mathcal{J}_{hk}$ is also determined by the operator hk on $\mathcal{M}_c$.

By (6) of (5.1-1), $(\mathcal{M},\mathcal{J}_{HK})$ is a T_1-space. By the results of (5.1-1) (3) the HK-closed sets in $\mathcal{M}$ are exactly the sets H(I) where I is an ideal in X and the HK-closed subsets of $\mathcal{M}_c$ are the sets h(I) where I is any ideal in X.

Our next result, (5.1-2), determines a base for the hull-kernel topology; H(x) and h(x) denote H([{x}]) and h([{x}]).

(5.1-2) A BASE OF CLOSED SETS FOR THE HULL-KERNEL TOPOLOGY (a) The sets H(x) are a base of closed sets for $\mathcal{J}_{HK}$ on $\mathcal{M}$. (b) The sets h(x) are a base of closed sets for $\mathcal{J}_{hk}$ on $\mathcal{M}_c$.

Proof (a) As previously noted in the discussion immediately following Definition 5.1-2, the closed sets of $(\mathcal{M},\mathcal{J}_{HK})$ are all of the form H(I) where I is an ideal in X. Since $H(I)=\bigcap_{x\in I}H(x)$, (a) follows. The proof of (b) is similar and is omitted. ∇

(4.12-3) has to do with when $\mathcal{M}_c$ was compact in the Gelfand topology $\mathcal{J}_G$, i.e. $\sigma(X',X)$-compact. (5.1-3) shows that $\mathcal{M}$ is always compact when it carries the hull-kernel topology.

(5.1-3) $\mathcal{M}$ IS HK-COMPACT $(\mathcal{M},\mathcal{J}_{HK})$ is a compact topological space.

Proof Let (F_μ) be a family of HK-closed subsets of $\mathcal{M}$ with empty intersection. Then $\bigcap_\mu F_\mu=\bigcap_\mu HK(F_\mu)=H([\bigcup_\mu K(F_\mu)])=\emptyset$. Thus $[\bigcup_\mu K(F_\mu)]=X$. Since $e\in X$ there must be indices $\mu_1,\ldots,\mu_n$, elements $x_i\in K(F_{\mu_i})$, and elements $y_1,\ldots,y_n\in X$ such that $e=\Sigma_i y_i x_i$. It now follows that $[\bigcup_{i=1}^n K(F_{\mu_i})]=X$ and that

$$\emptyset = H([\bigcup_{i=1}^{n} K(F_{\mu_i})]) = \bigcap_{i=1}^{n} HK(F_{\mu_i}) = \bigcap_{i=1}^{n} F_{\mu_i} . \quad \nabla$$

Definition 5.1-3 HULLS OF FINITE SETS For any finite subset $\{x_1,\ldots,x_n\}$ of the topological algebra X the $\mathcal{M}$-hull $H(x_1,\ldots,x_n)$ of $\{x_1,\ldots,x_n\}$ denotes the $\mathcal{M}$-hull in the sense of Definition 5.1-1 of the ideal $(x_1,\ldots,x_n)$ generated by $\{x_1,\ldots,x_n\}$ in X. The $\mathcal{M}_c$-hull $h(x_1,\ldots,x_n)$ of $\{x_1,\ldots,x_n\}$ is defined similarly.

We note that

$$\bigcap_i H(x_1) = H(x_1,\ldots,x_n) = \{M\in\mathcal{M} \mid x_i\in M,\ i=1,\ldots,n\}$$

and

$$\bigcap_i h(x_i) = h(x_1,\ldots,x_n) = \{M\in \mathcal{M}_c \mid x_i\in M,\ i=1,\ldots,n\}.$$

For many topological algebras X it is true that $\mathcal{M}=w(\mathcal{M}_c,\mathcal{L})$ where $\mathcal{L}$ is the lattice of hulls of finite subsets of X (see Theorem 5.4-1). We begin the approach to that result with (5.1-4) below.

(5.1-4) LATTICES OF HULLS OF FINITE SETS Let X be a topological algebra. (a) The lattice $\mathcal{L}=\{h(x_1,\ldots,x_n) \mid x_i \in X,\ n\in \underset{\sim}{N}\}$ is an $\alpha\beta$- lattice of closed subsets of $\mathcal{M}_c$ generating $\mathcal{J}_{hk}$. (b) The lattice $\mathcal{H}=\{H(x_1,\ldots,x_n) \mid x_i\in X,\ n\in\underset{\sim}{N}\}$ is an $\alpha\beta$-lattice of closed subsets of $\mathcal{M}$ generating $\mathcal{J}_{HK}$.

Proof We prove (a) only. In view of (5.1-2) it only remains to show that $\mathcal{L}$ is an $\alpha\beta$-lattice. To begin, since

$$h(x)\cup h(y) = h(xy) \text{ and } h(x)\cap h(y) = h(x,y)$$

it easily follows that $\mathcal{L}$ is a lattice.

To show that $\mathcal{L}$ is an α-lattice (Def. 3.3-1), let $h(x_1,\ldots,x_n)\in\mathcal{L}$ and $M\notin h(x_1,\ldots,x_n)$. There must be an x_j such that $x_j\notin M$. Moreover there must be some $z\in X$ and $m\in M$ such that $zx_j+m=e$. Thus $M\in h(m)$ but $h(m)\cap h(x_1,\ldots,x_n)=\emptyset$ for if $J\in h(m)\cap h(x_1,\ldots,x_n)$ then x_j, $m\in J$ which implies that $e\in J$ which is contradictory. It follows that $\mathcal{L}$ is an α-lattice. The proof that $\mathcal{L}$ is a β-lattice (Def. 3.3-1) is similar. ∇

5.2 Regular algebras and normality conditions

So far two topologies, the hull-kernel $\mathcal{J}_{HK}$ and the Gelfand $\mathcal{J}_G$, have been considered for the space $\mathcal{M}$ of maximal ideals of a topological algebra. Under certain conditions $\mathcal{J}_G$ is a compact topology for $\mathcal{M}$ as (4.12-3) shows and, as shown in (5.1-3), $\mathcal{J}_{HK}$ is always a compact topology for $\mathcal{M}$. These phenomena bear a strong resemblance to results of Chap. 1 dealing with maximal ideals of algebras $C(T,\underset{\sim}{F})$ of continuous functions: The maximal ideals of $C(T,\underset{\sim}{F})$ were (in 1-1 correspondence with) the Stone-Cech compactification βT of T (Theorem 1.4-1). By Example 4.10-2 the continuous homomorphisms of $C(T,\underset{\sim}{F},c)$ - or what amounts to the same thing, the closed maximal ideals $\mathcal{M}_c$ of $C(T,\underset{\sim}{F},c)$ - are just T for any completely regular Hausdorff space T. Thus in many cases $\mathcal{M}$ is the Stone-Cech compactification of $\mathcal{M}_c$. Section 5.4 is devoted to exploring when statements such as "$\mathcal{M}=\beta\mathcal{M}_c$" and "$\mathcal{M}=w(\mathcal{M}_c,\mathcal{L})$ for some $\mathcal{L}$" hold in various topological algebras and this section is devoted to preparing the ground for some of those results.

We begin by introducing the notion of regular algebra. We mention that aside from the role regularity will play in the theorems of Sec. 5.4, it also occupies a significant position in the theory of Banach algebras vis-à-vis continuous extendibility of homomorphisms of the Banach algebra X to superalgebras Y of X. If X is a regular Banach algebra and Y is any Banach algebra containing X then any homomorphism (perforce continuous if X is a Banach algebra) of X may be continuously extended to a homomorphism of Y. Equivalently, if X is regular, any maximal ideal of X may be embedded in a maximal ideal of Y (Naimark 1964, pp. 214 and 223).

For complete LMC algebras a maximal ideal is closed iff it is the kernel of a complex-valued homomorphism. Thus, for any closed maximal ideal M, X/M is topologically isomorphic to $\mathbb{C}$ and each $x\in X$ determines a unique scalar $x+M$. The points x in X may now be viewed as a family of maps $\hat{x}$ taking $\mathcal{M}_c$ into $\mathbb{C}$, namely $M \to x+M$. Viewed this way, if X separates points and closed subsets of $\mathcal{M}_c$ (appropriately topologized) then X is regular; if X separates disjoint closed sets in $\mathcal{M}_c$, X is normal. A reason for our interest in algebras which satisfy these "normality" conditions lies in the results of Sec. 5.4: For certain normal algebras, $\mathcal{M}$ may be realized as a Wallman compactification of $\mathcal{M}_c$.

<u>Definition 5.2-1</u> <u>REGULAR ALGEBRAS</u> A complex LMC algebra X is <u>regular</u> if for any $\mathcal{T}_G$-closed subset F of $\mathcal{M}_c$ and $M\notin F$ there exists $x\in X$ such that $\hat{x}(M)=\{1\}$ while $\hat{x}(F)=\{0\}$.

A large category of regular algebras is provided by (5.2-3). An equivalent formulation of regularity is given in (5.2-1) next.

<u>(5.2-1) REGULAR $\leftrightarrow \mathcal{T}_{hk}=\mathcal{T}_G$</u> The complex LMC algebra X is regular iff $\mathcal{T}_{hk}=\mathcal{T}_G$ on $\mathcal{M}_c$.

<u>Proof</u> We first show that $\mathcal{T}_G$ is generally finer than $\mathcal{T}_{hk}$.

If F is a $\mathcal{T}_{hk}$-closed subset of $\mathcal{M}_c$, then $F=h(k(F))$. Since $F=\bigcap_{x\in k(F)} h(x)=\bigcap_{x\in k(F)}\hat{x}^{-1}(0)$, then, since each $\hat{x}$ is continuous when $\mathcal{M}_c$ carries the weak-* $(=\mathcal{T}_G)$ topology, $\hat{x}^{-1}(0)$ is $\mathcal{T}_G$-closed for each $x\in X$. F is therefore a $\mathcal{T}_G$-closed subset of $\mathcal{M}_c$. It follows that $\mathcal{T}_{hk}\subset\mathcal{T}_G$.

Suppose now that X is regular, that F is closed in the Gelfand topology, and $M\notin F$. Then there is an $x\in X$ such that $\hat{x}(M)=\{1\}$ while $\hat{x}(F)=\{0\}$. Thus $x\notin M$ while $x\in M'$ for all $M'\in F$ so that $F\subset h(x)$ while $M\notin h(x)$. Thus for every $M\in CF$ there exists $x_M\in X$ such that $F\subset h(x_M)$ while $M\notin h(x_M)$. Consequently $F=\bigcap_{M\in CF} h(x_M)$ and F is $\mathcal{T}_{hk}$-closed.

Conversely if $\mathcal{T}_{hk}=\mathcal{T}_G$, let F be a $(\mathcal{T}_G=\mathcal{T}_{hk})$-closed subset of $\mathcal{M}_c$. Then $F=h(k(F))=\bigcap_{x\in k(F)} h(x)$ and for $M\notin F$ there exists $x_M\in k(F)$ such that $M\notin h(x_M)$ or, equivalently, $x_M\notin M$. On the other hand $x_M\in M'$ for each $M'\in F$. Consequently $\hat{x}_M(M)\neq 0$ while $\hat{x}_M(F)=\{0\}$. Now if we take $y_M=x_M/\hat{x}_M(M)$, it follows that $\hat{y}_M(F)=\{0\}$ while $\hat{y}_M(M)=\{1\}$ and X is therefore regular. ▽

A stronger separation condition than regularity - normality - is defined next. (5.2-2) shows it to be stronger than regularity.

Definition 5.2-2 NORMAL ALGEBRAS Let X be a complex LMC algebra. X is Gelfand normal (G-normal, weak-* normal) if for every pair F, K of disjoint $\mathcal{J}_G$-closed subsets of $\mathcal{M}_c$ there exists $x\in X$ such that $\hat{x}(F)=\{0\}$ while $\hat{x}(K)=\{1\}$. X is hull-kernel normal (hk-normal) if for every pair F, K of disjoint $\mathcal{J}_{hk}$-closed subsets of $\mathcal{M}_c$ there exists $x\in X$ such that $\hat{x}(F)=\{0\}$ while $\hat{x}(K)=\{1\}$.

(5.2-2) GELFAND NORMAL → REGULAR If the complex LMC algebra X is Gelfand normal then $\mathcal{J}_G$ is a normal topology on $\mathcal{M}_c$ and X is a regular algebra.

Proof The normality of $\mathcal{J}_G$ in the usual topological sense follows immediately from the Urysohn lemma characterization of normality and the fact that the functions $\hat{x}$ are all $\mathcal{J}_G$-continuous, $\mathcal{J}_G$ being the weakest topology for $\mathcal{M}_c$ making them so. Since $\mathcal{J}_G$ is just the topology induced by $\sigma(X^h,X)$ on X^h and weak-* topologies are generally Hausdorff, one-point subsets of $\mathcal{M}_c$ are $\mathcal{J}_G$-closed and the desired result follows. ∇

Having seen that Gelfand normality implies regularity, it is natural to inquire as to what relationship exists between hk-normality and regularity. It happens that regularity is not implied by hk-normality. In Sec. 5.6 it is shown that any Frechet algebra is hk-normal (Theorem 5.6-1); in Example 5.2-1 an example is given of a Frechet algebra which is not regular. Conversely, does regularity imply either version of normality? "No" is the answer to this question and this is discussed after (5.2-3).

The algebras $C(T,\underset{\sim}{C},c)$ of continuous functions are regular algebras when T is completely regular (and Hausdorff, as usual) and normal algebras when T is normal as the discussion below shows. For such algebras, with T a completely regular Hausdorff space, each $M\in\mathcal{M}_c$ is the kernel of a complex continuous homomorphism and $C(T,\underset{\sim}{C},c)^h=T^*$ by Example 4.10-2. Thus, for any $x\in C(T,\underset{\sim}{C},c)$ and $t\in T$, if M_t is the kernel of the evaluation map t^*, then

(*) $$\hat{x}(M_t) = t^*(x) = x(t).$$

As T is completely regular, its topology is the initial topology generated by $C(T,\underset{\sim}{C})$ on T, just as the topology of $\mathcal{M}_c$ is the initial topology generated by $\hat{X}=\{\hat{x}\,|\,x\in C(T,\underset{\sim}{C})\}$ on $\mathcal{M}_c$. Moreover the bijection $t\to M_t$ between T and $\mathcal{M}_c$ is actually a homeomorphism. Furthermore, separation of subsets of the completely regular space T by $C(T,\underset{\sim}{C})$ produces the same kind of separation of subsets of $\mathcal{M}_c$ by X due to

$$\begin{array}{ccc} C(T,\underset{\sim}{C}) & & \hat{X} \\ | & & | \\ T & \leftrightarrow & \mathcal{M}_c=\{M_t\,|\,t\in T\} \end{array}$$

the way (cf. (*)) functions in $C(T,\underset{\sim}{C},c)$ correspond to functions in X. In summary we have:

(5.2-3) REGULARITY AND NORMALITY OF $C(T,\underset{\sim}{C},c)$ The Hausdorff space T is completely regular or normal iff $C(T,\underset{\sim}{C},c)$ is a regular or normal algebra respectively.

The definition of hull-kernel normal does not suggest that $\mathcal{J}_{hk}$ need be normal topology since the functions $\hat{x}$ which separate the hk-closed subsets aren't necessarily hk-continuous. To see that regularity does not imply normality for algebras one need (in view of (5.2-3)) only consider a completely regular Hausdorff space T which is not normal. $C(T,\underset{\sim}{C},c)$ is then regular but not $\mathcal{J}_G$ $(=\mathcal{J}_{hk})$-normal.

Example 5.2-1 A NON-REGULAR FRECHET ALGEBRA OF ANALYTIC FUNCTIONS Let H be the algebra (with pointwise operations) of analytic functions on the open unit disc D of the complex plane with compact-open topology. Then

(a) $\mathcal{M}_c$ is homeomorphic to D,

(b) the proper hk-closed subsets of $\mathcal{M}_c$ are in 1-1 correspondence with the countable subsets of D having no limit point in D, and

(c) H is not regular.

Proof (a) By Example 4.10-3 the points of D are in 1-1 correspondence with $\mathcal{M}_c$. And because of this correspondence, Gelfand topology on $\mathcal{M}_c = H^h$ is the same as the initial topology determined by H on D. Since each $x \in H$ is continuous, the Euclidean topology of D must be finer than the initial topology determined by H on D. But the Euclidean topology is also the weakest topology for which the analytic map $\mu \to \mu$ is continuous, whence the initial topology determined by all the analytic functions on D is finer than the Euclidean.

(b) The basic hk-closed subset h(x) ((5.1-2)) corresponds to $x^{-1}(0)$ under the homeomorphism $t \to M_t$. If $h(x) \neq \mathcal{M}_c$, then $x \neq 0$ and we may invoke the analytic identity theorem* to conclude that $x^{-1}(0)$ is a countable subset of D having no limit point in D. As any proper closed subset of $\mathcal{M}_c$ is contained in some proper hull h(x), (b) follows.

(c) By (a) and (b) it follows that $\mathcal{J}_G \neq \mathcal{J}_{hk}$ so, by (5.2-1) H is not regular.∇

Since the algebra of the preceding example is not regular, it can't possibly be G-normal. On the other hand, as we have already mentioned, it

---

* If x is analytic on a domain E of $\underset{\sim}{C}$, then set $x^{-1}(0)$ of zeros of x does not have a limit point in E.

will be shown in Theorem 5.6-1 that all Frechet algebras are hk-normal, so that the algebra H is hk-normal. Thus hk-normality does not imply G-normality. However every G-normal algebra is regular by (5.2-2) so that $\mathcal{J}_G = \mathcal{J}_{hk}$ and hk-normality follows. In summary

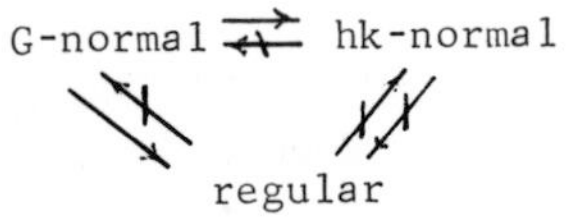

In the final result of this section the notions of G- and hk-normality are characterized in terms of how the closures in $\mathcal{M}$ of disjoint closed subsets of $\mathcal{M}_c$ behave. Prior to presenting that result we prove a prerequisite technicality.

<u>(5.2-4) SEPARATION IN $\mathcal{M}$</u> If E and F are disjoint subsets of $\mathcal{M}_c$ then the following are equivalent.

(a) $cl_{HK}E \cap cl_{HK}F = \emptyset$.

(b) For some $x \in X$, $\hat{x}(E)=\{1\}$ and $\hat{x}(F)=\{0\}$.

(c) $k(E)+k(F)=X$.

<u>Proof</u> (a)$\leftrightarrow$(c): As $cl_{HK}E \cap cl_{HK}F = H(k(E)) \cap H(k(F)) = H(k(E)+k(F))$ by (5.1-1) (7) and $H(k(E)+k(F))=\emptyset$ iff $k(E)+k(F)=X$, the equivalence follows.

(b)$\leftrightarrow$(c): If $x \in X$ is such that $\hat{x}(F)=\{0\}$ while $\hat{x}(E)=\{1\}$ it follows that $x \in k(F)$, $e-x \in k(E)$, and therefore $k(F)+k(E)=X$. Conversely if $X=k(F)+k(E)$, then there are $x \in k(F)$ and $y \in k(E)$ such that $e=x+y$. But then $\hat{x}(F)=\{0\}$ and $\hat{y}(E)=\{0\}$. Since $y=e-x$ it follows that $\hat{x}(E)=\{1\}$ and x satisfies condition (b). ∇

<u>(5.2-5) CLOSURES OF DISJOINT SETS AND NORMALITY</u> Let X be a complex LMC algebra. Then (a) disjoint $\mathcal{J}_G$-closed subsets of $\mathcal{M}_c$ have disjoint $\mathcal{J}_{HK}$-closures in $\mathcal{M}$ iff X is $\mathcal{J}_G$-normal, and (b) disjoint $\mathcal{J}_{hk}$-closed subsets of $\mathcal{M}_c$ have disjoint $\mathcal{J}_{HK}$-closures in $\mathcal{M}$ iff X is hk-normal.

<u>Proof</u> Since (a) and (b) are established in the same way, we only prove (a). If E and F are disjoint $\mathcal{J}_G$-closed subsets of $\mathcal{M}_c$ and X is G-normal, then there is some $x \in X$ satisfying (b) of (5.2-4). Thus, by (5.2-4)(a), $cl_{HK}E \cap cl_{HK}F = \emptyset$. The converse follows by simply reversing the steps. ∇

<u>5.3 Condition hH</u> There are always maximal ideals containing a given proper ideal, but it is not always true (in <u>topological</u> algebras) that there are closed maximal ideals containing a given ideal. In particular, there are dense ideals: The set I of $x \in C(\underset{\sim}{R}, \underset{\sim}{F}, c)$ which vanish outside some compact set G_x, for example, is a dense ideal in $C(\underset{\sim}{R}, \underset{\sim}{F}, c)$. Under certain conditions - special kinds of ideals or algebras - the existence of a closed

maximal ideal containing a given ideal can be guaranteed, as shown by (4.10-10) and (4.10-12).

The subject of this section is a condition, condition hH, which says that every finitely generated ideal is embeddable in a closed maximal ideal. We consider equivalent descriptions of the condition and mention classes of algebras, Frechet algebras and $C(T,\underset{\sim}{F},c)$ in particular, in which it holds. The reason for our interest in it is that in algebras in which condition hH holds, the space $\mathcal{M}$ of maximal ideals may be expressed as a Wallman compactification of $\mathcal{M}_c$, as shown in Sec. 5.4.

Definition 5.3-1 CONDITION hH A complex topological algebra X satisfies condition hH if each finitely generated proper ideal is contained in a closed maximal ideal.

The reason for calling this "condition hH" is contained in the following result.

(5.3-1) EQUIVALENTS OF hH In a complex topological algebra X the following statements are equivalent.

(a) X satisfies condition hH.

(b) With h and H as in Definition 5.1-1 and $x_1,\ldots,x_n \in X$ then $h(x_1,\ldots,x_n)=\emptyset \rightarrow H(x_1,\ldots,x_n)=\emptyset$.

(c) Letting "cl_{HK}" denote HK-closure in $\mathcal{M}$ (see Sec. 5.1), and $x_1,\ldots,x_n \in X$, $cl_{HK}h(x_1,\ldots,x_n)=H(x_1,\ldots,x_n)$.

Proof The equivalence of (a) and (b) is clear as is the fact that (c) → (a). To see that (a) → (c), we first note that

$$cl_{HK}h(x_1,\ldots,x_n) = Hkh(x_1,\ldots,x_n) \subset HKH(x_1,\ldots,x_n) = H(x_1,\ldots,x_n).$$

Suppose $M \notin Hkh(x_1,\ldots,x_n)$. To show that $M \notin H(x_1,\ldots,x_n)$ is equivalent to showing that some $x_i \notin M$. Clearly $kh(x_1,\ldots,x_n) \not\subset M$. Therefore $kh(x_1,\ldots,x_n)+M=X$, so there must be $z \in kh(x_1,\ldots,x_n)$ and $m \in M$ such that $z+m=e$. We contend that $h(x_1,\ldots,x_n,m)=\emptyset$. If $M' \in h(x_1,\ldots,x_n,m)$, then $M' \in h(x_1,\ldots,x_n)$ and therefore $z \in kh(x_1,\ldots,x_n) \subset M'$. Since m also belongs to M' however, $z+m=e \in M'$ which is a contradiction. Therefore $h(x_1,\ldots,x_n,m)=\emptyset$ from which it follows that $H(x_1,\ldots,x_n,m)=\emptyset$ by (b). Thus $I=(x_1,\ldots,x_n,m)=X$. Since $m \in M$, it follows that some $x_i \notin M$. ▽

Using (5.3-1) it is evident that:

(5.3-2) HK-CLOSURES OF h-HULLS If X satisfies condition hH, then $cl_{HK}h(x_1,\ldots,x_n)=\bigcap_{i=1}^{n} cl_{HK}h(x_i)$ for any $x_1,\ldots,x_n \in X$.

Some examples of algebras which satisfy condition hH follow.

(5.3-3) hH ALGEBRAS The class of hH algebras includes the complex Frechet algebras and the algebras $C(T,\underset{\sim}{C},c)$ where T is a completely regular Hausdorff space.

Proof The result about Frechet algebras has been established in (4.10-12). As for $C(T,\underset{\sim}{C},c)$ we show that if $I=(x_1,\ldots,x_n)$ is a proper ideal, then $\bigcap_{i=1}^{n} x_i^{-1}(0)\neq\emptyset$. If $\bigcap_{i=1}^{n} x_i^{-1}(0)=\emptyset$, then $x=\Sigma_i x_i\bar{x}_i=\Sigma_i |x_i|^2 \in I$ but never vanishes. ∇

5.4 $\mathcal{M}$ as a Wallman Compactification of $\mathcal{M}_c$ For the particular topological algebra $C(T,\underset{\sim}{C},c)$ the set $\mathcal{M}$ of maximal ideals is in 1-1 correspondence with the Stone-Cech compactification βT of T (Theorem 1.4-1). By Example 4.10-2, T is homeomorphic to the space $\mathcal{M}_c$ of closed maximal ideals for such algebras when $\mathcal{M}_c$ is endowed with the Gelfand (=hull-kernel) topology. Thus by just reflecting βT's topology over onto $\mathcal{M}$, we may say that $\mathcal{M}=\beta\mathcal{M}_c$. Letting $\mathcal{Z}$ denote the lattice of zero sets of T and using (3.4-3), which allows the convertibility of Stone-Cech compactifications into Wallman compactifications, we may say that $\mathcal{M}=w(\mathcal{M}_c,\mathcal{Z})$.

To what extent do results of this type for algebras of continuous functions extend to complex LMCH algebras? By (5.1-3) we know that $(\mathcal{M},\mathcal{J}_{HK})$ is a compact T_1-space containing $\mathcal{M}_c$. But is $\mathcal{M}$ a compactification of $\mathcal{M}_c$? i.e. is $\mathcal{M}_c$ dense in $\mathcal{M}$? For algebras that satisfy condition hH, the answer is yes by virtue of (5.3-1)(c) which asserts the equivalence of hH to the statement $cl_{HK}h(x_1,\ldots,x_n)=H(x_1,\ldots,x_n)$. We can go further and actually identify $\mathcal{M}$ as a Wallman compactification of $\mathcal{M}_c$ in the presence of hH and results of this type are the subject of this section.

As the lattice $\mathcal{H}$ of hulls $h(x_1,\ldots,x_n)$ of finite subsets $x_1,\ldots,x_n$ of a complex topological algebra X is an $\alpha\beta$-lattice of $\mathcal{J}_{hk}$-closed subsets of $\mathcal{M}_c$ by (5.1-4), it follows by (3.3-2) that the mapping

$$\varphi_H: \mathcal{M}_c \longrightarrow w(\mathcal{M}_c,\mathcal{H})$$
$$M \longrightarrow \mathcal{F}_M = \{A\in\mathcal{H} \mid M\in A\}$$

is a homeomorphism of $(\mathcal{M}_c,\mathcal{J}_{hk})$ onto a dense subspace of the compact Hausdorff space $w(\mathcal{M}_c,\mathcal{H})$, i.e. $w(\mathcal{M}_c,\mathcal{H})$ is a Hausdorff compactification of $\mathcal{M}_c$. We shall see that if X satisfies condition hH then φ_H is extendible to a homeomorphism of $(\mathcal{M},\mathcal{J}_{HK})$ onto $w(\mathcal{M}_c,\mathcal{H})$.

Theorem 5.4-1 $\mathcal{M}=w(\mathcal{M}_c,\mathcal{H})$ If X satisfies condition hH then there is a homeomorphism σ taking $(\mathcal{M},\mathcal{J}_{HK})$ onto $w(\mathcal{M}_c,\mathcal{H})$ such that $\sigma|_{\mathcal{M}_c}=\varphi_H$.

Proof For each $M\in\mathcal{M}$ we define $\sigma(M)=\{B\in\mathcal{H} \mid M\in cl_{HK}B\}$.

We claim that $\sigma(M)$ is an $\mathcal{H}$-filter. To see this suppose $h(x_1,\ldots,x_n)$

and $h(y_1,\ldots,y_m)$ belong to $\sigma(M)$. Since X satisfies condition hH, (5.3-1)(c) may be invoked and it follows that $M\in cl_{HK}h(x_1,\ldots,x_n)=H(x_1,\ldots,x_n)$ and $M\in cl_{HK}h(y_1,\ldots,y_m)=H(y_1,\ldots,y_m)$. Hence $x_1,\ldots,x_n$ and $y_1,\ldots,y_m$ belong to M so

$$M\in H(x_1,\ldots,x_n,\ y_1,\ldots,y_m) = cl_{HK}h(x_1,\ldots,x_n,\ y_1,\ldots,y_m)$$
$$= cl_{HK}h(x_1,\ldots,x_n)\cap h(y_1,\ldots,y_m).$$

Thus $\sigma(M)$ is stable with respect to the formation of finite intersections. As supersets of $\sigma(M)$ in $\mathcal{H}$ clearly belong to $\sigma(M)$, $\sigma(M)$ is an $\mathcal{H}$-filter.

To see that $\sigma(M)$ is an $\mathcal{H}$-ultrafilter, suppose that $h(x_1,\ldots,x_n)\notin\sigma(M)$ or equivalently that

$$M\notin cl_{HK}h(x_1,\ldots,x_n) = H(x_1,\ldots,x_n).$$

Then for some i, $x_i\notin M$ so, since M is a maximal ideal, there must be some $y\in X$ and $m\in M$ such that $x_iy+m=e$. Since $m\in M$, $M\in cl_{HK}h(m)=H(m)$ and therefore $h(m)\in\sigma(M)$. But $h(m)\cap h(x_1,\ldots,x_n)=\emptyset$ for if $M'\in h(m)\cap h(x_1,\ldots,x_n)$ then x_i and m each belong to M' and the contradictory conclusion that $x_iy+m=e\in M'$ follows. Hence $\sigma(M)$ is an $\mathcal{H}$-ultrafilter.

Next we show that σ is bijective. As for onto-ness, let $\mathcal{F}\in w(\mathcal{M}_c,\mathcal{H})$. Since the sets of $\mathcal{F}$ have the finite intersection property and $\mathcal{M}$ is a compact T_1-space by (5.1-3) it follows that

$$\bigcap_{h(x_1,\ldots,x_n)\in\mathcal{F}} H(x_1,\ldots,x_n) = \bigcap_{h(x_1,\ldots,x_n)\in\mathcal{F}} cl_{HK}h(x_1,\ldots,x_n) \neq \emptyset$$

We contend that there is some unique $M\in\mathcal{M}$ such that

$$\bigcap_{h(x_1,\ldots,x_n)\in\mathcal{F}} H(x_1,\ldots,x_n) = \{M\} \text{ and } \sigma(M) = \mathcal{F}.$$

If $M\in\bigcap_{h(x_1,\ldots,x_n)\in\mathcal{F}} H(x_1,\ldots,x_n)$ and $M'\neq M$ then there exists $m\in M$ such that $m\notin M'$. Hence $M\in cl_{HK}h(m)=H(m)$ while $M'\notin H(m)$. Thus, to prove that $\bigcap_{h(x_1,\ldots,x_n)\in\mathcal{F}} H(x_1,\ldots,x_n)=\{M\}$, it suffices to show that $h(m)\in\mathcal{F}$. To this end let $h(x_1,\ldots,x_n)\in\mathcal{F}$. Then, as $M\in H(x_1,\ldots,x_n,m)=cl_{HK}h(x_1,\ldots,x_n,m)$,

$$h(x_1,\ldots,x_n,m) = h(x_1,\ldots,x_n)\cap h(m) \neq \emptyset.$$

Thus, since $\mathcal{F}$ is an ultrafilter, it follows that $h(m)\in\mathcal{F}$. To show that $\sigma(M)=\mathcal{F}$ we observe that $\mathcal{F}\subset\sigma(M)$ follows from the definition of $\sigma(M)$ and the fact that $M\in cl_{HK}h(x_1,\ldots,x_n)$ for any $h(x_1,\ldots,x_n)\in\mathcal{F}$. But both $\mathcal{F}$ and $\sigma(M)$ are $\mathcal{H}$-ultrafilters; hence $\mathcal{F}=\sigma(M)$.

To see that σ is 1-1, consider distinct elements $M_1, M_2 \in \mathcal{M}$. Selecting $m_1 \in M_1$ such that $m_1 \notin M_2$ it follows that $M_1 \in H(m_1) = cl_{HK}h(m_1)$ while $M_2 \notin H(m_1) = cl_{HK}h(m_1)$. Thus $h(m_1) \in \sigma(M_1)$ and $h(m_1) \notin \sigma(M_2)$ so that $\sigma(M_1) \neq \sigma(M_2)$.

Having shown that σ is bijective, next we contend that $\sigma|_{\mathcal{M}_c} = \varphi_H$. Indeed if $M \in \mathcal{M}_c$ then $M \in H(x_1,\ldots,x_n)$ iff $M \in h(x_1,\ldots,x_n)$. Thus $\sigma(M) = \{h(x_1,\ldots,x_n) \mid M \in H(x_1,\ldots,x_n)\} = \{h(x_1,\ldots,x_n) \mid M \in h(x_1,\ldots,x_n)\} = \varphi_H(M)$.

To complete the proof we need only show that σ is bicontinuous. To prove this it suffices to show that

$$\sigma(H(x)) = \{\mathcal{F} \in w(\mathcal{M}_c, \mathcal{H}) \mid h(x) \in \mathcal{F}\} = \mathcal{B}_{h(x)} \quad \text{for each } x \in X.$$

To this end let $M \in H(x)$. Then, as $H(x) = cl_{HK}h(x)$ by condition hH, $h(x) \in \sigma(M) = \{h(x_1,\ldots,x_n) \mid M \in cl_{HK}h(x_1,\ldots,x_n)\}$. Thus $\sigma(M) \in \mathcal{B}_{h(x)}$ and $\sigma(H(x)) \subset \mathcal{B}_{h(x)}$. On the other hand if $\mathcal{F} \in \mathcal{B}_{h(x)}$ we know that $\mathcal{F} = \sigma(M)$ for some M. As $h(x) \in \mathcal{F}$, it follows that $M \in cl_{HK}h(x) = H(x)$ and $\mathcal{B}_{h(x)} \subset \sigma(H(x))$, that is that basic closed subsets of $\mathcal{M}$ are taken onto basic closed subsets of $w(\mathcal{M}_c, \mathcal{H})$ and conversely. ∇

As we already know from previous considerations (see Sec. 3.6) $\mathcal{H}$ may be replaced in the previous theorem by any lattice which is equivalent (in the sense of Def. 3.6-1) to $\mathcal{H}$.

In Theorem 3.6-1 we established the fact that two α-lattices $\mathcal{L}$ and $\mathcal{L}'$ defined on a set T, for which it is true that $\mathcal{L}' \subset \mathcal{L}$, are equivalent whenever two arbitrary disjoint sets $A, B \in \mathcal{L}$ have disjoint closures in the Wallman space defined by the lattice $\mathcal{L}'$. Since $\mathcal{C}_{hk}$, the lattice of all hk-closed subsets of the T_1-space $\mathcal{M}_c$, is an $\alpha\beta$-lattice by Example 3.3-2 and the $\alpha\beta$-lattice $\mathcal{H} \subset \mathcal{C}_{hk}$, we may apply the previously mentioned result to conclude that $\mathcal{H}$ may be replaced in Theorem 5.4-1 by $\mathcal{C}_{hk}$ iff the closures in $w(\mathcal{M}_c, \mathcal{H})$ of any disjoint pair of sets $A, B \in \mathcal{C}_{hk}$ are disjoint. Since $w(\mathcal{M}_c, \mathcal{H}) = (\mathcal{M}, \mathcal{T}_{HK})$ this is equivalent to requiring that for $A, B \in \mathcal{C}_{hk}$

$$(*) \qquad A \cap B = \emptyset \rightarrow cl_{HK}A \cap cl_{HK}B = \emptyset.$$

By (5.2-5)(b), (*) holds iff X is hk-normal. Summarizing the above discussion we state (for the LMC algebra X):

<u>Theorem 5.4-2</u> <u>$\mathcal{M} = w(\mathcal{M}_c, \mathcal{C}_{hk})$</u> Let X satisfy condition hH and $\mathcal{C}_{hk}$ denote all hk-closed subsets of $\mathcal{M}_c$. Then $\mathcal{M}$ and $w(\mathcal{M}_c, \mathcal{C}_{hk})$ are homeomorphic via a map which reduces to $\varphi_{\mathcal{C}_{hk}} : \mathcal{M}_c \rightarrow w(\mathcal{M}_c, \mathcal{C}_{hk})$, $M \rightarrow \mathcal{F}_M = \{A \in \mathcal{C}_{hk} \mid M \in A\}$ on $\mathcal{M}_c$ iff X is hk-normal.

Going one step further, if we consider only regular algebras that satisfy condition hH then $\mathcal{H}$ may be replaced by the lattice $\mathcal{C}_G$ of all Gelfand -

closed subsets of $\mathcal{M}_c$ iff X is Gelfand normal for then $\mathcal{J}_{hk}=\mathcal{J}_G$ and Gelfand normality is equivalent to hk-normality. Furthermore, in this case, $w(\mathcal{M}_c, \mathcal{J}_G)=\beta\mathcal{M}_c$ by (3.4-3). Thus:

Theorem 5.4-3 WHEN IS $\mathcal{M}$ A STONE-CECH COMPACTIFICATION? Let the regular algebra X satisfy condition hH and $\mathcal{C}_G$ denote the $\alpha\beta$-lattice of all Gelfand closed subsets of $\mathcal{M}_c$. Then $\mathcal{M}$ and $\beta\mathcal{M}_c=w(\mathcal{M}_c,\mathcal{C}_G)$ are homeomorphic via a map that reduces to

$$\varphi_{\mathcal{C}_G}: \mathcal{M}_c \to w(\mathcal{M}_c,\mathcal{C}_G)$$
$$M \to \mathcal{F}_M = \{A\in\mathcal{C}_G \mid M\in A\}$$

on $\mathcal{M}_c$.

5.5 The X-repletion of $\mathcal{M}_c$

As shown in Chap. 1 the maximal ideals of algebras $C(T,\underset{\sim}{R})$ of continuous functions were characterized by βT. Those maximal ideals in $C(T,\underset{\sim}{R})$ which were kernels of (real-valued) homomorphisms were characterized by the repletion υT of T ((1.6-1)). For certain topological algebras X the closed maximal ideals $\mathcal{M}_c$ behave analogously to T in that $\mathcal{M}=\beta\mathcal{M}_c$ as discussed in Sec. 5.4. The question naturally arises: Can those maximal ideals of X which are kernels of homomorphisms be characterized as some sort of repletion of $\mathcal{M}_c$? Briefly, the answer is yes (for suitable algebras). We introduce such a repletion of $\mathcal{M}_c$ in Def. 5.5-1 and show how it characterizes homomorphisms in Theorem 5.5-1.

Letting X be a complex Gelfand-normal algebra satsfying condition hH, we may say that $\mathcal{M}=\beta\mathcal{M}_c$ by Theorem 5.4-3. Letting $\underset{\sim}{C}\cup\{\infty\}$ denote the one-point compactification of $\underset{\sim}{C}$, each Gelfand extension $\hat{x}$ on $(\mathcal{M}_c,\mathcal{J}_G=\mathcal{J}_{hk})$ has a unique continuous extension $\hat{x}^*:\beta\mathcal{M}_c\to\underset{\sim}{C}\cup\{\infty\}$. By the results of the preceding section we may say that $\mathcal{M}=\beta\mathcal{M}_c=w(\mathcal{M}_c,\mathcal{H})$ where $\mathcal{H}$ is the collection of all hulls $h(x_1,\ldots,x_n)$ of finite subsets $\{x_1,\ldots,x_n\}$ of X. We can now define the X-repletion of $\mathcal{M}_c$.

Definition 5.5-1 THE X-REPLETION OF $\mathcal{M}_c$ With X a complex Gelfand-normal algebra and notation as above the X-repletion of $\mathcal{M}_c$ is $\upsilon_X(\mathcal{M}_c)=\{\mathcal{F}\in w(\mathcal{M}_c,\mathcal{H}) \mid \hat{x}^*(\mathcal{F})\in\underset{\sim}{C}$ for all $x\in X\}$.

As we are assuming that X is normal, the results of Theorems 5.4-1, 2, and 3 permit us to represent $\mathcal{M}$ in a number of ways. In Th. 5.5-1 we have chosen to use the representation of Theorem 5.4-1 for reasons that will emerge in the proof of Theorem 5.5-1.

Theorem 5.5-1 THE HOMOMORPHISMS OF A GELFAND-NORMAL ALGEBRA Let X be a complex Gelfand-normal algebra satisfying condition hH. Then $M\in\mathcal{M}$ is the kernel of a complex homomorphism of X iff $\sigma(M)\in\upsilon_X(\mathcal{M}_c)$ where (cf. Theorem

5.4-1) $\sigma(M)=\{B\in\mathcal{H}\,|\,M\in cl_{HK}B\}$.

Proof Suppose first that $\sigma(M)=\mathcal{F}\in\upsilon_X(\mathcal{M}_c)$ and let $I=\{x\in X\,|\,\hat{x}^*(\mathcal{F})=0\}$. Clearly I is an ideal in X. Once it has been shown that I=M it will follow that the map f_M sending x into $\hat{x}^*(\mathcal{F})$ is a complex homomorphism with kernel M. In other words if $\sigma(M)=\mathcal{F}\in\upsilon_X(\mathcal{M}_c)$ then M is the kernel of a homomorphism of X. To prove that I=M, it suffices to show that $M\subset I$ since I is a proper ideal ($\hat{e}^*(\mathcal{F})=1$ for all $\mathcal{F}\in w(\mathcal{M}_c,\mathcal{H})$). For $x\in M$ then $M\in H(x)=cl_{HK}h(x)$. Since M is the limit of a net (M_μ) in h(x), then, with φ_H as in Theorem 5.4-1, $\sigma(M)=\mathcal{F}$ is the limit of the net $\sigma(M_\mu)=\varphi_H(M_\mu)=\mathcal{F}_{M_\mu}=\{B\in\mathcal{H}\,|\,M_\mu\in cl_{HK}B\}$. Carrying the function $\hat{x}$ on $\mathcal{M}_c$ over to $\varphi_H(\mathcal{M}_c)$ and continuing to write it as $\hat{x}$, we see that as $M_\mu\in h(x)$, $\hat{x}(M_\mu)=\hat{x}(\mathcal{F}_{M_\mu})=0\rightarrow\hat{x}^*(\mathcal{F})=0$. Thus $M\subset I$ and half the proof is done.

Conversely if M is the kernel of a nontrivial complex homomorphism of X and $\sigma(M)=\mathcal{F}$, then for each $x\in X$ there is some $\mu\in\underset{\sim}{C}$ such $x-\mu e\in M$. Thus $M\in H(x-\mu e)=cl_{HK}h(x-\mu e)$ and once again $\mathcal{F}=\sigma(M)$ is the limit of a net $\varphi_H(M_\alpha)=\mathcal{F}_{M_\alpha}$ where $M_\alpha\in h(x-\mu e)$. Therefore $(\hat{x}-\mu\hat{e})(M)=0\rightarrow(\hat{x}-\mu\hat{e}^*)(\mathcal{F})=0$. Thus $\hat{x}^*(\mathcal{F})=\mu$ and $\mathcal{F}\in\upsilon_X(\mathcal{M}_c)$. ∇

5.6 Frechet algebras It is well-known that for Banach algebras regularity and (Gelfand) normality are equivalent (Naimark 1964, p. 224). We demonstrate the generalization of this result to Frechet algebras in this section. Thus, as any Frechet algebra satisfies condition hH ((4.10-12)), it follows in light of Theorem 5.4-3 that for Frechet algebras $\mathcal{M}=\beta\mathcal{M}_c$.

(5.6-1) Let I and J be closed ideals in the Frechet algebra X. Then, if $h(I)\cap h(J)=\emptyset$, $I+J=X$ $(H(I)\cap H(J)=\emptyset)$.

Proof Let (p_n) be an increasing sequence of multiplicative seminorms (i.e. $p_n\leq p_{n+1}$ for each n) generating the topology on X. For each n let X_n denote the factor algebras generated by p_n (Sec. 4.5) and let κ_n be the canonical homomorphism of X into X_n. We recall that $X^h=\bigcup_n\kappa_n'(X_n^h)$ by (4.10-7). Furthermore X^h and $\mathcal{M}_c$ are in 1-1 correspondence under the pairing $h\longleftrightarrow\ker h$ [(4.10-4)]. Since all maximal ideals of a Banach algebra are closed [(4.10-1)], a 1-1 correspondence also exists between X_n^h and the collection $\mathcal{M}_n$ of all maximal ideals of X_n. Thus, identifying $\mathcal{M}_c$ with X^h and $\mathcal{M}_n$ with X_n^h, we have $\mathcal{M}_c=\bigcup_n\kappa_n'(\mathcal{M}_n)$.

Note that neither h(I) nor h(J) can be empty by (4.10-10). Moreover we claim that h(I) and h(J) meet all the $\kappa_n'(\mathcal{M}_n)$ past some n. Indeed h(I) meets $\mathcal{M}_c=\bigcup_n\kappa_n'(\mathcal{M}_n)$ and therefore $\kappa_n'(\mathcal{M}_n)$ for some n. This means that there is a complex homomorphism $f_n\in X_n^h$ such that the continuous homomorphism $f=f_n\cdot\kappa_n$ vanishes on I.

Since for any $m > n$ we may write $f=f_n \cdot \kappa_n = f_n \cdot h_{nm} \cdot \kappa_m$, where h_{nm} is the continuous map (see Theorem 4.6-1)

$$h_{nm}: X_m \longrightarrow X_n$$

$$\kappa_m(x) = x+p_m^{-1}(0) \longrightarrow x+p_n^{-1}(0) = \kappa_n(x)$$

it follows that $f \in \kappa_m'(X_m^h)$. Thus $h(I) \cap \kappa_m'(\mathcal{M}_m) \neq \emptyset$ for $m \geq n$. As a similar conclusion holds for $h(J)$, there is no loss of generality in assuming that $h(I)$ and $h(J)$ meet all the $\kappa_n'(\mathcal{M}_n)$.

Since $h(I) \cap h(J) = \emptyset$, it follows that $h(\overline{\kappa_n(I)}) \cap h(\overline{\kappa_n(J)}) = \emptyset$ for all n. Indeed if this is not so, i.e. if for some $M \in \mathcal{M}_n$, $\overline{\kappa_n(I)} + \overline{\kappa_n(J)} \subset M = \ker f$ where f is a nontrivial homomorphism of the Banach algebra X_n onto $\underset{\sim}{C}$, then $I+J \subset \ker f \cdot \kappa_n \in \mathcal{M}_c$ and therefore $h(I) \cap h(J) \neq \emptyset$ which is contradictory. Hence $\overline{\kappa_n(I)} + \overline{\kappa_n(J)}$ (which is an ideal in X_n is not contained in any maximal ideal in X_n and so the ideal $\overline{\kappa_n(I) + \kappa_n(J)}$ coincides with X_n. Also we note that by Theorem 4.6-1, I and J are the projective limits of the algebras $(\overline{\kappa_n(I)})$ and $(\overline{\kappa_n(J)})$. Having recalled these basic facts, we begin the main body of the proof. We will construct $x \in I$, $y \in J$ such that $x+y=e$.

Let (μ_n) be any sequence of positive numbers such that $\Sigma_n \mu_n$ converges. Since $\overline{\kappa_o(I)} + \overline{\kappa_o(J)} = X_o$, we choose $x_o \in \overline{\kappa_o(I)}$, $y_o \in \overline{\kappa_o(J)}$ such that $x_o + y_o = \kappa_o(e)$. Next choose $x_1' \in \overline{\kappa_1(I)}$, $y_1' \in \overline{\kappa_1(J)}$ such that $x_1' + y_1' = \kappa_1(e)$, and then choose $z_1 \in \kappa_1(I)$ and $w_1 \in \kappa_1(J)$ such that

$$\text{(1)} \qquad \|h_{o1}(z_1) - x_o\|_o < \min(\mu_1/3,\ \mu_1/3\max(\|x_1\|_1, \|y_1\|_1))$$

and

$$\text{(1')} \qquad \|h_{o1}(w_1) - y_o\|_o < \min(\mu_1/3,\ \mu_1/3\max(\|x_1'\|_1, \|y_1'\|_1)).$$

This can be done because $h_{o1}(X_1)$ is dense in X_o [recall that $h_{o1}(X/p_1^{-1}(0)) = X/p_o^{-1}(0)$ so that $h_{o1}(X_1) \supset X/p_o^{-1}(0)$ and that X_o is the completion of $X/p_o^{-1}(0)$].

Let

$$x_1 = z_1 + x_1'(\kappa_1(e) - z_1 - w_1)$$

and

$$y_1 = w_1 + y_1'(\kappa_1(e) - z_1 - w_1)$$

Clearly $x_1 + y_1 = \kappa_1(e)$. If we apply h_{o1} to x_1 and note that $h_{o1}(\kappa_1(e)) = \kappa_o(e) = x_o + y_o$, then it follows from (1) and (1') that

$$\|h_{o1}(x_1) - x_o\|_o < \mu_1 \quad \text{and} \quad \|h_{o1}(y_1) - y_o\|_o < \mu_1 .$$

By induction we may construct sequences (x_n) and (y_n), $x_n \epsilon X_n$, $y_n \epsilon Y_n$, such that $x_n + y_n = \varkappa_n(e)$ and

(2) $\|h_{n-1,n}(x_n) - x_{n-1}\|_{n-1} < \mu_n$ and $\|h_{n-1,n}(y_n) - y_{n-1}\|_{n-1} < \mu_n$.

If we apply the norm-decreasing homomorphism $h_{n-2,n-1}$ to the vectors in (2) we obtain

$$\|h_{n-2,n}(x_n) - h_{n-2,n-1}(x_{n-1})\|_{n-2} < \mu_n \quad \text{and} \quad \|h_{n-2,n}(y_n) - h_{n-2,n-1}(y_{n-1})\|_{n-2} < \mu_n .$$

Continuing this process we obtain

(3) $\|h_{kn}(x_n) - h_{k,n-1}(x_{n-1})\|_k < \mu_n$ and $\|h_{kn}(y_n) - h_{k,n-1}(y_{n-1})\|_k < u_n$

for all $k < n$.

In (3) fix k and for $n \geq k$ define two sequences $(x_k(n))$ and $(y_k(n))$ from X_k by taking

(4) $x_k(n) = h_{kn}(x_n)$ and $y_k(n) = h_{kn}(y_n)$

for $n \geq k$.   Since $x_n + y_n = \varkappa_n(e)$, it follows that

(5) $x_k(n) + y_k(n) = h_{kn}(\varkappa_n(e)) = \varkappa_k(e)$

for all $n \geq k$.   The sequences $(x_k(n))$ and $(y_k(n))$ satisfy conditions (6), and (7) below.

(6) $h_{k,k+1}(x_{k+1}(n)) = x_k(n)$ and $h_{k,k+1}(y_{k+1}(n)) = y_k(n)$ $(n \geq k+1)$.

Proof of (6)   Recall that $h_{k,k+1} \cdot h_{k+1,n} = h_{kn}$.   Thus

$$h_{k,k+1}(x_{k+1}(n)) = h_{k,k+1}(h_{k+1,n}(x_n)) = h_{k,n}(x_n) = x_k(n) .$$

The third relationship, (7) below, together with the fact that $\Sigma_n \mu_n$ converges will imply that, for any fixed k, $(x_k(n))$ and $(y_k(n))$ are Cauchy sequences in $\overline{\varkappa_k(I)}$ and $\overline{\varkappa_k(J)}$ respectively.

(7) $\|x_k(n) - x_k(n+p)\|_k < \mu_{n+1} + \mu_{n+2} + \dots + \mu_{n+p}$

and

$$\|y_k(n) - y_k(n+p)\|_k < \mu_{n+1} + \dots + \mu_{n+p}$$

for any positive integer p and $n \geq k$.

Proof of (7) We verify the result for p=1 and p=2.

(a) p=1 $\|x_k(n) - x_k(n+1)\|_k = \|h_{kn}(x_n) - h_{k,n+1}(x_{n+1})\|_k < \mu_{n+1}$ by (3).

(b) p=2 $\|x_k(n) - x_k(n+2)\|_k = \|x_k(n) - x_k(n+1) + x_k(n+1) - x_k(n+2)\|_k$

$$\leq \|x_k(n) - x_k(n+1)\|_k + \|x_k(n+1) - x_k(n+2)\|_k < \mu_{n+1} + \mu_{n+2} .$$

The proof is completed by putting statements (5), (6), and (7) together. By (7) there exist , for each k, vectors $x(k)\in\overline{\kappa_k(I)}$ and $y(k)\in\overline{\kappa_k(J)}$ such that

(9) $x_k(n) \overset{n}{\to} x(k)$ and $y_k(n) \overset{n}{\to} y(k)$.

By (6) $h_{k,k+1}(x_{k+1}(n))=x_k(n)$ for $n \geq k+1$. Furthermore by the continuity of $h_{k,k+1}$ and (9) it follows that

$$h_{k,k+1}(x_{k+1}(n)) \overset{n}{\to} h_{k,k+1}(x(k+1)).$$

On the other hand $x_k(n) \overset{n}{\to} x(k)$. Thus we obtain

(10) $h_{k,k+1}(x(k+1)) = x(k)$ and $h_{k,k+1}(y(k+1)) = y(k)$.

As I and J are the projective limits of $(\overline{\kappa_k(I)})$ and $(\overline{\kappa_k(J)})$ respectively, there exist $x\in I$ and $y\in J$ such that

(11) $\kappa_k(x) = x(k)$ and $\kappa_k(y) = y(k)$ for all k.

By (5) $x_k(n)=y_k(n)=\kappa_k(e)$ so, applying (9), we obtain $x(k)+y(k)=\kappa_k(e)$ for all k. Applying (11) we obtain

(12) $\kappa_k(x+y) = \kappa_k(e)$ for all k.

Thus, as X is a LMCH algebra, x+y=e. As $x\in I$ and $y\in J$, it follows that I+J= X. ∇

Theorem 5.6-1 FRECHET → hk-NORMAL If X is a Frechet algebra then X is hull-kernel normal.

Proof Let F and G be disjoint hk-closed subsets of $\mathcal{M}_c$. By (5.2-5)(b) it suffices to show that F and G have disjoint HK-closures in $\mathcal{M}$. Let I=k(F) and J=k(G). As $h(I)\cap h(J)=F\cap G=\emptyset$, by (5.6-1), I+J=X. Thus, using (5.1-1)(7),

$$\emptyset = H(I+J) = H(k(F) + k(G)) = Hk(F) \cap Hk(G) = cl_{HK}F \cap cl_{HK}G. \quad \nabla$$

Theorem 5.6-2 REGULARITY → NORMALITY If X is a regular Frechet algebra, then X is Gelfand normal and $\mathcal{M} = \beta \mathcal{M}_c$.

Proof As X is regular, the Gelfand and hull-kernel topologies coincide by (5.2-1). As X is hk-normal, by the previous theorem, X is Gelfand normal. To prove that $\mathcal{M} = \beta \mathcal{M}_c$ note that X satisfies condition hH by (4.10-12) and apply Theorem 5.4-3. ∇

Exercises 5

The eight exercises presented here concern themselves with continuity of Banach algebra-valued homomorphisms on complex Banach algebras as well as continuity of derivations on Banach algebras and certain topological algebras. The principal results are 3(c), 4(b), 5(c), 5(d), 7(d) and 8(e).

5.1 BOUNDEDNESS OF HOMOMORPHISMS ON IDEMPOTENTS (Curtis and Bade 1960)

In this exercise X and Y are complex commutative Banach algebras with identity and H is a homomorphism from X into Y. We show in (c) that H is "bounded on the idempotents" of X - that is, that there exists $M > 0$ such that

$$\|H(p)\| \leq M\|p\|^2$$

for every idempotent $p \in X$. These results can be proved without assuming that identity is present in Y or that Y is complete through use of the usual procedures of completion and adjunction of the identity.

(a) Let (a_n) and (b_n) be sequences of nonzero elements in X satisfying

(1) $a_n b_n = a_n$ for all n

(2) $b_n b_m = 0$ for $m \neq n$.

Then $\sup \|H(a_n)\|/\|a_n\|\|b_n\| \neq \infty$.

<u>Hint</u>: With no loss of generality assume that $\|a_n\|=1$. Clearly by (1) it follows that $\|b_n\| \geq 1$. Let us assume that on the contrary $\sup\|H(a_n)\|/\|b_n\|=\infty$. Select distinct elements c_{ij} from among a_n and d_{ij} from among b_n such that $\|H(c_{ij})\| \geq 4^{i+j}\|d_{ij}\|$ where d_{ij} corresponds to c_{ij} through (1). Define $x_i = \sum_{j=1}^{\infty} 2^{-j} c_{ij}$ and consequently $d_{ij}x_i = 2^{-j}c_{ij}$, $H(d_{ij}) \neq 0$. Let $y = \sum_{i=1}^{\infty} d_{ij}/2^i\|d_{ij}\|$. Then show that $x_i y = c_{ij}/2^{i+j}\|c_{ij}\|$ and $\|H(y)\|\|H(x_i)\| \geq \|H(x_i y)\| > 2^{i+j} > 2^i\|H(x_i)\|$; hence $\|H(y)\| > 2^i$ for any i which is a contradiction.

(b) Let H be a homomorphism of X into Y. If (x_n) is a sequence of orthogonal idempotents in X, then for some $M > 0$, $\|H(x_n)\| \leq M\|x_n\|^2$ for each $n \in \underset{\sim}{N}$.

(c) Let H be a homomorphism of X into Y and let I be the set of idempotents of X. Then there exists $M > 0$ such that $\|H(x)\| \leq M\|x\|^2$ for all $x \in I$. Hence if I is bounded, H(I) is bounded.

<u>Hint</u>: We may assume that if e is the identity of X, then H(e) is the identity of Y for H(e) is the identity of $\overline{H(X)}$ and we may take $\overline{H(X)}=Y$.

We say that $z \leq x$ if $xz=z$. Let $I_1=\{x \in I \mid \sup\|H(z)\|/\|z\|^2=\infty\}$.

We assume that $e \in I_1$. If $x \in I_1$ and $z \leq x$ then z or $x-z$ is in I_1 for otherwise $\|H(w)\| \leq K\|w\|^2$ for all $w \leq z$ and $w \leq x-z$. Then if $v \leq x$, $v=vz+v(x-z)$ so $\|H(v)\| \leq K(\|vz\|^2+\|(x-z)v\|^2) \leq K\|v\|^2(\|z\|^2+\|x-z\|^2)$ and this contradicts $x \in I_1$.

Let $x_1 \in I_1$ and choose $z_1 \leq x_1$ such that $\|H(z_1)\|/\|z_1\|^2 > 16\|z_1\|^4(2+2\|H(x_1)\|/\|x_1\|^2)$. Then $\|H(x_1-z_1)\|/\|x_1-z_1\|^2 > [\|H(z_1)\|-\|H(x_1)\|]/[\|x_1\|\|z_1\|+\|z_1\|]^2 \geq [\|H(z_1)\|/4\|x_1\|^2\|z_1\|^2]-[\|H(x_1)\|/\|x_1\|^2] > 4\|x_1\|^2[2+\|H(x_1)\|/\|x_1\|^2]$. We generate in this way a sequence of idempotents (x_n) in I_1 such that $x_{n+1} \leq x_n$ and $\|H(x_n)\|/\|x_n\|^2 > 4\|x_{n-1}\|^2[n+\|H(x_{n-1})\|/\|x_{n-1}\|^2]$. Letting $p_n = x_n-x_{n+1}$ we have $p_np_k=0$ $(n \neq k)$ and $\|H(p_n)\|/\|p_n\|^2 \geq [\|H(p_{n+1})\|/4\|p_n\|^2\|p_{n+1}\|^2]-[\|H(p_n)\|/\|p_n\|^2] > n+1$ which contradicts (b).

<u>5.2 HOMOMORPHISMS OF REGULAR ALGEBRAS (Curtis and Bade 1960)</u> In this exercise X is a complex commutative regular semisimple Banach algebra with identity, Y is a complex commutative Banach algebra with identity, and H a homomorphism from X into Y. It is established that there exists a finite set F of points in $\mathcal{M}$ (the maximal ideals of X) such that if V is any neighborhood of F, H is continuous on the ideal $I(V)=\{x \in X \mid \psi x(V)=\{0\}\}$. Thus if F is empty, then H is continuous. It turns out that this occurs if $X= C(\mathcal{M},\underset{\sim}{C},c)$ and Y is semisimple as shown in the next exercise.

We denote by $\mathcal{S}$ the family of open sets $E \subset \mathcal{M}$ such that

$$m_E = \sup\|H(x)\|/\|x\|\|z\| < \infty$$

whenever the supports of ψx and ψz lie in E and $xz=x$. In (a) - (e) it is shown that the union of all sets in $\mathcal{S}$ is also a set in $\mathcal{S}$ and that the complement F of this set is finite. F will be the desired finite set discussed in the introduction.

(a) If (E_n) is any sequence of disjoint open sets in $\mathcal{M}$, then E_n belongs to $\mathcal{S}$ for sufficiently large n.

<u>Hint</u>: If (a) is false then there exists a sequence (E_n) of disjoint open sets in X and x_n, $z_n \in X$ such that the supports of ψx_n and ψz_n lie in E_n, $x_nz_n=x_n$, $\|x_n\|=1$, and $\|H(x_n)\| > n\|z_n\|$. But this contradicts 5.1(a).

(b) Let E_1 and $E_2 \in \mathcal{S}$. If G is an open set such that $\overline{G} \subset E_2$, then $E_1 \cup G \in \mathcal{S}$.

<u>Hint</u>: Choose x_1 such that ψx_1 is 1 on a neighborhood of CE_2 and 0 on a neighborhood of $\overline{G}$. Let $x_2=e-x_1$. We may find z_1, $z_2 \in X$ such that

$$z_1x_1 = x_1 \qquad \operatorname{supp} \psi z_1 \cap \overline{G} = \emptyset$$

$$z_2x_2 = x_2 \qquad \operatorname{supp} \psi z_2 \cap CE_2 = \emptyset$$

Let $H=E_1\cup G$ and suppose supp $\psi g\subset H$, supp $\psi h\subset H$, and $gh=g$. Then

$$\text{supp } \psi(gx_i)\subset E_i,$$

$$\text{supp } \psi(hz_i)\subset E_i,$$

$gx_i = ghx_iz_i = gx_ihz_i$ and $\|H(g)\| \le \|H(gx_1)\| + \|Hgx_2\| \le m_{E_1}\|gx_1\|\|hz_1\| + m_{E_2}\|gx_2\|\|hz_2\| \le (m_{E_1}\|x_1\|\|z_1\| + m_{E_2}\|x_2\|\|z_2)\ \|g\|\|h\|$

(c) If E_1, $E_2\in\mathcal{S}$ and G is open with $\bar{G}\subset E_1\cup E_2$ then $G\in\mathcal{S}$.

Hint: Let $F=CE_1\cap\bar{G}$ and U be open such that $F\subset U\subset\bar{U}\subset E_2$. Then $G\subset E_1\cup U\in\mathcal{S}$.

(d) If E_1, $E_2\in\mathcal{S}$ then $E_1\cup E_2\in\mathcal{S}$.

Hint: Suppose $E_1\cup E_2\notin\mathcal{S}$. If $F\subset E_1\cup E_2$ and F is closed, then $G=E_1\cup E_2 - F\notin\mathcal{S}$, for if $F\subset E_1\cup E_2$ then choose open sets U, V such that $F\subset V\subset\bar{V}\subset U\subset\bar{U}\subset E_1\cup E_2$. Thus $U\in\mathcal{S}$ by (c). If $G\in\mathcal{S}$ then $E_1\cup E_2=G\cup V\in\mathcal{S}$ by (b) contrary to assumption.

Returning to the assumption $E_1\cup E_2\notin\mathcal{S}$ we can find z_1, x_1 such that $x_1=x_1z_1$, supp $\psi z_1\subset E_1\cup E_2$ and $\|H(x_1)\| > \|x_1\|\|z_1\|$. Pick an open set U_1 such that supp $\psi z_1\subset U_1\subset\bar{U}_1\subset E_1\cup E_2$. Then $G_2=E_1\cup E_2 - \bar{U}_1\notin\mathcal{S}$ and there exists g_2, h_2 such that $g_2h_2=g_2$ and supp $\psi h_2\subset G_2$ with $\|H(g_2)\| \ge 2\|g_2\|\|h_2\|$. We generate sequences (g_n), (h_n) such that $g_nh_n=g_n$ and $\|H(g_n)\| \ge n\|g_n\|\|h_n\|$. The supports of ψh_n lie in disjoint open sets. This produces a contradiction of 5-1(a).

(e) $\mathcal{S}$ is closed under arbitrary unions.

(f) There exists a finite set F of points and $m > 0$ such that

$$\|H(x)\| \le m\|x\|\|z\|$$

for all $x, z\in X$ such that supp ψx, supp $\psi z\subset CF$ and $xz=x$. Consequently, if F is empty, then $\mathbf{H}$ is continuous.

Hint: Let $F=C(\cup\mathcal{S})$. If F is infinite, then a sequence of points in F can be separated by disjoint open sets. Hence F is finite.

The set F of (f) will be referred to as the singularity set of H.

(g) If V is any neighborhood of the singularity set of H, then the restriction of H to $I(V)=\{x\in X\,|\hat{x}(V)=\{0\}\}$ is continuous and $\|H(x)\| \le \|x\|\|h\|$ for all $x\in I(V)$ where h is such that $\psi h(CV)=\{1\}$ and ψh is zero in some neighborhood of F.

(h) Let X be a complex commutative regular semisimple Banach algebra with identity under $\|\ \|$ and let $\|\ \|_1$ be any other norm on X making X a normed algebra. Then there is a finite set $F\subset\mathcal{M}=X^h$ such that if G is any open set in $\mathcal{M}$ with $\bar{G}\cap F=\emptyset$, then for some $m_G > 0$, $\|x\|_1 \leq m_G\|x\|$ for any $x\in X$ such that supp $\psi x\subset G$. Conversely if G is an open set such that $\|x\|_1 \leq m_G\|x\|$ for all $x\in X$ such that supp $\psi x\subset G$, then $G\cap F=\emptyset$.

5.3 HOMOMORPHISMS OF $C(T,\underset{\sim}{C},c)$ (Curtis and Bade 1960) In this exercise we show that ((c)) any homomorphism of $C(T,\underset{\sim}{C},c)$, T a compact Hausdorff space, into a complex semisimple Banach algebra is continuous. It is not known if the result remains true if semisimplicity is removed.

In the parts to follow, H denotes a homomorphism from $C(T,\underset{\sim}{C},c)$ into a Banach algebra Y, F the singularity set (Exercise 2(f)) of H, $R(F)$ the dense subalgebra of $C(T,\underset{\sim}{C},c)$ consisting of functions which are constant in some neighborhood of each point of F, and $I(F)$ the ideal of functions which vanish in some neighborhood of F (the neighborhood is not fixed).

(a) H is continuous on $R(F)$.

Hint: Apply (5.2)(f) to $C(T,\underset{\sim}{C},c)$ and show that for some $M > 0$

$$\|H(x)\| \leq M\|x\|$$

for all $x\in I(F)$.

If $F=\{t_1,\ldots,t_n\}$ let $e_i\in C(T,\underset{\sim}{C})$ be such that $e_ie_j=0$ $(i\neq j)$, $0 \leq e_i(t) \leq 1$, for all $t\in T$, and $e_i(t_i)=1$ in some neighborhood of t_i. Then consider $x - \sum_{i=1}^{n} x(t_i)e_i\in I(F)$.

We denote by K the unique continuous homomorphism of $C(T,\underset{\sim}{C},c)$ which agrees with H on $R(F)$. Define $\lambda=H-K$; λ and K are the continuous and singular parts of H respectively. We reserve M for a constant with the property $\|K(x)\| \leq M\|x\|$ for any $x\in C(T,\underset{\sim}{C})$. We assume with no loss of generality that $\mathrm{cl}H(C(T,\underset{\sim}{C}))=Y$.

(b) The range of K is closed in Y and $K(C(T,\underset{\sim}{C}))\cap R=\{0\}$ where R is the radical of Y.

Hint: Let $I=K^{-1}(\{0\})$. Since I is a closed ideal there exists a closed set $S\subset T$ such that $I=\{x\in C(T,\underset{\sim}{C}) \mid x(S)=\{0\}\}$ (Exercise 4.3). If $C(T,\underset{\sim}{C},c)/I$ carries the norm

$$\|x + I\| = \inf\{\|z\| \mid z\in x + I\}$$

then $C(T,\underset{\sim}{C},c)/I$ is isometrically isomorphic to $C(X,\underset{\sim}{C},c)$ and $\|x+I\|=\sup_{t\in S}|x(t)|$. We may also norm, $C(T,\underset{\sim}{C})/I$ be defining $|x+I|=\|K(x)\|$. By fundamental results of Banach algebra theory (Kaplansky 1949, Theorem 6.2) $\|x+I\|=\sup_{t\in S}|x(t)| \leq |x+I|=\|K(x)\|$. As K is continuous, if $y\in x+I$, $|x+I|=|y+I|=\|K(y)\| \leq M\|y\|$. Thus $|x+I| \leq M\inf\{\|y\| \mid y\in x+I\}=M\|x+I\|$. Hence $|x+I|$ and $\|x+I\|$ are equivalent norms on $C(T,\underset{\sim}{C})/I$. To show that the range of K is closed, let $y \in Y$ and $y = \lim K(x_n)$. Since $\|x_m - x_n + I\| \le \|K(x_m - x_n)\| \to 0$, there exists $x_o \in C(T,\underset{\sim}{C})$ such that $\|x_o - x_n + I\| \to 0$. Since $\|K(x_o) - K(x_n)\| \le M\|x_o - x_n + I\| \to 0$, $y = K(x_o)$.

By the previous argument $K(C(T,\underset{\sim}{C}))$ is topologically isomorphic to $C(S,\underset{\sim}{C},c)$. Since in any Banach algebra $x \in R$ iff $\lim \|x^n\|^{1/n} = 0$, it follows that $K(C(T,\underset{\sim}{C})) \cap R = \{0\}$.

(c) $\lambda(C(T,\underset{\sim}{C})) \subset R$. Hence if Y is semisimple, $\lambda = 0$ and $H = K$. Thus H is continuous.

Hint: Let $\emptyset$ be a nontrivial multiplicative (hence continuous) linear functional on $\mathrm{cl}H(C(T,\underset{\sim}{C}))$. Then the well-defined functionals $\emptyset_H$ and $\emptyset_K$ defined by $\emptyset_H(x) = \emptyset(H(x))$ and $\emptyset_K(x) = \emptyset(K(x))$ are multiplicative, hence continuous, linear functionals on $C(T,\underset{\sim}{C},c)$. Since they coincide on $R(F)$, $\emptyset_H = \emptyset_K$. Thus $\emptyset(\lambda(x)) = 0$ for all $x \in C(T,\underset{\sim}{C})$ and all $\emptyset$.

(d) $\mathrm{cl}H(C(T,\underset{\sim}{C}))$ is the topological direct sum of $K(C(T,\underset{\sim}{C}))$ and R.

Hint: Surely $H(C(T,\underset{\sim}{C})) \subset K(C(T,\underset{\sim}{C})) + R$ so that $\mathrm{cl}H(C(T,\underset{\sim}{C})) \subset K(C(T,\underset{\sim}{C})) + R$.

Let $y = \lim H(x_n)$. Since $K(C(T,\underset{\sim}{C}))$ is closed, $\|H(x_n - x_m)\| \ge r_\sigma(H(x_n - x_m)) = r_\sigma(K(x_n - x_m))$. Denoting by r_K the spectral radius of a vector in $K(C(T,\underset{\sim}{C}))$, $r_\sigma(K(x_n - x_m)) = r_K(K(x_n - x_m)) \ge M^{-1}\|K(x_n - x_m)\|$. Thus there exists $x_o \in C(T,\underset{\sim}{C})$ such that $K(x_o) = \lim K(x_n)$. If $z = y - K(x_o)$ then $z = \lim \lambda(x_n)$ and $z \in R$. Hence

$$\mathrm{cl}H(C(T,\underset{\sim}{C})) = K(C(T,\underset{\sim}{C})) + R$$

The topologies of $H(C(T,\underset{\sim}{C}))$ and the topological direct sum agree because $K(C(T,\underset{\sim}{C}))$ and R are closed subspaces of a Banach space (Horvath 1966, p. 122).

<u>5.4 $\underset{\sim}{C}^\infty$ IS NOT A BANACH ALGEBRA</u> Let $\underset{\sim}{C}^\infty$ denote the complex algebra of all complex-valued infinitely differentiable functions on an open interval. There exists no norm under which $\underset{\sim}{C}^\infty$ is a Banach algebra.

Hint: Suppose there is such a norm. Let $Dx = dx/dt$ for each $x \in \underset{\sim}{C}^\infty$. Consider the bounded linear transformations L_n where

$$L_n(x) = (x(t_o + 1/n) - x(t_o))/1/n.$$

Since $\lim L_n(x) = x'(t_o)$, then $A_{t_o}(x) = x'(t_o)$ is continuous for each t_o. Applying the closed graph theorem, D is bounded and by (4.11-3), $D = 0$. But this is clearly a contradiction.

<u>5.5 DISCONTINUOUS HOMOMORPHISMS (Curtis and Bade 1960)</u> In this exercise we deal with the question of existence of discontinuous homomorphisms on a complex (commutative) Banach algebra (with identity) taking values in a (commutative) Banach algebra (with identity). In showing that they exist, it follows therefore that the singularity set of Exercise 5.2 is not always

empty.

The basic result (b) presented in this exercise is that if there is a maximal ideal M of an algebra X such that $M^2 \neq M$ and M^2 is dense in M, then there is a discontinuous homomorphism of X into a Banach algebra Y. Examples are given in (c) and (d).

If $X=C(T,\underset{\sim}{C},c)$, T a compact Hausdorff space, then $M^2=M$ for all maximal ideals in X. Consequently the question of existence of discontinuous homomorphisms remains open in this case although Sinclair (1969) has recently proved that existence of a discontinuous homomorphism of any C^*-algebra implies existence of a discontinuous homomorphism of $C([0,1],\underset{\sim}{C},c)$.

(a) Let X be a commutative Banach algebra with identity e and M a maximal ideal. Let I be an ideal such that $I \subset M$ properly and I is dense in M. Then there exists a discontinuous linear functional on X which vanishes on I.

Hint: Let $x \in M-I$ and f be a linear functional which vanishes on I, vanishes on e, and such that $f(x)=I$.

(b) Let X be a commutative Banach algebra with identity and M be a maximal ideal in X such that $M^2 \subset M$ and M^2 is dense in M. Show that there exists a discontinuous homomorphism of X into an algebra Y.

Hint: Let $Y=M \oplus \{\lambda y \,|\, \lambda \in \underset{\sim}{C}\}$ where operations in Y are taken to be pointwise addition and $(m+\lambda y)(m'+\lambda' y)=mm'$ so that the symbol y satisfies $y^2=0$, $yM=\{0\}$. Define $\|x+\lambda y\|=\|x\|+|\lambda|$ and

$$H:M \longrightarrow M \oplus \{\lambda y \,|\, \lambda \in \underset{\sim}{C}\}$$

$$x \longrightarrow x + f(x)y$$

where f is the discontinuous linear functional of (a). Then H is a discontinuous homomorphism of M into Y. H can be extended to X after the identity is adjoined to Y.

(c) Let X be the algebra ℓ_p, $1 \leq p < \infty$. Show that an identity can be adjoined to X forming an algebra X and a discontinuous homomorphism of X exists.

Hint: ℓ_p is a maximal ideal in X and $\ell_p^{\,2}=\ell_{p/2}$ which does not equal ℓ_p but is dense in ℓ_p.

(d) Let $X=\mathcal{D}^1$ be the algebra of continuously differentiable functions on [0,1] with norm

$$\|x\| = \sup_{t\in[0,1]}|x(t)| + \sup_{t\in[0,1]}|x'(t)|$$

Show there is a discontinuous homomorphism of $\mathcal{D}^1$.

Hint: The structure space (space of maximal ideals of $\mathcal{D}^1$) is [0,1] (see Gelfand, Raikov, and Shilov 1964, p. 23). Let $M=\{x\in\mathcal{D}^1 \mid x(0)=0\}$ and $N=\{x\in\mathcal{D}^1 \mid x'(0)=0\}$. If $y\in N$ and (p_n) is a sequence of polynomials from M converging uniformly to y', then $q_n(t)=\int_0^t p_n(s)ds$ is a sequence of polynomials in M^2 with $q_n \to y$ in $\mathcal{D}^1$. Thus M^2 is dense in M but not closed as it is readily seen that y''(0) exists for every y in M^2.

5.6 POINT DERIVATIONS AND DERIVATIONS INTO LARGER ALGEBRAS (Singer and Wermer 1955 A point derivation of a complex commutative Banach algebra of X with identity is a linear functional d_Φ associated with a multiplicative linear functional Φ such that $d_\Phi(x_1x_2)=\Phi(x_2)d_\Phi(x_1)+\Phi(x_1)d_\Phi(x_2)$. This notion is used to explore conditions under which there exists a bounded derivation of X into a superalgebra Y.

(a) If there exists a nontrivial bounded point derivation d_Φ of X then there exists an extension Banach algebra Y of X and a nonzero bounded derivation D of X into Y. If X is semisimple, Y is semisimple.

Hint: Let $Y=\{(x,\lambda) \mid x\in X, \lambda\in \underset{\sim}{C}\}$ where $(x_1,\lambda_1)(x_2,\lambda_2)=(x_1x_2,\lambda_1\lambda_2)$ and $\|(x,\lambda)\|=\max(\|x\|, |\lambda|)$. Consider the map

$$D:X^* \longrightarrow Y$$

$$(x,0) \longrightarrow (x,\Phi(x))$$

where X^* is the isometric isomorphic image of X in Y.

(b) If Φ is a continuous homomorphism and $M_\Phi=\Phi^{-1}(\{0\})$, a nonzero (bounded) point derivation of X exists iff $M_\Phi^2\neq M_\Phi$ $(\overline{M_\Phi^2}\neq \overline{M_\Phi})$.

Hint: Let f be a nontrivial linear functional annihilating M_Φ^2 and e. As any $x\in X$ can be written $x=x'+\Phi(x)e$ with $x'\in M_\Phi$, then f can be shown to be the desired point derivation of X associated with Φ.

(c) If $X\subset Y$ and if D is a nonzero bounded derivation of X into Y but D(X) is not in the radical of Y, then there exists a nonzero bounded point derivation of X.

Hint: By the condition given above there exists a nontrivial homomorphism Φ whose restriction to D(X) is not zero. Define d_Φ on X by $d_\Phi(x)=\Phi(D(x))$.

(d) Assume that X is semisimple. Then X has no nontrivial (continuous derivations into a semisimple commutative extension Y of X iff $M^2=M$ $(\overline{M^2}=\overline{M})$ for all maximal ideals $M\subset X$.

(e) If T is a compact Hausdorff space, then $C(T,\underset{\sim}{C},c)$ has no nontrivial derivations into any semisimple commutative extension Y.

5.7 DERIVATIONS OF COMMUTATIVE REGULAR SEMISIMPLE BANACH ALGEBRAS (Curtis 1961) In this exercise it is shown that if the Banach algebra S is regular and semisimple, then the "boundedness" requirement of (4.11-3) can be dropped; that is, any derivation of X into itself is bounded and therefore trivial. Moreover B.E. Johnson (1969) has shown that the "regularity" assumption can be dropped.

Let X be a complex commutative regular semisimple Banach algebra with identity. A 'derivation' D from X into $B(\mathcal{M},\underset{\sim}{C},c)$ (bounded complex-valued functions on $\mathcal{M}$ with pointwise operations and sup norm) is a linear transformation such that

$$Dxy = xDy + (Dx)y$$

where $\hat{x}:\mathcal{M} \to \underset{\sim}{C}$ is the Gelfand map determined by x.

(a) If x is an idempotent in X, then $Dx=0$.

(b) If k_M denotes the characteristic function of $\{M\}$, $M \in \mathcal{M}$ and $k_M \in \hat{X}$, then letting $k_M = \hat{x}_M$ with $x_M \in X$, $Dx_M x=0$ for all $x \in X$.

<u>Hint</u>: $x=m_M+\lambda e$ for some $m_M \in M$. As X is semisimple, $x_M x = x_M m_M + \lambda x_M = \lambda x_m$ and x_M is idempotent.

(c) Let D be a derivation of X into $B(\mathcal{M},\underset{\sim}{C},c)$. Then there exists a finite subset F of $\mathcal{M}$ and a continuous derivation D_1 of X into $B(\mathcal{M},\underset{\sim}{C},c)$ such that if $D_2=D-D_1$, then $D_2x(M)=0$ for all $x \in X$ and $M \notin F$. If $M \in F$ then $f_M(x)=D_2x(M)$ is an unbounded point derivation. If $Dx \in C(\mathcal{M},\underset{\sim}{C},c)$ for all $x \in X$ then $F=\emptyset$ and D is continuous.

<u>Hint</u>: Consider a new norm on X defined by $\|x\|_1=\|x\|+\sup_{M \in \mathcal{M}} |Dx(M)|$. Let F be the singularity set (5.2(f)) of $\|\ \|_1$. The fact that the linear functional $f_M(x)=Dx(M)$ is bounded iff $M \notin F$ is established as follows.

If $M \notin F$ then choose a neighborhood V of F and a vector $x \in X$ such that $\hat{x}_M(M)=1$ while $\hat{x}_M(V)=\{0\}$. Take $I(V)=\{x \in X \mid \hat{x}(V)=\{0\}\}$. Let W be an open subset of $\mathcal{M}$ such that $W \cap F=\emptyset$ and if $x \in I(V)$ then the support of $\hat{x}$ lies in W. If $x_n \in X$ and $x_n \to 0$ then $x_n x \in I(V)$. Hence $D(x_n x_M) \to 0$. But $Dx_n \hat{x}_M(M)=Dx_n(M)+\hat{x}_n(M)Dx_M(M) \to 0$ and $\hat{x}_n(M) \to 0$ implies $Dx_n(M)=f_M(x_n) \to 0$. Thus letting $H=\{M \mid f_M$ is bounded$\}$ we see that $CF \subset H$ and H is open. By the principle of uniform boundedness there exists $k > 0$ such that $|Dx(M)| \leq k\|x\|$ for all $x \in X$ and all $M \in H$.

To show that $H \subset F$ suppose $M_o \in H \cap F$. Let G be an open set such that $M_o \in G \subset H$ and let $y \in X$ be such that $\hat{y}(G)=1$ and $\hat{y}(\mathcal{M}-H)=\{0\}$. If $x \in X$ and the support of $\hat{x}$ is contained in G, then $xy=x$. Thus

$$Dx = \hat{y}Dx + \hat{x}Dy \quad \text{and}$$

$$\sup_{M\in\mathcal{M}}|\hat{x}(M)| \le \sup_{M\in\mathcal{M}}|\hat{y}(M)Dx(M) + \|x\|\sup_{M\in\mathcal{M}}|Dy(M)| \le (k \sup_{M\in\mathcal{M}}|\hat{y}(M)| + \sup_{M\in\mathcal{M}}|Dy(M)|)\|x\|.$$

Then $\|x\|_1 \le K\|x\|$ and this contradicts the properties of F.

If $F\neq\emptyset$ and D is unbounded, define D_1 by

$$D_1x(M) = \begin{cases} Dx(M) & M\notin F \\ 0 & M\in F \end{cases}$$

Applying the principle of uniform boundedness, D_1 is bounded. If $D_2=D-D_1$, D_2 can readily be shown to have the desired properties.

Now show that $\overline{\mathcal{M}-F}=\mathcal{M}$. For if M is isolated in $\mathcal{M}$ then let $x_M\in X$ be such that $\hat{x}_M=k_M$. Then it follows from (b) that $Dx_Mx=0$ for all $x\in X$. Thus $Dx_M=-\hat{x}(M)Dx_M(M)=0$ and $M\in F$. Now if $D(X)\subset C(\mathcal{M},\underset{\sim}{C},c)$ then $\sup_{M\in\mathcal{M}}|Dx(M)| \le k\|x\|$ and D is continuous.

(d) If X is a commutative regular semisimple Banach algebra with identity and D is any derivation of X into itself, then D is continuous and therefore trivial.

Hint: Consider the derivation $D=\psi D$ of X into $C(\mathcal{M},C,c)$ where $\psi x=\hat{x}$ and apply the closed graph theorem.

5.8 DERIVATIONS OF REGULAR SEMISIMPLE FRECHET ALGEBRAS (Rosenfeld 1966)

Let X be a commutative regular semisimple Frechet algebra with identity. We show that a derivation $D:X \to C(\mathcal{M},\underset{\sim}{C},c)$ is continuous ((e)). From this it follows that any derivation $D:X \to X$ is continuous by essentially the same application of the closed graph theorem as in Exercise 5.7(d). Using these results we show that for a certain algebra the derivations can be completely determined. It is worth noting that unlike the situation when X is a Banach algebra, the derivations need not be trivial.

(a) Let H be a homomorphism from X into a seminormed algebra Y (i.e. denoting the seminorm by p, $p(x,y) \le p(x)p(y)$, x, $y\in Y$). Let $\{p_n | n\in\underset{\sim}{N}\}$ be a family of multiplicative seminorms generating the topology on X such that $p_n \le p_{n+1}$ for each $n\in\underset{\sim}{N}$. Then if (x_n) and (y_n) are sequences in X such that $x_ny_n=x_n$ and $y_ny_m=0$ for $n\neq m$, it is not possible for $p(Hx_n) > np_n(x_n)p_n(y_n)$ for all $n \ge 1$.

Hint: Modify Exercise 5.1(a).

(b) If X is a commutative Banach algebra with identity and e_1 is an idempotent ($e_1^2=e_1$) belonging to the radical, then $e_1=0$.

Hint: As e_1 belongs to the radical, $\lim \|e_1^n\|^{1/n}=0$. Now use the fact that $e_1^{2n}=e_1^n$ for any $n\in\underset{\sim}{N}$.

(c) If X is a commutative Banach algebra with identity and e_i (i=1,2) are idempotents with e_1-e_2 belonging to the radical, then $e_1=e_2$.

Hint: $(e_1-e_2)^2$ is idempotent and therefore equal to 0. Thus $e_1^2+e_2^2=2e_1e_2$. Now $(e-e_1)^2=e-e_1$; $e_1-e_2=2e_2(e_1-e)$, and therefore $0=(e_1-e_2)^2=4e_2^2(e_1-e)^2=4e_2(e-e_1)=-2(e_1-e_2)$.

(d) If X is a commutative Banach algebra with identity and M is isolated in $\mathcal{M}$, then there exists an idempotent $e_1 \in X$ such that $\hat{e}_1$ is the characteristic function of $\{M\}$.

Hint: This is a result due to Shilov (Rickart 1960, p. 168).

(e) If D is a derivation from X into $C(\mathcal{M}, \underset{\sim}{C}, c)$ then D is continuous. Hence any derivation of X into itself is continuous.

Hint: Let the factor algebras of X be denoted by (X_n) and the maximal ideals of X_n by $\mathcal{M}_n$ we identify $\mathcal{M}$ and $\bigcup \mathcal{M}_n$ by (4.10-7). To prove the result it suffices to prove the following three statements.

(1) For each $n > 0$ there exists $F_n \subset \mathcal{M}_n$ such that F_n is finite and if $x_n \to 0$ then $Dx_n \to 0$ uniformly on $cl(\mathcal{M}_n - F_n)$.

(2) If $M \in \mathcal{M}$ and M is isolated, then $Dx(M)=0$ for all $x \in X$.

(3) If $M \in \mathcal{M}_n$ and M is isolated in $\mathcal{M}_m$ for all $m \geq n$ then M is isolated in $\mathcal{M}$.

Once these statements have been established then it follows that $Dx_k(M) \to 0$ for every $M \in \mathcal{M}$ whenever $x_k \to 0$. Statement (1) then implies that (Dx_k) converges to 0 uniformly on every compact set of $\mathcal{M}$. (If K is compact in $\mathcal{M}$, then $K \subset \mathcal{M}_n$ for some n by (4.10-7) since X is barreled.)

To prove (1) fix $n \in \underset{\sim}{N}$ and let Y be the seminormed algebra X supplied with seminorm

$$p(x) = p_n(x) + \sup_{M \in \mathcal{M}_n} |Dx(M)|$$

Let $F_n = \{M \in \mathcal{M}_n \mid x \to f_M(x) = Dx(M)$ is not continuous$\}$. Since X is a Frechet space and for each $x \in X$ $|f_M(x)| \leq \sup_{M' \in \mathcal{M}_n} |Dx(M')| = k_x$, $Dx_k \to 0$ uniformly on $\mathcal{M}_n - F$. If F_n is not finite then there exists a sequence (M_k) in F_n and a sequence of mutually disjoint neighborhoods V_k of M_k for each k. Since X is regular there are sequences (y_k) and (z_k) such that $\hat{y}_k(M_k)=1$, $y_k z_k = y_k$ and $z_k z_p = 0$ whenever $k \neq p$. Since f_{M_k} is a discontinuous linear functional there exists $x_k \in X$, $k \in \underset{\sim}{N}$, such that

$$|f_{M_k}(x_k)| = |Dx_k(M_k)| > k\, p_k(x_k)\, p_k(y_k)\, p_k(z_k).$$

As X is a Frechet algebra, f_{M_k} restricted to any closed maximal ideal is discontinuous. Thus we may choose x_k from M_k. Let $g_k = x_k y_k$ and $h_k = z_k$. Then

$$p(g_k) \geq \sup_{M \in \mathcal{M}_n} |Dg_k(M)| \geq |Dg_k(M_k)| =$$

$$|\hat{x}_k(M_k)\ Dy_k(M_k) + \hat{y}_k(M_k)\ Dx_k(M_k)| = |Dx_k(M_k)|$$

$$> kp_k(x_k)\ p_k(y_k)\ p_k(h_k) \geq k\,p_k(g_k)\ p_k(h_k) \text{ for each } k.$$

But this contradicts (a).

We next outline the proof of (3). Suppose $M \in \mathcal{M}_n$ and M is isolated in $\mathcal{M}_m$ for $m \geq n$. Then by part (d) there exists for each $m \geq n$ an idempotent $e_m \in X_m$ such that $\hat{e}_m(M)=1$ while $\hat{e}_m(M')=0$ for all $M' \neq M$ with $M' \in \mathcal{M}_m$. Now show that if $n \leq r \leq s$, with h_{rs} as in Theorem 4.6-1, it follows that $\widehat{h_{rs}(e_s)} = \hat{e}_r$ and therefore $h_{rs}(e_s)=e_r$. Thus there exists an idempotent $e \in X$ such that, with κ_n as in (4.10-7), $\kappa_n e = e_n$ for all n. Clearly $\hat{e}(M)=1$ while $\hat{e}(M')=0$ for all $M' \in \mathcal{M}$ with $M' \neq M$.

We use (3) to sketch (2). Let $e \in X$ be an idempotent satisfying the requirements of (3). Then by Exercise 5.7(b) $Dxe'=0$ for all $x \in X$ and $Dxe'(M)=0=x(M)De'(M)+e'(M)Dx(M)=Dx(M)$.

(f) Let $C^\infty(\underset{\sim}{R})$ be the algebra of all infinitely differentiable complex-valued functions on the real line $\underset{\sim}{R}$. Then for each $y \in C^\infty(\underset{\sim}{R})$ there exists a derivation D of $C^\infty(\underset{\sim}{R})$ into itself such that for all $x \in C^\infty(\underset{\sim}{R})$, $Dx(t)=x'(t)y(t)$ and conversely.

<u>Hint</u>: Let $C^\infty(\underset{\sim}{R})$ carry the topology defined by the seminorms

$$p_{n,k}(x) = \sup\{ |x^{(k)}(t)| \mid t \in [-n,n]\}$$

for all $n,k \geq 0$. The polynomial functions are dense in $C^\infty(\underset{\sim}{R})$ and for each polynomial p, $Dp(t)=p'(t)D(t)$. As D is continuous, the result follows.

SIX
LB-Algebras

WE HAVE SEEN that many of the features of Banach algebras are carried over to the larger class of LMCH Q-algebras, e.g., openness of the set of units, continuity of inversion, continuity of homomorphisms, compactness of the spectrum, etc. The central notion of this and the next two sections - the LB-algebra - provides another class of structures containing the Banach algebras in which some of the important properties carry over, only this time by way of a more algebraic approach. While many of the LMCH Q-algebras, namely the complete ones, are projective limits in the TVS sense of their factor algebras (Theorem 4.6-1), an LB-algebra is an algebraic inductive limit of a system of Banach algebras and continuous unital isomorphisms. The LB-algebra itself need not be supplied with a topology, however.

<u>6.1 Definition and Examples</u> In this section we define an LB-algebra and present some examples.

<u>Definition 6.1-1 BOUND STRUCTURES AND LB-ALGEBRAS*</u> Let X be a complex algebra. A <u>bound structure</u> for X is a non-empty collection $\mathcal{B}$ of absolutely m-convex subsets B of X containing e satisfying the following stability condition :

For each pair B_1, $B_2 \in \mathcal{B}$ there exists a set $B_3 \in \mathcal{B}$ and a scalar $\lambda > 0$ such that $B_1 \cup B_2 \subset \lambda B_3$.

The pair $(X, \mathcal{B})$ is referred to as a <u>bound algebra</u> and it is said to be <u>complete</u> provided each of the subalgebras $X(B)=\{\alpha x \mid \alpha \in \mathbb{C},\ x \in B\}$ is a Banach algebra with respect to the gauge p_B of B:

$$p_B(x) = \inf\{a > 0 \mid x \in aB\} \quad (x \in B,\ B \in \mathcal{B}).$$

X is an <u>LB-algebra</u> (<u>pseudo-Banach algebra</u>) <u>with respect to the bound structure $\mathcal{B}$</u> if $(X, \mathcal{B})$ is a complete bound algebra and $X = \bigcup_{B \in \mathcal{B}} X(B)$. If $\mathcal{B}$ is understood then we simply say that X is an <u>LB-algebra (pseudo-Banach algebra)</u>. ("LB-algebra" is used in loose analogy with "LF-space", each LB-algebra being an inductive limit (not necessarily strict) of Banach algebras.)

Certainly every Banach algebra is an LB-algebra with respect to the bound structure $\mathcal{B}$ consisting solely of the closed unit ball. In the remainder of this section we present two examples of LB-algebras which are not

---
* First considered in Allan, Dales, McClure 1971.

generally Banach algebras.

Example 6.1-1 p-BANACH ALGEBRAS Let X be a complex algebra equipped with a p-norm, $0 \leq p \leq 1$, i.e. a real-valued function $\| \ \|$ defined on X such that

(i) $\|x\| \geq 0$ for each $x \in X$ and $\|x\|=0$ iff $x=0$,

(ii) $\|\lambda\| = |\lambda|^p \|x\|$ for each $\lambda \in \underset{\sim}{C}$ and $x \in X$,

(iii) $\|x+y\| \leq \|x\|+\|y\|$ for each pair $x, y \in X$.

If, in addition, $\|xy\| \leq \|x\| \ \|y\|$ for each pair $x, y \in X$ and $\|e\|=1$, then $\| \ \|$ is an algebra p-norm and X a p-normed algebra*. It is a p-Banach algebra if X is complete with respect to the metric $d(x,y)=\|x-y\|$, $x, y \in X$.

A p-Banach algebra X is an LB-algebra with respect to a certain bound structure $\mathcal{B}$ which we now describe. If $x_1,\ldots,x_n$ are elements of X with p-norm less than one, and $M(x_i)$ denotes the collection of all monomials in $x_1,\ldots,x_n$ (i.e. elements of the form $x_1^{i_1}\ldots x_n^{i_n}$ where $i_1,\ldots,i_n$ are non-negative integers and $x_k^o=e$) then $\mathcal{B}$ consists of all sets of the form

$$B(x_i) = cl(M(x_i)_{bc}).$$

Clearly each of the sets $B(x_i)$ is absolutely-convex and contains e. Since $\|xy\| \leq \|x\| \ \|y\|$ for each pair $x, y \in X$ it follows just as in the Banach algebra case that multiplication is (jointly) continuous. Using this and the fact that the sets $M(x_i)_{bc}$ are multiplicative we obtain

$$(B(x_i))^2 = (cl((M(x_i)_{bc}))^2 \subset cl((M(x_i)_{bc}{}^2)$$

$$\subset cl(M(x_i)_{bc}) = B(x_i)$$

Observing that $B(x_i) \cup B(y_j) \subset B(x_i,y_j)$ we conclude that $\mathcal{B}$ is a bound structure for X. To show that $\mathcal{B}$ is complete it is first established that each $B=B(x_i)$ is bounded in the metric space X. Indeed, if $b=\Sigma\lambda(i_1,\ldots,i_n)x_1^{i_1}\ldots x_n^{i_n} \in M(x_i)_{bc}$ where the sum is taken over some finite collection of n-tuples $(i_1,\ldots,i_n)$ and $\Sigma|\lambda(i_1,\ldots,i_n)| \leq 1$, then

$$\|b\| \leq \Sigma \ |\lambda(i_1,\ldots,i_n)|^p \|x_1\|^{i_1} \ldots \|x_n\|^{i_n}$$

If we choose $1 > r > 0$ such that $\|x_k\| < r$ for each $k=1,\ldots,n$ then it follows from the above inequality that

$$\|b\| \leq \Sigma \ \|x_1\|^{i_1} \ldots \|x_n\|^{i_n} \leq (\sum_{i_1=0}^{\infty} \|x_i\|^{i_1}) \ldots (\sum_{i_n=0}^{\infty} \|x_n\|^{i_n}) < (\frac{1}{1-r})^n$$

* For more on p-normed algebras see Zelazko 1965, Chapter I.

so each $M(x_i)_{bc}$ is bounded. Since the closure of a bounded set in a metric space is bounded, $B=B(x_i)$ is also bounded. Next we claim that p_B induces a stronger topology on $X(B)$ than the relative topology from X thereby implying that p_B is a norm. Since B is bounded there exists an $\epsilon > 0$ such that $B \subset S_\epsilon(0) \cap X(B) = \{x \in X(B) \mid \|x\| < \epsilon\}$. Thus $S_\epsilon(0) \cap X(B)$ is a neighborhood of 0 in the topology induced by p_B on $X(B)$, and our contention follows.

Now suppose that (x_n) is a Cauchy sequence in $X(B)$ with the topology induced by p_B. Then (x_n) is certainly Cauchy in X, a complete metric space, so $x_n \to x$ for some $x \in X$. Given $\epsilon > 0$ there is an index $N > 0$ such that $x_n - x_m \in \epsilon B$ for all $n, m \geq N$. It is easy to see that ϵB is closed in X so $x_n - x = \lim_m (x_n - x_m) \in \epsilon B$. Hence $x \in X(B)$, $p_B(x_n - x) \leq \epsilon$ whenever $n \geq N$, and we conclude that $X(B)$ is a Banach algebra.

The only remaining thing to note is that for each $x \in X$, $x \in X(B(\lambda x))$, where $\lambda \in \underset{\sim}{C}$ has the property that $\|\lambda x\| < 1$. Thus $X = \bigcup_{B \in \mathcal{B}} X(B)$ and X is an LB-algebra.

<u>Example 6.1-2 X-HOLOMORPHIC FUNCTIONS</u> Let X be a uniform algebra on the compact Hausdorff space T (Def. 4.12-4). A function $y \in C(T, \underset{\sim}{C})$ is <u>X-holomorphic at $t \in T$</u> if there is a neighborhood U of t such that y can be approximated uniformly on U by elements of X, i.e. for each $\epsilon > 0$ there exists some $x \in X$ such that $\sup |(y-x)(U)| < \epsilon$. The function $y \in C(T, \underset{\sim}{C})$ is <u>X-holomorphic on T</u> if it is X-holomorphic at each $t \in T$. The collection of all X-holomorphic functions on T, denoted by $H(T,X)$, clearly forms a complex algebra which contains X as a subalgebra. It is our contention that $H(T,X)$ is an LB-algebra.

Let $\mathcal{U} = (U_j)_{j=1}^n$ be a finite open cover of T and $H(T,\mathcal{U},X)$ be the collection of all $y \in C(T, \underset{\sim}{C})$ such that y can be uniformly approximated on U_j by elements of X for each $j=1,\ldots,n$. To see that $H(T,\mathcal{U},X)$ is a uniform algebra on T it suffices to show that it is closed in $C(T,\underset{\sim}{C},c)$. To this end suppose that $z \in cl(H(T,\mathcal{U},X)$. Then for each $\epsilon > 0$ there is a $y \in H(T,\mathcal{U},X)$ such that $\sup |(z-y)(T)| < \epsilon/2$. Since $y \in H(T,\mathcal{U},X)$ then for each i there is an $x_i \in X$ such that $\sup |(y-x_i)(U_i)| < \epsilon/2$. Hence, $\sup |(z-x_i)(U_i)| < \epsilon$ and it follows that $z \in H(T,\mathcal{U},X)$.

Next we claim that $H(T,X) = \bigcup H(T,\mathcal{U},X)$ where $\mathcal{U}$ runs through the collection of all finite open covers of T. Indeed if $y \in H(T,X)$ then for each $t \in T$ there is a neighborhood U_t of t on which y can be uniformly approximated on U_t by elements of X. Since $T = \bigcup_{t \in T} U_t$ and T is compact, there is a finite subcover, $U_1,\ldots,U_n$ of T. It easily follows that $y \in H(T,\mathcal{U},X)$ where $\mathcal{U} = (U_i)$.

It only remains to show that the collection of closed unit balls $B_{\mathcal{U}}$ of $H(T,\mathcal{U},X)$ constitutes a bound structure for X. Clearly each $B_{\mathcal{U}}$ is absolutely m-convex and contains the identity. If $\mathcal{U}_1$ and $\mathcal{U}_2$ are finite open coverings of T then so is $\mathcal{U}_{12}=\{U_1\cap U_2 \mid U_1\in\mathcal{U}_1, U_2\in\mathcal{U}_2\}$. Furthermore, because each element of $\mathcal{U}_{12}$ is contained in elements of $\mathcal{U}_1$ and $\mathcal{U}_2$, $H(T,\mathcal{U}_1,X)$ and $H(T,\mathcal{U}_2,X)$ are subalgebras of $H(T,\mathcal{U}_{12},X)$. Thus $B_{\mathcal{U}_1}\cup B_{\mathcal{U}_2} \subset B_{\mathcal{U}_{12}}$ and this concludes the proof. In general $H(T,X)$ is not closed in $C(T,\underset{\sim}{C},c)$ [Rickart 1966] so $H(T,X)$ is not generally a Banach algebra.

Neither of the algebras discussed above were endowed with topological structure. A condition under which an LMCH algebra is LB-algebra is considered in Sec. 6.3.

6.2 Some Properties of LB-Algebras In this section we indicate some of the important features of Banach algebras that are carried over to LB-algebras. After characterizing LB-algebras as inductive limits of Banach algebras we prove such things as the compactness of the class of non-trivial complex-valued homomorphisms in the Gelfand topology, that $\underset{\sim}{C}$ is the only LB-algebra which is a field, and that every maximal ideal is the kernel of a complex homomorphism.

Suppose that X is an LB-algebra with respect to the bound structure $\mathcal{B}$. We wish to show that X may be viewed as a sort of increasing limit of the algebras X(B). First we note that the set $\mathcal{B}$ is directed by the ordering: $B \leq B'$ iff there is a $\lambda > 0$ such that $B\subset\lambda B'$. It is clear that X(B) is a subalgebra of X(B') and the injection map $I_{B'B}:X(B) \to X(B')$ is a continuous unital algebra isomorphism (i.e. a continuous algebra isomorphism that maps the identity of the domain into the identity of the range), whenever $B \leq B'$. Furthermore X, being the union of the X(B) is the limit in the ordinary set-theoretic sense of the increasing net $(X(B))_{B\in\mathcal{B}}$ (recall that a net of sets $(S_m)_{m\in M}$ converges to S if $S=\bigcap_{m\in M}\bigcup_{t>m} S_t=\bigcup_{m\in M}\bigcap_{t>m} S_t$). The net of algebras $(X(B))_{B\in\mathcal{B}}$ together with the collection of continuous unital isomorphisms $I_{B'B}$, defined whenever $B \leq B'$, is a particular example of an "**inductive** system" with X being the "inductive limit" of the system.

More generally an <u>inductive system of linear spaces</u> is a net of vector spaces (all over the same field), $(Z_t)_{t\in M}$ and a collection of linear maps $I_{ts}:Z_s \to Z_t$, one defined for each pair (t,s) where $s \leq t$, such that I_{tt} is the identity map for each t and $I_{ut}\cdot I_{ts}=I_{us}$ for all $s \leq t \leq u$. The <u>inductive limit</u> of the inductive system is the vector space $\sum_{t\in M}Z_t/N$, where $\sum_{t\in M}Z_t$ is the direct sum of the Z_t's and N the subspace of the direct sum consisting of all elements (z_t) with the property that $\sum_{s\leq t}I_{ts}(z_s)=0$ in Z_t for each t large enough so that $z_s \neq 0$ implies that $s \leq t$.*

We leave it to the reader to verify that the LB-algebra X is in fact the inductive limit of the system given by the net $(X(B))_{B\in\mathcal{B}}$ and the continuous unital (algebra) isomorphisms $I_{B'B}$ (the injection of X(B) into X(B')) for $B \leq B'$.

It is natural to ask if an inductive limit of a system of complex Banach algebras and continuous unital isomorphisms equipped with pointwise multiplication, is an LB-algebra. The answer to this question as we shall now show, is yes. Let $(X_t)_{t\in M}$ be a net of Banach algebras and the maps

$$I_{ts}:X_s \to X_t$$

be continuous unital (algebra) isomorphisms. The inductive limit of this system is $X=\sum_{s\in M} X_s/N$ where $(x_s)\in N$ iff $\sum_{s \leq t} I_{ts}(x_s)=0$ whenever t is so large that $x_s \neq 0$ implies $s \leq t$. Consider the map

$$I_s:X_s \longrightarrow \sum_{t\in M} X_t \longrightarrow X$$

$$x_s \longrightarrow <x_s> \longrightarrow <x_s> + N$$

where the entry of $<x_s>$ indexed by s is specified to be x_s and all the rest are 0. I_s is easily seen to be an algebra isomorphism so we can transfer the norm of X_s to $I_s(X_s)$, i.e. $\|<x_s> + N\|_s=\|x_s\|_s$, whereby

---

* One can motivate the definition of the inductive limit in the following way. The idea is to construct a vector space Z containing the Z_t's as subspaces (more precisely isomorphic images of the Z_t's) such that Z_s is a subspace of Z_t whenever $s \leq t$ and Z is the increasing limit in the set-theoretic sense of the Z_t's. The direct sum $\sum_{t\in M} Z_t$ certainly contains isomorphic images of the Z_t's but the other desired properties are lacking. Thus we are led to consider a quotient space of the direct sum, $\sum_{t\in M} Z_t/N$, and the following candidates for each index s for the required isomorphism

$$I_s:Z_s \to \Sigma Z_t \to \Sigma Z_t/N$$

$$z_s \to <z_s> \to <z_s> + N$$

where $<z_s>$ is the tuple with zero entries for $t\neq s$ and z_s in the entry indexed by s. It readily follows that each I_s is an isomorphism, $I_s(Z_s)$ is a subspace of $I_t(Z_t)$ for $s \leq t$ provided N is defined as above ($<I_{ts}x_s> - <x_s> \in N$ for each $x_s \in X_s$) and that $\Sigma Z_t/N = \bigcup_{s\in M} I_s(Z_s)$.

$I_s(X_s)$ becomes a Banach algebra. Suppose that $s \leq t$; since $< I_{ts}x_s > - < x_s > \in N$ it follows that $< x_s > + N = < I_{ts}x_s > + N$ for each $x_s \in X_s$. With the aid of this equality it follows that $I_s(X_s)$ is a subalgebra of $I_t(X_t)$ whenever $s \leq t$. Furthermore the continuity of I_{ts} implies that the norm $\| \|_t$ of $I_t(X_t)$ restricted to $I_s(X_s)$ is "weaker" than $\| \|_s$: There exists a scalar $\lambda_{st} > 0$ such that $\| \|_t \leq \lambda_{ts} \| \|_s$ whenever $s \leq t$. Let B_s be the closed unit ball of $I_s(X_s)$ and $\mathcal{B}=\{B_s\}_{s\in M}$. We claim that $\mathcal{B}$ is a complete bound structure for X. The elements of $\mathcal{B}$ are certainly absolutely m-convex and each contains the identity $< e_s > + N$. Moreover, if $s \leq t$ then, since $\| \|_t \leq \lambda_{ts}\| \|_s$ it follows that $B_s \subset \lambda_{ts}B_t$. Hence if B_s, $B_r \in \mathcal{B}$ we can choose $t \in M$ such that $r, s \leq t$ and obtain the conclusion that $B_r \cup B_s \subset \max(\lambda_{tr},\lambda_{ts})B_t$. Since $X(B_s)=I_s(X_s)$ is a Banach algebra, $(X,\mathcal{B})$ is a complete bound algebra.

It remains to show that $X=\bigcup X(B_s)$. To this end consider an arbitrary element $(x_s)+N \in X$. There are only a finite number of indices s such that $x_s \neq 0$. Denoting these indices by $s_1,\ldots,s_n$ we see that $(x_s)+N=\sum_{i=1}^{n} < x_{s_i} > + N \in I_t(X_t)=X(B_t)$ for any $t \geq s_1,\ldots,s_n$.

We summarize these results in:

<u>Theorem 6.2-1 AN LB-ALGEBRA IS AN INDUCTIVE LIMIT OF BANACH ALGEBRAS</u> A complex algebra X is an LB-algebra with respect to some bound structure iff X is the inductive limit of an inductive system of complex Banach algebras and continuous unital isomorphisms.

Since any complex Banach algebra is a Q-algebra each non-trivial complex-valued homomorphism of a Banach algebra is continuous [(4.10-5)].

<u>(6.2-1) HOMOMORPHISMS OF AN LB-ALGEBRA ARE "BOUNDED"</u> If X is an LB-algebra with respect to the bound structure $\mathcal{B}$ then each non-trivial complex-valued homomorphism h maps the elements of $\mathcal{B}$ into bounded subsets of $\underset{\sim}{C}$.

<u>Proof</u> If $B \in \mathcal{B}$ then, since $e \in B$, h restricted to the Banach algebra X(B) is a non-trivial homomorphism of X(B). As such it is continuous. Thus it takes B, a bounded subset of X(B), into a bounded set of complex numbers. ∇

In view of the fact that the bound structure $\mathcal{B}$ associated with an LB-algebra X need not arise from a topology, it doesn't make sense in general to ask when a non-trivial homomorphism h is continuous. Even in the event that the elements of $\mathcal{B}$ are bounded in some compatible topology on X, it need not follow that each h be continuous (see Example 6.3-2).

Recall that if X is a complex Banach algebra then X^h carrying its Gelfand topology, i.e. $\sigma(X^h,X)$ is a compact Hausdorff space. This property is carried over to the LB-algebra case, i.e. if X is an LB-algebra, the collection of all non-trivial complex-valued homomorphisms H(X) with the

$\sigma(H(X),X)$ topology (the Gelfand topology) is a non-empty compact Hausdorff space. We will prove this by first showing that $H(X)$ is a "topological projective limit" (a notion which is defined below) of non-empty compact Hausdorff spaces followed by a demonstration of the fact that a topological projective limit of non-empty compact Hausdorff spaces is a non-empty compact Hausdorff space.

Theorem 6.2-2 H(X) IS A NON-EMPTY COMPACT HAUSDORFF SPACE Let $H(X)$ be the collection of all non-trivial complex-valued homomorphisms of the LB-algebra X with bound structure $\mathcal{B}$. Then:

(a) $H(X)$ with its Gelfand topology is homeomorphic to the topological projective limit (defined in the proof) of the non-empty compact Hausdorff spaces $X(B)^h$ and continuous restriction maps $R_{BB'}$ with $B, B' \in \mathcal{B}$ and $B \leq B'$;

(b) $H(X)$ with its Gelfand topology is a non-empty compact Hausdorff space.

Proof First we note that $X(B)^h$ is not empty because $\chi(B)$ is a Banach algebra.

Suppose now that $h \in H(X)$ and $B \in \mathcal{B}$, the bound structure associated with X. Let $h_B = h|_{X(B)}$ and consider the element $(h_B)_{B\in\mathcal{B}}$ of the product $\prod_{B\in\mathcal{B}} X(B)^h$. It is clear that if $B \leq B'$ (where $\leq$ is as in the proof of Theorem 6.2-1) which implies that $X(B) \subset X(B')$, then h_B is the restriction of $h_{B'}$ to $X(B)$. Thus the element $h \in H(X)$ determines an element $(h_B) \in \prod_{B\in\mathcal{B}} X(B)^h$ with the property that $h_B = R_{BB'}(h_{B'})$ where $R_{BB'}$ is the restriction map which restricts elements of $X(B')^h$ to $X(B)$ when $B \leq B'$. Conversely if $(h_B) \in \prod_{B\in\mathcal{B}} X(B)^h$ and $R_{BB'}(h_{B'}) = h_B$ whenever $B \leq B'$ we can define $h: X \to \underset{\sim}{C}$ via the rule $h(x) = h_B(x)$ for $x \in X(B)$. It is clear that h is a well-defined non-trivial complex homomorphism of X. Thus we have established a correspondence (which is evidently 1-1) between $H(X)$ and the subset of the product consisting of elements (h_B) for which $R_{BB'}(h_{B'}) = h_B$ whenever $B \leq B'$. We also observe that since $X(B) \subset X(B')$, each such $R_{BB'}$, is continuous when $X(B')^h$ and $X(B)^h$ carry their Gelfand topologies. This subset of the product $\prod_{B\in\mathcal{B}} X(B)^h$, equipped with the relative product topology, is referred to as the "topological projective limit" of the "topological projective system" consisting of the net of topological spaces $(X(B)^h)_{B\in\mathcal{B}}$ and the continuous restrictions maps $R_{BB'}$ defined for each pair (B,B') such that $B \leq B'$.

More generally, a "topological projective system" (Definition 4.6-1) is a set $(Z_m)_{m\in M}$ of topological spaces indexed by a preordered set M and a collection of continuous maps $R_{mt}: Z_t \to Z_m$, one defined for each pair (m,t)

for which $m \leq t$, such that R_{mm} is the identity map for each $m \in M$ and $R_{ms} \circ R_{st} = R_{mt}$ for all $m \leq s \leq t$. The topological projective limit Z of this system is defined to be the topological subspace of all elements $(z_m) \in \prod_{m \in M} Z_m$ such that $R_{mt}(z_t) = z_m$ for all $m \leq t$.*

Suppose that each Z_m is a non-empty compact Hausdorff space; we claim that Z is also a non-empty compact Hausdorff space. That Z is a compact Hausdorff space will follow from having shown it to be closed in the compact Hausdorff product $\prod_{m \in M} Z_m$. Indeed, if $(w_i)_{i \in I}$ is a net of elements $w_i = (z_{mi}) \in Z$ converging to the element $z = (z_m) \in \prod_{m \in M} Z_m$ then, denoting the continuous projection of the product onto Z_m by pr_m, we see that $(z_{mi})_{i \in I} = (pr_m(w_i))_{i \in I}$ converges to z_m in Z_m. Since $(z_{mi}) \in Z$, then for each $i \in I$ we have $R_{mt}(z_{ti}) = z_{mi}$ whenever $m \leq t$. Since each such R_{mt} is continuous it follows that $R_{mt}(z_t) = z_m$ for each pair (m,t) for which $m \leq t$. Hence $(z_m) \in Z$ and Z is closed.

To see that Z is non-empty it is enough to show that Z is the intersection of a family of closed subsets of the compact spaces $\prod_{m \in M} Z_m$ having the finite intersection property. Fix $t \in M$ and let Y_t be the set of all $(z_m) \in \prod_{m \in M} Z_m$ with the property that $R_{mt}(z_t) = z_m$ for all $m \leq t$. That Y_t is closed follows in the same way as the closedness of Z. Now if we choose some z_t from the non-empty space Z_t and set $z_m = R_{mt}(z_t)$ for each $m \leq t$ then, by arbitrarily assigning values to all other z_m's, we obtain a member of Y_t. Hence each $Y_t \neq \emptyset$. Suppose that $m_1, \ldots, m_n \in M$; then by choosing $t \geq m_1, \ldots, m_n$ it follows that $Y_t \subset \bigcap_{j=1}^{n} Y_{m_j}$. Thus the collection $\{Y_t \mid t \in M\}$ has the finite intersection property and $Z = \bigcap_{t \in M} Y_t \neq \emptyset$.

Returning to the situation at hand we recall that the correspondence $T: h \longleftrightarrow (h_B)_{B \in \mathcal{B}}$ is a 1-1 mapping of H(X) onto the topological projective limit Z of the compact Hausdorff spaces $X(B)^h$. Next we claim that this correspondence establishes a homeomorphism when H(X) carries the Gelfand topology. A typical subbasic neighborhood of the point $(h_B)_{B \in \mathcal{B}}$ in Z is a set of the form

* If one takes the intuitive viewpoint that the mappings R_{mt} act to "reduce the size" of the Z_t's then the projective limit may be considered as a "decreasing" limit of the net $(Z_m)_{m \in M}$. This viewpoint is especially appealing in the case when the projective limit of the spaces of homomorphisms $X(B)^h$ is formed, for then $B \leq B'$ implies that $X(B) \subset X(B')$ and, therefore, $X(B')^h$ "$\subset$" $X(B)^h$.

$$< V(h_{B_o}, x_o, \epsilon) > = \{(f_B)\epsilon Z \mid f_{B_o} \epsilon V(h_{B_o}, x_o, \epsilon)\}$$

where B_o is fixed in $\mathcal{B}$ and $V(h_{B_o}, x_o, \epsilon) = \{h \epsilon X(B_o)^h \mid |h(x_o) - h_{B_o}(x_o)| < \epsilon\}$, with $x_o \epsilon X(B_o)$, and $\epsilon > 0$, is a typical subbasic neighborhood of h_{B_o} in $X(B_o)^h$. Now it is clear that $T(V(h, x_o, \epsilon)) \subset < V(h_{B_o}, x_o, \epsilon) >$, from which continuity follows. T is bicontinuous since it is a 1-1 continuous mapping of a compact Hausdorff space into a Hausdorff space. ∇

As an immediate consequence of the fact that $H(X) \neq \emptyset$ we obtain a result for LB-algebras similar to the Gelfand-Mazur theorem [Theorem 4.9-1].

<u>(6.2-2) IF THE LB-ALGEBRA X IS A FIELD, X"="C</u> If X is an LB-algebra and a field then X is algebraically isomorphic to $\mathbb{C}$.

<u>Proof</u> By Theorem 6.2-2 (b) there exists a non-trivial homomorphism $h: X \to \mathbb{C}$. If we assume that h is not 1-1 then there exists a non-zero x in the field X, such that $h(x)=0$. Thus, for each $y \epsilon X$,

$$h(y) = h(ye) = h(yx^{-1} \cdot x) = h(yx^{-1})\, h(x) = 0$$

which contradicts the non-triviality of h. Thus h is 1-1 and X is algebraically isomorphic to $\mathbb{C}$. ∇

Recall that in any Banach algebra the maximal ideals are in 1-1 correspondence with the non-trivial complex homomorphisms. It is always the case that the kernel of a non-trivial (complex) homomorphism of an algebra is a maximal ideal. On the other hand if M is a maximal ideal in an algebra X then X/M is certainly a field. Hence if we can show that X/M is an LB-algebra whenever X is an LB-algebra and M is a maximal ideal, it will follow by (6.2-2) that M is the kernel of the complex homomorphism

$$\begin{aligned} h: X &\longrightarrow X/M \longrightarrow \mathbb{C} \\ x &\longrightarrow x+M \longrightarrow \lambda \end{aligned}$$

where $x+M \to \lambda$ is an isomorphism between X/M and $\mathbb{C}$.

To do this we prove a more general result:

<u>(6.2-3) QUOTIENTS OF LB-ALGEBRAS</u> Let X be an LB-algebra with respect to the bound structure $\mathcal{B}$ and I be an ideal of X with the property that $I \cap X(B)$ is a closed ideal in X(B) for each $B \epsilon \mathcal{B}$. Then the algebra X/I is an LB-algebra with respect to the bound structure $\mathcal{B}/I = \{B+I \mid B \epsilon \mathcal{B}\}$. In particular X/M is a LB-algebra for any maximal ideal M of X.

<u>Proof</u> First we note that by some routine considerations $\mathcal{B}/I$ is a bound structure on X/I. It is also clear that for each $B \epsilon \mathcal{B}$ the subalgebra of X/I generated by B+I is just $X(B)+I = \{x+I \mid x \epsilon X(B)\}$ and $X/I = \bigcup_{B \epsilon \mathcal{B}} X(B)+I$. Thus it

remains to show that each algebra X(B)+I is a Banach algebra with respect to the gauge of B+I. To this end suppose that $x\in X(B)$ and consider

$$p_{B+I}(x+I) = \inf\{\lambda > 0 \mid x + I\in\lambda(B + I)\}$$
$$= \inf \bigcup_{z\in I\cap X(B)} \{\lambda > 0 \mid x + z\in\lambda B\}$$
$$= \inf_{z\in I\cap X(B)} \inf\{\lambda > 0 \mid x + z\in\lambda B\}$$
$$= \inf_{z\in I\cap X(B)} p_B(x + z) = \dot{p}_B(x + I\cap X(B)).$$

Thus the mapping

$$X(B) + I \longrightarrow X(B)/I\cap X(B)$$
$$x + I \longrightarrow x + I\cap X(B)$$

is seen to be an onto "isometric" isomorphism when X(B)+I carries the semi-norm p_{B+I} and $X(B)/I\cap X(B)$ the semi-norm $\dot{p}_B$ induced by p_B. Since $I\cap X(B)$ is closed in X(B), $\dot{p}_B$ is a norm which, because X(B) is complete in the norm p_B, renders $X(B)/I\cap X(B)$ a Banach algebra. Hence each algebra X(B)+I is a Banach algebra and X/I is a LB-algebra.

Now if M is a maximal ideal in X then we claim that each ideal $M\cap X(B)$ is closed, in the Banach algebra X(B). Suppose that this isn't the case, i.e. there exists a $B\in\mathcal{B}$ such that $M\cap X(B)$ is not closed in X(B). Then there exists a sequence $(m_k)\subset M\cap X(B)$ convergent in X(B) to some $x\notin M$. Since M is maximal in X elements $y\in X$ and $m\in M$ exist such that e=yx+m. Choose $B'\in\mathcal{B}$ such that $X(B')\supset X(B)$ and $y\in X(B')$. As the injection mapping $I_{B'B}$ taking X(B) into X(B') is continuous by Theorem 6.2-1 it follows that $ym_k+m \to yx+m=e$ in X(B'). Since a Banach algebra is a Q-algebra [(4.8-2)] we may choose k so large that $ym_k+m\in M$ belongs to the open set of units. However $y\, m_k+m\in M$ and M, being proper, can contain no invertible elements. This contradiction implies that each $M\cap X(B)$ is closed in X(B) thereby concluding the proof.∇

As an immediate consequence of this result and the remarks preceding it we have

<u>(6.2-5) IN AN LB-ALGEBRA H(X) "=" $\mathcal{M}$</u> Each maximal ideal of a LB-algebra is the kernel of some $h\in H(X)$.

In light of the last result it is evident that whenever M is a maximal ideal in an LB-algebra X each element $x\in X$ can be written uniquely in the form $x=\lambda e+m$ where $\lambda\in\underset{\sim}{C}$ and $m\in M$. Thus M is a subspace of X of codimension one consisting of singular elements. Gleason 1967 (cf. Beckenstein, Narici, and Bachman 1971) proved the converse for Banach algebras, i.e. any subspace of a Banach algebra consisting solely of singular elements and having co-dimension one is a maximal ideal. We conclude this section with a presentation of Gleason's result followed by an extension of it to LB-algebras.

(6.2-6) IN AN LB-ALGEBRA, $M \cap Q = \emptyset$ AND COD(M) = 1 IMPLIES $M \in \mathcal{M}$ If M is a subspace of the LB-algebra X of codimension one, consisting of singular elements then M is a maximal ideal.

Proof First suppose that X is a complex Banach algebra. The subspace M, being of codimension one, must be either dense or closed in X. Since X is Q-algebra (4.8-2), Q, the set of units in X, is non-empty and open; so M, containing only singular elements, must be closed. If we let h be that linear functional on X having M as its null space and mapping e into 1, then since M is closed, h is continuous. Now it suffices to show that h is a homomorphism. Furthermore, the equation

$$xy = \frac{(x+y)^2 - x^2 - y^2}{2}$$

together with the linearity of h reduces the problem to showing that $h(x^2) = (h(x))^2$ for each $x \in X$.

To this end consider the function

$$\exp(\lambda x) = \sum_{n=0}^{\infty} \frac{\lambda^n x^n}{n!}.$$

Since $\|x^n\| \leq \|x\|^n$ in a normed algebra it follows that the defining series is absolutely convergent and, therefore, convergent in the Banach algebra X for each $\lambda \in \underset{\sim}{C}$. Thus, by the continuity of h,

$$h(\exp(\lambda x)) = \sum_{n=0}^{\infty} \frac{\lambda^n h(x^n)}{n!} \qquad \lambda \in \underset{\sim}{C},$$

and $h(\exp(\lambda x))$ is an entire function with no zeros. Furthermore, setting $M(r) = \sup_{|\lambda|=r} |h(\exp(\lambda x))|$ we obtain

$$\text{order } (h(\exp(\lambda x))) = \overline{\lim_{r \to \infty}} \frac{\ln\ln M(r)}{\ln r}^{*}$$

$$\leq \overline{\lim_{r \to \infty}} \frac{\ln\ln \|h\| \sum_{n=0}^{\infty} \frac{(r\|x\|)^n}{n!}}{\ln r}$$

* The notion of the order of an entire function may be found in Markushevich (1965), Vol. 2, p. 251.

$$= \overline{\lim_{r \to \infty}} \frac{\ln(\ln\|h\| + r\|x\|)}{\ln r}$$

$$\leq \lim_{r \to \infty} \frac{\ln\ln\|h\| + \ln r + \ln\|x\|}{\ln r}$$

$$= 1.$$

Thus, by a weak version of Hadamard's factorization theorem*, the fact that $h(\exp(\lambda x))$ never assumes the value zero, and, $h(\exp(\lambda x))=1$, it follows that

$$h(\exp(\lambda x)) = e^{\alpha\lambda} = \sum_{n=0}^{\infty} \frac{\alpha^n \lambda^n}{n!}$$

for some $\alpha \in \mathbb{C}$. Now by the identity theorem for power series**, $h(x^n)=\alpha^n$ for each $n \geq 0$; so

$$h(x^2) = \alpha^2 = (h(x))^2$$

for each $x \in X$.

Next suppose that X is an LB-algebra with respect to the bound structure $\mathcal{B}$ and M is linear subspace of X as in the hypothesis. Since the codimension of M is one, we may write each $x \in X$ uniquely in the form $x=\lambda e+m$ where $\lambda \in \mathbb{C}$, and $m \in M$. Now if $x \in X(B)$, $(B \in \mathcal{B})$, then, since $e \in B$, $m \in X(B)$, and it follows that the codimension of $M \cap X(B)$ is one in $X(B)$. As the elements of M are all singular so are the elements of $M \cap X(B)$ singular in $X(B)$. Hence by the result established above we conclude that $M \cap X(B)$ is a maximal ideal in $X(B)$. It remains to show that M is an ideal in X. To this end let $x \in M$ and $y \in X$ and choose $B \in \mathcal{B}$ such that $x, y \in X(B)$. Then $x \in M \cap X(B)$ and $xy \in M \cap X(B) \subset M$. ∇

Section 6.3-1. Complete LMC LB-Algebras The concept of an LB-algebra is primarily algebraic - it need not be a topological algebra to begin with, nor do we have to add topological structure to it for its salient properties (see Sec. 6.2). However, it is the case that we can completely characterize complete LMC LB-algebras, i.e. those complete-LMCH-algebras that

* (Markushevich (1965), Vol 2, p. 266) if the entire function $f(\lambda)$ of order ρ never assumes the value $w \in \mathbb{C}$ then ρ is an integer and $f(\lambda)$ is of the form $f(\lambda)=w+e^{p(\lambda)}$ where $p(\lambda)$ is a polynomial of degree ρ.

**(Markushevich (1965), Vol. 1, p. 352). If the complex power series $\sum_{n=0}^{\infty} a_n(z-z_o)^n$ agree on a bounded infinite set of complex numbers then $a_n=b_n$ for each $n \geq 0$.

are also LB-algebras. A bound structure of a Banach algebra consists of the unit ball, a topologically bounded closed absolutely m-convex set containing the identity; a natural bound structure to consider in an LMC algebra is the collection $\mathcal{B}_n$ of all absolutely m-convex closed bounded sets which contain the identity. We shall show in the main theorem of this section (Theorem 6.3-1) that a complete LMC algebra is an LB-algebra with respect to this bound structure whenever the space of non-trivial complex-valued homomorphism is compact in its Gelfand topology. It follows from this and the fact that the space of non-trivial complex-valued homeomorphisms is always compact for an LB-algebra (Theorem 6.2-2) that if a complete LMC algebra is an LB-algebra with a bound structure $\mathcal{B}$ then it must be an LB-algebra with respect to $\mathcal{B}_n$. It is also established that the Frechet LB-algebras are precisely the Frechet Q-algebras (Theorem 6.3-2).

Definition 6.3-1. THE NATURAL BOUND STRUCTURE Let X be a complex topological algebra. Then the associated natural bound structure, denoted by $\mathcal{B}_n$, is the collection of all subsets $B\subset X$ such that

(a) B is absolutely m-convex and $e\in B$,

(b) B is closed and bounded.

The collection $X_o = \bigcup_{B\in\mathcal{B}_n} X(B)$ where $X(B)=\{\alpha y \mid \alpha\in\mathbb{C},\ y\in B\}$ is referred to as the bounded elements of X and is a subalgebra of X. If $x\in X$ while $x\notin X_o$ then x is an unbounded element.

Prior to presenting examples we characterize the elements of X_o.

(6.3-1) A CHARACTERIZATION OF X_o The element of $x\in X$ is a bounded element (i.e. $x\in X_o$) iff there exists a scalar λ such that $\{(\lambda x)^n \mid n\in\mathbb{N}\}$ is a bounded set.

Proof If $x\in X_o$ then $x\in X(B)$ for some $B\in\mathcal{B}_n$. Since $X(B)=\{\alpha y \mid \alpha\in\mathbb{C},\ y\in B\}$, $x=\alpha y$ for some $\alpha\in\mathbb{C}$ and $y\in B$. If $x=0$ then $\{(\lambda x)^n \mid n\in\mathbb{N}\}$ is certainly a bounded subset of X. If $x\neq 0$ then $y=\alpha^{-1}x\in B$, and, since B is multiplicative, $y^n\in B$ for any n. Thus $\{(\alpha^{-1}x)^n \mid n\in\mathbb{N}\}\subset B$ and is therefore bounded.

Conversely, suppose that $\{(\lambda x)^n \mid n\in\mathbb{N}\}$ is bounded. It is clear that $S=(\lambda x)^n \mid n\in\mathbb{N}\}\cup\{e\}$ is multiplicative and bounded. Since multiplicativity and boundedness are preserved when forming the closed absolute convex hull of a set, $B=\mathrm{cl}(S_{bc})\in\mathcal{B}_n$. Thus $x\in X(B)\subset X_o$. ∇

Certainly all elements of a normed algebra are bounded as can be seen directly from the definition of X_o or from the characterization given in (6.3-1). In our next example we characterize the elements of $C(T,\mathbb{C},c)$ that are bounded. As might be expected they are just the set of uniformly bounded complex-valued functions on T.

Example 6.3-1. C(T,$\underset{\sim}{C}$,c) "=" UNIFORMLY BOUNDED FUNCTIONS Let $x \in C(T,\underset{\sim}{C},c)$; we claim that s is a bounded element iff it is uniformly bounded on T. Indeed if x is uniformly bounded on T by the number $M > 0$, i.e. $\sup|x(T)| \leq M$, then the function $(1/M)x$ and all of its positive powers are uniformly bounded by 1. Thus it follows that each function $((1/Mx)^n$, n=1,2,..., is bounded by 1 on each compact set K. Since the seminorms $p_K(y)=\sup|y(K)|, y \in C(T,\underset{\sim}{C},c)$, generate the compact-open topology, we see that $\{((1/M)x)^n | n \in \underset{\sim}{N}\}$ is a bounded set in $C(T,\underset{\sim}{C},c)$. Conversely, suppose that x is not uniformly bounded on T; then neither is the function λx, for any $\lambda \neq 0$. Thus there exists a compact set K on which λx assumes a modulus larger than 1. It follows that the set of numbers $\{p_K((\lambda x)^n) | n \in \underset{\sim}{N}\}$ is unbounded and $\{(\lambda x)^n | n \in \underset{\sim}{N}\}$ is not a bounded set. Since λ was arbitrary, $x \notin C(T,\underset{\sim}{C},c)_o$ by (6.3-1) and the proof is complete. ∇

As a result of the preceding example we see that $C(T,\underset{\sim}{C},c)$ possesses only bounded elements iff T is pseudo-compact (Sec. 1.5). We shall make further use of this example at the end of this section when we prove that a completely regular Hausdorff hemi-compact k-space (Secs. 2.1 and 2.2) is pseudo-compact iff it is compact.

In order for X to be an LB-algebra with respect to $\mathcal{B}_n$ it is necessary that each X(B) be a Banach algebra with respect to the gauge p_B of B. In our next result we present a sufficient condition on X for this to be the case.

(6.3-2) EACH X(B), $B \in \mathcal{B}_n$, IS COMPLETE IF X IS SEQUENTIALLY COMPLETE If X is a sequentially complete locally convex Hausdorff algebra then each X(B), $B \in \mathcal{B}_n$, is a Banach algebra with respect to the gauge p_B, i.e. $(X, \mathcal{B}_n)$ is a complete bound algebra.

Proof Since each $B \in \mathcal{B}_n$ is a topologically bounded set in X, no such B can contain a non-trivial subspace; hence each p_B is a norm. Next we contend that the norm topology defined by p_B on X(B) is stronger than the topology induced by X. To see that this is so let $U \in V(0)$. Then there is a scalar $a > 0$ such that $aB \subset U$. Now suppose that $x \in aVp_B$, i.e. $p_B(x)=\inf\{\alpha > 0 | x \in \alpha B\} < a$. Since B is balanced, $x \in aB \subset U$. Thus $Vp_B \subset U \subset X(B)$.

Consequently we see that any sequence $(x_n) \subset X(B)$ which is Cauchy in the norm p_B is also Cauchy in the induced topology on X(B). By the sequential completeness of X, (x_n) has a limit $x \in X$. It remains to show that $p_B(x_n - x) \to 0$ and $x \in X(B)$. Since (x_n) is Cauchy in the norm p_B for each $\delta > 0$ there is an index $N > 0$ such that $p_B(x_n - x_m) < \delta$ whenever n, $m \geq N$. Thus $x_n - x_m \in \delta B$ for n, $m \geq N$ and, noting that B is closed in X, it follows, that $x_n - x = \lim_m (x_n - x_m) \in \delta B$

for $n \geq N$. Hence $x \in x_N + \delta B \subset X(B)$ and $p_B(x_n - x) \leq \delta$ for all $n \geq N$. ∇

The next result, the central theorem of the section, provides a characterization of complete LMCH LB-algebras.

Theorem 6.3-1. THE COMPLETE LMCH ALGEBRA X IS AN LB-ALGEBRA IFF H(X) IS COMPACT Let X be a complete LMCH complex algebra. If either X^h or $H(X)$ is compact in its Gelfand topology then X is an LB-algebra with respect to $\mathcal{B}_n$.

Remark Theorem 6.2-2 shows H(X) to be Gelfand-compact if X is an LB-algebra. Thus the compactness condition on H(X) is necessary as well as sufficient.

Proof We already know by the previous result that $(X, \mathcal{B}_n)$ is a complete bound algebra. Thus it only remains to show that $X = X_o = \bigcup_{B \in \mathcal{B}_n} X(B)$. Furthermore, as a result of the characterization of X_o given in (6.3-1) and the fact that a subset of a LMCH algebra is bounded iff it is weakly bounded, it suffices to show that for each $x \in X$ a scalar $\lambda \in \underset{\sim}{C}$ exists such that $\{x'((\lambda x)^n) n \in \underset{\sim}{N}\}$ is a bounded set of complex numbers for each $x' \in X'$. In order to do this we make use of the analyticity of the resolvent function of a element in an LMCH algebra to construct a power series having $-x'(x^{n-1})$ as its n-th coefficient. We then apply the Cauchy-Hadamard formula* for the radius of convergence to prove that the set $\{x'((\lambda x)^n) \mid n \in \underset{\sim}{N}\}^*$ is bounded.

We begin the proof by observing that in the complete LMCH algebra X, the spectrum of each element $x \in X$ is obtained as either $\psi x(H(X))$ or $\psi x(X^h)$ ((4.10-8)), and is therefore compact. Hence $\rho(x)$, the resolvent set, is open, and by (4.8-5) and (4.8-6), the resolvent function $r_x(\lambda) = (x - \lambda e)^{-1}$ is analytic on $\rho(x)$. Furthermore, again by the compactness of $\sigma(x)$, there is an $r > 0$ such that $\{\lambda \in \underset{\sim}{C} \mid |\lambda| > r\} \subset \rho(x)$. Thus, setting $\epsilon = 1/r$, we can define the function s at λ by

$$s(\lambda) = \begin{cases} r_x(1/\lambda) & \text{for } 0 < |\lambda| < \epsilon \\ 0 & \text{for } \lambda = 0 \end{cases}$$

which is clearly analytic for $0 < |\lambda| < \epsilon$. Moreover we claim that s is analytic $S_\epsilon(0) = \{\lambda \mid |\lambda| < \epsilon\}$ and the n-th derivative of s at λ is given by

(*) $$s^{(n)}(\lambda) = (-1)^n\, n!\, x^{n-1}(\lambda x - e)^{-(n+1)}$$

for $\lambda \in S_\epsilon(0)$ and $n = 1, 2, \ldots$. First consider the case $n = 1$. Since $s(\lambda) = (x - (1/\lambda)e)^{-1} = \lambda(\lambda x - e)^{-1}$ for $0 < |\lambda| < \epsilon$ and inversion is continuous in any LMC algebra [(4.8-6)]

* The Cauchy-Hadamard formula for the radius of convergence of $\Sigma a_n \lambda^n$ is $1/\limsup \sqrt[n]{|a_n|}$. (Markushevich (1965), Vol. 1, p. 344.

$$s^{(1)}(0) = \lim_{\lambda \to 0} \frac{s(\lambda) - s(0)}{\lambda - 0} = \lim_{\lambda \to 0} (\lambda x-e)^{-1} = -e$$

which agrees with (*) for $\lambda=0$ and $n=1$. Next suppose that $0 < |\mu|, |\lambda| < \epsilon$ and consider

$$s^{(1)}(\mu) = \lim_{\lambda \to \mu} \frac{s(\lambda) - s(\mu)}{\lambda - \mu}$$

$$= \lim_{\lambda \to \mu} \frac{r_x(1/\lambda) - r_x(1/\mu)}{\lambda - \mu}$$

$$= \lim_{\lambda \to \mu} \frac{(1/\lambda - 1/\mu)\ r_x(1/\lambda)\ r_x(1/\mu)}{\lambda - \mu} \quad \text{of}$$

$$= (-1/\mu^2)\ (r_x(1/\mu))^2 = -(\mu x - e)^{-2}\ .$$

Hence we have established (*) for $n=1$. Proceeding by induction we assume that (*) is valid for the integers $1,\ldots,n$ and then for $\lambda, \mu \in S_\epsilon(0)$, we consider

$$s^{(n)}(\lambda)-s^{(n)}(\mu) = (-1)^n\ n!\ x^{n-1}\ [(\lambda x-e)^{-(n+1)} - (\mu x-e)^{-(n+1)}]$$

$$= (-1)^n\ n!\ x^{n-1}(\lambda x-e)^{-(n+1)}(\mu x-e)^{-(n+1)}\ [(\mu x-e)^{n+1} - (\lambda x-e)^{n+1}]$$

$$= (-1)^n\ n!\ x^{n-1}(\lambda x-e)^{-(n+1)}(\mu x-e)^{-(n+1)}\ [(\mu-\lambda)x][\sum_{k=0}^{n}(\mu x-e)^k(\lambda x-e)^{n-1}]$$

Therefore,

$$\frac{s^{(n)}(\lambda)-s^{(n)}(\mu)}{\lambda - \mu} = (-1)^{n+1} n!x^n(\lambda x-e)^{-(n+1)}(\mu x-e)^{-(n+1)}[\sum_{k=0}^{n}(\mu x-e)^k(\lambda x-e)^{n-1}]$$

and, by the continuity of inversion, addition and multiplication, it follows that by taking the limit as $\lambda \to \mu$ that

$$s^{(n+1)}(\mu) = (-1)^{n+1}(n+1)!x^n(\mu x-e)^{-(n+2)}$$

for each $\mu \in S_\epsilon(0)$.

Thus having established (*), it follows that for $x' \in X'$, the complex-valued function $x' \cdot s$ is analytic on $S_\epsilon(0)$, and, by induction together with a routine computation,

$$(x' \cdot s)^{(n)}(\mu) = (-1)^n n! x'(x^{n-1}(\mu x-e)^{-(n+1)})$$

for each $\mu \in S_\epsilon(0)$ and $n=1,2,\ldots$. By the analyticity of $x' \cdot s$ on $S_\epsilon(0)$ the Taylor series of $x' \cdot s$ centered at $\mu=0$ converges in $S_\epsilon(0)$. Since

$(x'.s)^{(n)}(0)/n!=-x'(x^{n-1})$ for $n \geq 1$, we may apply the Cauchy-Hadamard formula to obtain

$$\limsup |x'(x^{n-1})|^{1/n} \leq 1/\epsilon = r$$

which implies

$$\limsup |\tfrac{\epsilon}{2} x'((\tfrac{\epsilon}{2} x)^{n-1})|^{1/n} \leq \tfrac{1}{2}$$

Thus for all but at most a finite number of values of n,

$$|\tfrac{\epsilon}{2} x'(\tfrac{\epsilon}{2} x)^{n-1})|^{1/n} \leq 1$$

and the final conclusion that the set $\{x'((\frac{\epsilon}{2} x)^n) n \in \underset{\sim}{N}\}$ is bounded follows. ▽

Suppose now that X is not an LB-algebra with respect to $\mathcal{B}_n$. Is it possible for X to be an LB-algebra with respect to some other bound structure? The answer is no, in view of our previous result and the fact that an LB-algebra has a Gelfand-compact set of non-trivial homomorphisms.

Any complex homomorphism of a Banach algebra is bounded or equivalently continuous - as mentioned prior to (Theorem 6.2-2). Boundedness cannot possibly be equivalent to continuity for LB-algebras since they don't carry a topology. Suppose however that $(X,\mathcal{B})$ is an LB-algebra and that a compatible topology $\mathcal{T}$ for X exists such that each $B \in \mathcal{B}$ is $\mathcal{T}$-bounded. By (6.2-1) any complex homomorphism of X maps elements of $\mathcal{B}$ into bounded subsets of the complex plane. Since, algebraically, X is structurally related to Banach algebras (as an inductive limit) and, since a compatible topology for X is linked to this algebraic structure it is natural to ask if all homomorphisms of X onto $\underset{\sim}{C}$ are $\mathcal{T}$-continuous. As shown below, this needn't happen.

Example 6.3-2. A DISCONTINUOUS HOMOMORPHISM Let T be a compact Hausdorff topological space with a non-isolated point t_o (e.g. $T=[0,1]$ and $t_o=0$). Since $C(T,\underset{\sim}{C},c)$ is a Banach algebra each element of $H(C(T,\underset{\sim}{C},c))$ is continuous and, moreover, by Example 4.10-2 $H(C(T,\underset{\sim}{C},c))=T^*$ (the evaluation maps). Denoting by $\mathcal{T}$ the weakest topology of $C(T,\underset{\sim}{C})$ for which the homomorphisms $T^*-\{t_o^*\}$ are continuous, it follows that $\mathcal{T}$ is weaker than the compact-open topology. Thus the elements of the natural bound structure are all $\mathcal{T}$-bounded while, by Example 4.10-1, t_o^* is not continuous with respect to $\mathcal{T}$.

As the above example illustrates not all homomorphisms of an LB-algebra $(X,\mathcal{B})$ are $\mathcal{T}$-continuous even tho $\mathcal{T}$ is a compatible topology in which the elements of $\mathcal{B}$ are $\mathcal{T}$-bounded. Reversing direction somewhat, starting with a Frechet algebra $(X,\mathcal{T})$, if $\mathcal{B}$ is the natural bound structure determined by $\mathcal{T}$ and $(X,\mathcal{B})$ is an LB-algebra then all complex homomorphisms of X must be continuous as shown by Theorem 6.3-2: The main thrust of Theorem 6.3-2 is that in the class of Frechet algebras, being an LB-algebra is equivalent to being a Q-algebra.

Theorem 6.3-2. FOR FRECHET ALGEBRAS, LB $\leftrightarrow$ Q If X is a Frechet algebra then the following are equivalent:

(a) X is an LB-algebra.

(b) X is a Q-algebra - hence all homomorphisms are continuous.

(c) X^h is compact in its Gelfand topology.

(d) $r_\sigma(x) < \infty$ for each $x \in X$.

Proof (a) $\to$ (d): By Theorem 6.2-2, H(X) is compact in its Gelfand topology. Since each singular element lies in a maximal ideal and each maximal ideal is the kernel of some complex homomorphism [(6.2-5)], it follows that $\sigma(x)=\{h(x) \mid h \in H(X)\}=(\psi x)(H(X))$ - where ψ is as in Def. 4.12-1 - and is therefore a bounded set of complex numbers for each $x \in X$.

(b) $\to$ (c): We know that a Frechet algebra, being nonmeager, is barreled. In (4.12-3) we established the equivalence of (b) and (c) for barreled complete complex LMCH algebras.

(d) $\to$ (b): Recall that X is a Q-algebra iff the set $U(\sigma)=\{x \in X \mid r_\sigma(x) \leq 1\}$ has non-empty interior (4.8-3). In a complete complex LMCH algebra $U(\sigma)=(X^h)^o$ (4.12-2). Since the polar of a set is weakly closed and absolutely convex and a $\sigma(X,X')$ - closed absolutely convex set is closed in any topology of the dual pair, $U(\sigma)$ is a closed subset of X. Now by (d), we have that $X=\bigcup_{n \in \underset{\sim}{N}} nU(\sigma)$, a countable union of closed sets, and, so by the Baire category theorem, $U(\sigma)$ has non-empty interior.

(c) $\to$ (a): By Theorem 6.3-1. ∇

In Chap. 1 it was shown that for replete spaces, compactness and pseudo-compactness are equivalent ((1.5-3)). As any completely regular Hausdorff hemicompact k-space is replete (Th. 1.5-3) such a space is compact whenever it is pseudo-compact. An alternate proof of this result can be given using the ideas developed in this chapter: More precisely, the previous theorem and the characterization of the bounded elements of $C(T,\underset{\sim}{C},c)$ given in Example 6.3-1. Suppose that T is a completely regular Hausdorff hemicompact k-space which is not compact. We know from Example 4.10-2 that because T is completely regular $C(T,\underset{\sim}{C},c)^h=T^*$. Since the Gelfand topology of $C(T,\underset{\sim}{C},c)^h$ is the weakest with respect to which each function ψx, where

$(\psi x)(t^*)=x(t)$ for $x\in C(T,\underset{\sim}{C})$ and $t\in T$ is continuous, the original topology of T is stronger than the Gelfand topology. To see that the two topologies are in fact the same, let U be a proper open subset of T in the original topology. In that T is completely regular, it follows that corresponding to each $t\in U$ there exists a continuous real-valued function x_t such that $x_t(t)=0$ and $x_t(CU)=\{1\}$. Hence the Gelfand neighborhood of t, $x_t^{-1}(S_{\frac{1}{2}}(0))$, is contained in U and U is open in the Gelfand topology. Having established that the Gelfand topology of T coincides with the original non-compact topology, it follows by the previous theorem ((a) and (c)) that the Frechet algebra $C(T,\underset{\sim}{C},c)$ is not an LB-algebra. Now by virtue of the fact that $X=C(T,\underset{\sim}{C},c)$ is complete, each X(B), for $B\in\mathcal{B}_n$, is a Banach algebra by (6.3-2). Consequently it must be that $C(T,\underset{\sim}{C},c)\neq C(T,\underset{\sim}{C},c)_o$. Thus by Example 6.3-1 there are continuous unbounded complex-valued functions defined on T which precludes the possibility of T being pseudo-compact.

We are now in a position to give specific examples of complete LMCH algebras that are not LB-algebras. If T is any non-compact completely regular Hausdorff hemicompact k-space, e.g. $T=\underset{\sim}{R}$, then $C(T,\underset{\sim}{C},c)$ is a complete LMCH algebra that is not an LB-algebra.

REFERENCES

Alexandrov, A.D.

1940. Additive set functions in abstract spaces, Mat. Sb. (N.S.) 8(50), pp. 307-348.

1941. Additive set functions in abstract spaces, Mat. Sb. (N.S.) 9(51), pp. 563-628.

1943. Additive set functions in abstract spaces, Mat. Sb. (N.S.) 13(55), pp. 169-238.

Allan, G.R.

1965. A spectral theory for locally convex algebras, Proc. Lond. Math. Soc. (3), 15, pp. 399-421.

1967. On a class of locally convex algebras, Proc. Lond. Math. Soc. (3), 17, pp. 91-114.

Allan, G.R., Dales, H.G., and McClure, J.P.

1971. Pseudo-Banach algebras, Studia Math., 40, pp. 55-69.

Alo, R., and Shapiro, H.

1968a. A note on compactifications and semi-normal spaces, J. Austral. Math. Soc., 8, pp. 102-108.

1968b. Normal bases and compactifications, Math. Ann., 175, pp. 337-340.

Arens, R.

1946a. A topology for spaces of linear transformations, Ann. of Math., 47, pp. 480-495.

1946b. The space L^{ω} and convex topological rings, Bull. Amer. Math. Soc., 52, pp. 931-935.

1947a. Linear topological division algebras, Bull. Amer. Math. Soc., 53, pp. 623-630.

1947b. Representations of *-algebras, Duke Math. J., 14, pp. 269-282.

1952. A generalization of normed rings, Pacific J. Math., 2, pp. 455-471.

1958a. Inverse-producing extensions of normed algebras, Trans. Amer. Math. Soc., 88, pp. 536-548.

1958b. The maximal ideals of certain function algebras, Pacific J. Math., 8, pp. 641-648.

1958c. Dense inverse limit rings, Michigan Math. J., 5, 169-182.

Bachman, G., Beckenstein, E., and Narici, L.

1971a. Spectral continuity and permanent sets in topological algebras, Atti. Accad. Naz. Lincei Rend. Cl. Sci. Fis. Mat. Natur. Serie VIII, vol. L, fasc. 3, pp. 277-283.

1971b. Function algebras over valued fields and measures I, Atti. Accad. Naz. Lincei Rend. Cl. Sci. Fis. Mat. Natur. Serie VIII, vol. LI, fasc. 5, pp. 293-300.

1972. Function algebras over valued fields and measures II, Atti. Accad. Naz. Lincei Rend. Cl. Sci. Fis. Mat. Natur. Serie VIII, vol. LII, fasc. 2, pp. 78-83.

1973. Function algebras over valued fields, Pacific J. Math., 44, pp. 45-58.

Bachman, G., Beckenstein, E., Narici, L., and Warner, S.

1975. Rings of continuous functions with values in a topological field, to appear.

Bagley, R., and Yang, J.

1966. On k-spaces and function spaces, Proc. Amer. Math. Soc., 17, pp. 703-705.

Bagley, R., Connell, E., and McKnight, J.

1958. On properties characterizing pseudocompact spaces, Proc. Amer. Math. Soc., 9, pp. 500-506.

Banach, S.

1932. Théorie des operations lineaires, Chelsea, New York.

Banaschewski, B.

1955. Uber nulldimensionale Räume, Math. Nachr., 13, pp. 129-140.

Beckenstein, E., Bachman, G., and Narici, L.

1972a. Function algebras over valued fields and measures III, Atti. Accad. Naz. Lincei Rend. Cl. Sci. Fis. Mat. Natur. Serie VIII, vol. LII, fasc. 6, pp. 498-503.

1972b. Function algebras over valued fields and measures IV, Atti. Accad. Naz. Lincei Rend. Cl. Sci. Fis. Mat. Natur. Serie VIII, vol. LIII, fasc. 5, pp. 349-358.

1973. Topological algebras of continuous functions over valued fields, Studia Math., T. XLVIII, 119-127.

Birtel, F.

1965. Function algebras. Proceedings of the International Symposium on Function Algebras, Tulane University, edited by F. Birtel, Scott-Foresman, Chicago.

Bourbaki, N.

1953a. Sur certain espaces vectoriels topologiques, Ann. Inst. Fourier, 2, pp. 5-16.

1953b. Espaces vectoriels topologiques, Chapitres I et II, Hermann, Paris.

1955. Espaces vectoriels topologiques, Chapitres III-V, Hermann, Paris.

1964. Algèbre commutative, Hermann, Paris.

Brooks, R.M.

1967a. On Wallman compactifications, Fund. Math., 60, pp. 157-173.

1967b. Boundaries for locally m-convex algebras, Duke Math. J., 34, pp. 103-116.

1968a. Partitions of unity in F-algebras, Math. Ann., 177, pp. 265-272.

1968b. The structure space of a commutative locally m-convex algebra, Pacific J. Math., 25, pp. 443-454.

1968c. On the spectrum of finitely-generated locally m-convex algebras, Studia Math., 29, pp. 143-150.

1968d. On the invariance of the spectrum in locally m-convex algebras, Can. J. Math., 20, pp. 658-664.

1968e. On commutative locally m-convex algebras, Duke Math. J., 35, pp. 257-268.

1970. On singly generated locally m-convex algebras, Duke Math. J., 38, pp. 529-536.

1971a. On Arens' "Weierstrass Products" paper, Math. Ann., 190, pp. 329-337.

1971b. On representing F-algebras, Pacific J. Math., 39, pp. 51-69.

1971c. Boundaries for natural systems, Indiana Univ. Math. J., 20, pp. 865-875.

1972. Boundaries for natural systems II, Math. Scand., 30, pp. 281-289.

Browder, A.

1969a. Rational approximation and function algebras, Benjamin, New York.

1969b. Introduction to function algebras, Benjamin, New York.

Cech, E.

1937. On bicompact spaces, Ann. of Math., 38, pp. 823-844.

Cochran, A.

1973. Inductive limits of A-convex algebras, Proc. Amer. Math. Soc., 37, pp. 489-496.

Cochran, A., Keown, R., and Williams, C.R.

1970. On a class of topological algebras, Pacific J. Math., 34, pp. 17-25.

Comfort, W.

1967. A nonpseudocompact product space whose finite subproducts are pseudocompact, Math. Ann., 170, pp. 41-44.

1968. On the Hewitt realcompactification of a product space, Trans. Amer. Math. Soc., 131, pp. 107-118.

Correl, E., and Henriksen, M.

1956. On rings of continuous functions with values in a division ring, Proc. Amer. Math. Soc., 7, pp. 194-198.

Curtis, P.C.

1961. Derivations of commutative Banach algebras, Bull. Amer. Math. Soc., 67, pp. 271-273.

Curtis, P.C., and Bade, W.G.

1960. Homomorphisms of commutative Banach algebras, Amer. J. Math., pp. 589-607.

deMarco, G.

1972. On the countably generated z-ideals of C(X), Proc. Amer. Math. Soc., 31, no. 2, pp. 574-576.

Dietrich, W.E.

1970. On the ideal structure of C(X), Trans. Amer. Math. Soc., 152, pp. 61-77.

1969a. A note on the ideal structure of C(X), Proc. Amer. Math. Soc., 23, pp. 174-178.

1969b. The maximal ideals space of the topological algebra C(X,E), Math. Ann., 183, pp. 201-212.

Dieudonné, J.

1953. Recent developments in the theory of locally convex spaces, Bull. Amer. Math. Soc., 59, pp. 495-512.

1970. Treatise on analysis, Academic Press, New York.

Dugundji, J.

1966. Topology, Allyn and Bacon, Boston.

Dunford, N., and Schwartz, J.

1958. Linear operators I, Wiley, New York.

Engelking, R., and Mrówka, S.

1958. On E-compact spaces, Bull. Acad. Pol. Sci. Ser. Math. Astro Phys., 6, pp. 429-436.

Frolik, Z.

1960. The topological product of two pseudocompact spaces, Czechoslovak Math. J., 10, pp. 339-349.

Fleischer, I.

1971. Generalized compactifications for initial topologies, Rev. Roum. Math. Pures et Appl., Tome XVI, no. 8, pp. 1185-1191.

Gamelim, T.

1969. Uniform algebras, Prentice-Hall, Englewood Cliffs, New Jersey.

Gelfand, I.

1939. On normed rings, Dokl. Akad. Nauk. SSSR, 23, pp. 430-432.

1941. Normierte ringe, Mat. Sbornik, 9, pp. 3-24.

Gelfand, I., and Kolmogoroff, A.

1939. On rings of continuous functions on a topological space, Doklady, 22, pp. 11-15.

Gillman, L. and Jerison, M.

1960. Rings of continuous functions, van Nostrand, New York.

Gillman, L., and Henriksen, M.

1954. Concerning rings of continuous functions, Trans. Amer. Math. Soc., 77, pp. 340-362.

1956. Rings of continuous functions in which every finitely generated ideal is principal, Trans. Amer. Math. Soc., 82, pp. 366-391.

Gleason, A.

1967. A characterization of maximal ideals, J. d'Analyse Math., 19, pp. 171-172.

Glicksberg, I.

1959. Stone-Cech compactifications of products, Trans. Amer. Math. Soc., 90, pp. 369-382.

Goldhaber, J. and Wolk, E.

1954. Maximal ideals in rings of bounded continuous functions, Duke Math. J., 21, pp. 565-569.

Gordon, H.

1970. Compactifications defined by means of generalized ultrafilters, Ann. Mat. Pura ed App., pp. 15-23.

Gelfand, I., Raikov, D., and Silov, G.

1964. Commutative normed rings, Chelsea, New York.

Guennebaud, B.

1967. Algèbres localement convexes sur les corps values, Bull. Sci. Math. Fr. 2e serie, 91, pp. 75-96.

Gulik, F.

1970. Systems of derivations, Trans. Amer. Math. Soc., 149, pp. 1-24.

Hamburger, P.

1971a. On internal characterizations of complete regularity and Wallman-type compactifications, General topology and its relations to modern analysis and algebra III, Proceedings of the third Prague topology symposium, pp. 171-172.

1971b. On Wallman-type, regular Wallman-type and z-compactifications, Per. Math. Hung., I(4), pp. 303-309.

1972a. On k-compactifications and realcompactifications, Acta Math. Acad. Sci. Hung., 23, (1-2), pp. 255-262.

1972b. A general method to give internal characterizations of completely regular and Tychonoff spaces, Acta Math. Acad. Sci. Hung., 23 (3-4), pp. 479-494.

Henriksen, M.

1956. On the equivalence of the ring, lattice, and semigroup of continuous functions, Proc. Amer. Math. Soc., 7, pp. 959-960.

Henriksen, M., and Isbell, J.

1957. Local connectivity in the Stone-Cech compactification, Israel J. Math., 1, pp. 574-582.

Hermes, H.

1955. Einführung in die Verbandstheorie, Berlin.

Hewitt, E.

1948. Rings of real-valued continuous functions I, Trans. Amer. Math. Soc., 64, pp. 54-99.

1950. Linear functionals on spaces of continuous functions, Fund. Math., 37, pp. 161-189.

Hoffman, K.

1962. Banach spaces of analytic functions, Prentice-Hall, Englewood Cliffs, New Jersey.

Horvath, J.

1966. Topological vector spaces and distributions, Addison-Wesley, Reading, Massachusetts.

Husain, T.

1965. The open mapping and closed graph theorems in topological vector spaces, Oxford University Press, London.

Ingleton, A.W.

1952. The Hahn-Banach theorem for nonarchimedean fields, Proc. Cambridge Philos. Soc., 48, pp. 41-45.

Johnson, B.E.

1969. Continuity of derivations on commutative algebras, Amer. J. Math., 91, pp. 1-10.

Johnson, K.G.

1964. $B(S,\Sigma)$-algebras, Proc. Amer. Math. Soc., 15, pp. 247-251.

Kakutani, S.

1940. Concrete representations of abstract (M)-spaces, Ann. of Math., 42, pp. 994-1024.

Kaplansky, I.

1947a. Lattices of continuous functions, Bull. Amer. Math. Soc., 53, pp. 617-623.

1947b. Topological rings, Amer. J. Math., 69, pp. 153-183.

1948. Topological rings, Bull. Amer. Math. Soc., 45, pp. 809-826.

1949. Normed algebras, Duke Math. J., 16, pp. 399-418.

Kelley, J.

1955. General topology, van Nostrand, New York.

Kennison, J.F.

1965. Reflective functors in general topology and elsewhere, Trans. Amer. Math. Soc., 118, pp. 303-315.

Kohls, C.

1958a. Prime ideals in rings of continuous functions, Illinois J. Math., 2, pp. 505-536.

1958b. Prime ideals in rings of continuous functions II, Duke Math. J., 25, pp. 447-458.

Köthe, G.

1969. Topological vector spaces I, Springer-Verlag, New York.

Kowalsky, H.

1955. Stonesche Körper und ein überdeckungsatz, Math. Nachr., 14, pp. 57-64.

Krein, M. and Krein, S.

On an inner characterization of the set of all continuous functions defined on a bicompact Hausdorff space, C.R. (Doklady) Academie des Sciences de l'URSS, 27, pp. 427-430.

Kuczma, M.E.

1958. On a problem of E. Michael concerning topological divisors of 0, Coll. Math., 19, pp. 295-299.

Leibowitz, G.

1970. Lectures on complex function algebras, Scott, Foresman, Glenview, Illinois.

Levitz, K.

1972. A characterization of general z.p.I. rings, Proc. Amer. Math. Soc., 32, pp. 376-380.

Markushevich, A.I.

1965. Theory of functions of a complex variable I, II, Prentice-Hall, Englewood Cliffs, New Jersey.

1967. Theory of functions of a complex variable III, Prentice-Hall, Englewood Cliffs, New Jersey.

Mackey, G.

1946. On convex topological linear spaces, Trans. Amer. Math. Soc., 60, pp. 519-537.

Mazur, S.

1938. Sur les anneaux lineaires, C.R. Acad. Sci. Paris, 207, pp. 1025-1027.

1952. On continuous mappings on cartesian products, Fund. Math., 39, pp. 229-238.

Mayers, S.B.

1949. Spaces of continuous functions, Bull. Amer. Math. Soc., 55, pp. 402-407.

Michael, E.A.

1952. Locally multiplicatively-convex topological algebras, Mem. Amer. Math. Soc., no. 11.

Miller, J.B.

1970. Higher derivations on Banach algebras, Amer. J. Math., XCII, no. 2, pp. 301-331.

Morris, P., and Wulbert, D.

1967. Functional representations of topological algebras, Pacific J. Math., 22, pp. 323-337.

Mrowka, S.

1958. On the union of Q-spaces, Bull. Acad. Polon. Sci. Ser. Sci. Math., Astra et Phys., 6, pp. 365-368.

Nachbin, L.

1954. Topological vector spaces of continuous functions, Proc. Nat. Acad. Sci. U.S.A., 40, pp. 471-474.

Naimark, M.

1960. Normed rings, P. Noordhoff, Ltd., Groningen, The Netherlands.

Nanzetta, P., and Plank, D.

1972. Closed ideals in C(X), Proc. Amer. Math. Soc., 35, pp. 601-606.

Narici, L., Beckenstein, E., and Bachman, G.

1971. Functional Analysis and valuation theory, Marcel Dekker, New York.

1972a. Some recent developments on repletions and Stone-Cech compactifications of 0-dimensional spaces, Topo 72 - general topology and its applications, Second Pittsburgh International Conference, Springer-Verlag, Berlin.

1972b. Locally convex algebras of continuous functions over valued fields, Proc. Int. Coll. Analyse Rio de Janeiro, Hermann, Paris.

Nielson, R., and Sloyer, C.

1970. Ideals of semicontinuous functions and compactifications of T_1 spaces, Math. Ann., 187, pp. 329-331.

Noble, N.

1967. k-spaces and some generalizations, Dissertation, University of Rochester.

Nobeling, G.

1954. Grundlagen der analytischen topologie, Berlin.

Piacun, N., and Li Pi Si.

1973. Wallman compactifications on E-completely regular spaces, Pacific J. Math., 45, no. 1, pp. 321-326.

Pierce, R.S.

1961. Rings of integer-valued continuous functions, Trans. Amer. Math. Soc., 100, 371-394.

Ptak, V.

1953. On complete topological linear spaces, Czech. Math. J., 78, pp. 301-364.

Rickart, C.

1960. General theory of Banach algebras, van Nostrand, Princeton, New Jersey.

1966. The maximal ideals space functions locally approximable in a function algebra, Proc. Amer. Math. Soc., 7, pp. 1320-1326.

Rosenfeld, M.

1966. Commutative F-algebra, Pacific J. Math., 16, pp. 159-166.

Samuel, P.

1948. Ultrafilters and compactifications of uniform spaces, Trans. Amer. Math. Soc., 64, pp. 100-132.

Shirota, T.

1952a. A class of topological spaces, Osaka Math. J., 4, pp. 23-40.

1952b. A generalization of a theorem of Kaplansky, Osaka Math. J., 4, pp. 121-132.

1954. On locally convex vector spaces of continuous functions, Proc. Japan Acad., 30, pp. 294-298.

Silov, G.

1940. On the extension of maximal ideals, Dokl. Akad. Nauk. SSSR, 29, pp. 83-84.

Sinclair, A.M.

1969. Continuous derivations on Banach algebras, Proc. Amer. Math. Soc., pp. 166-170.

Singer, I.M., and Wermer, J.

1955. Derivations on commutative algebras, Math. Ann., 129, pp. 260-264.

Steiner, E.F.

1968. Wallman spaces and compactifications, Fund. Math., 61, pp. 295-304.

Stephenson, R.M., Jr.

1968. Pseudocompact spaces, Trans. Amer. Math. Soc., 134, no. 3, pp.

437-448.

Stone, M.

1936. The theory of representations for Boolean algebras, Trans. Amer. Math. Soc., 40, pp. 37-111.

1937. Applications of Boolean rings to general topology, Trans. Amer. Math. Soc., 41, pp. 375-481.

1941. A general theory of spectra II, Proc. Nat. Acad. Sci. U.S.A., 27, pp. 83-87.

Stout, E.L.

1971. The theory of uniform algebras, Bogden and Quigley, Tarrytown-on-Hudson, New York.

Suffel, C., Beckenstein, E., and Narici, L.

1974. A note on permanently singular elements in topological algebras, Coll. Math., XXXI, no. 1, to appear.

van der Slot, J.

1968. Some properties related to compactness, Mathematical Centre tracts 19, Mathematical Centre 49, 2nd Boerhaavestraat, Amsterdam, The Netherlands.

Varadarajan, V.

1965. Measures on topological spaces, Amer. Math. Soc. Translations, Series 2, vol. 48, pp. 161-228.

Wallman, H.

1938. Lattices and topological spaces, Ann. of Math., 39, pp. 112-126.

Warner, S.

1955. Weak locally multiplicatively-convex algebras, Pacific J. Math., 5, pp. 1025-1032.

1957. Weakly topologized algebras, Proc. Amer. Math. Soc., 8, pp. 314-316.

1958. The topology of compact convergence on continuous function spaces, Duke Math. J., 25, pp. 265-282.

Weil, A.

1937. Sur les espaces uniformes et sur la topologie generale, Actualites Scientifique et Industrielles, no. 551, Hermann, Paris.

Wielandt, H.

1949. Über die Unbeschränktheit der Operatoren der Quantenmechanik, Math. Ann., 121, 21 (1949-1950).

Williamson, J.H.

1954. On topologizing the field C(t), Proc. Amer. Math. Soc., 5, pp. 729-734.

Zelazko, W.

1965. Metric generalizations of Banach algebras, Rozprawy̌ Math., Warsaw 47.

1971. On permanently singular elements in commutative m-convex locally convex algebras, Studia Math., XXXVII, pp. 181-190.

INDEX OF SYMBOLS

INDEX